Small
Appliance
Servicing Guide

Small Appliance Servicing Guide

MOTOR-DRIVEN AND RESISTANCE-HEATED APPLIANCES

Edited by
ROBERT SCHARFF

McGRAW-HILL BOOK COMPANY
NEW YORK
ST. LOUIS
SAN FRANCISCO

AUCKLAND
DÜSSELDORF
JOHANNESBURG
KUALA LUMPUR
LONDON

MEXICO
MONTREAL
NEW DELHI
PANAMA
PARIS

SÃO PAULO
SINGAPORE
SYDNEY
TOKYO
TORONTO

Library of Congress Cataloging in Publication Data
Main entry under title:

Small appliance servicing guide.

   (Practical appliance servicing and repair course;
book 2)
   Includes index.
   1. Household appliances, Electric—Maintenance
and repair.   I. Scharff, Robert.   II. Series.
TK7018.S56   643'.6   76-2670
ISBN 0-07-055142-1

1234567890   HDHD   785432109876

The editors for this book were Tyler G. Hicks and Lester Strong,
the designer was Edward J. Fox, and the production supervisor
was Teresa F. Leaden.

It was printed and bound by Halliday Lithograph Corporation.

Contents

CONTENTS

Preface

The purpose of this book is to give the service technician the information needed to maintain and repair small appliances—the motor-type as well as the heat-resistance ones. It is not necessary to break down the subject matter any further in the Preface; the extent of the completeness of the information can be found in the Contents, Index, and, of course, the text itself. The book is not a treatise on *appliance* engineering; rather it is intended to tell the service technician how to maintain and service small appliances.

This book—as well as all the others in the series—was prepared and written with the full cooperation of the Association of Home Appliance Manufacturers (AHAM). While AHAM and all its members were most cooperative in the preparation of this book, we would especially like to thank the following for their outstanding help: General Electric Company; Sunbeam Corporation; Schick Incorporated; Hamilton Beach Division of Scovill Corporation; Remington Division of Sperry Rand Corporation; Sessions Clock Company; Cory Manufacturing Company; Dominion Electric Company; West Bend Company; Rival Manufacturing Company; Oster Corporation; McGraw-Edison Company; Hoover Company; Arvin Industries Incorporated; Salton, Incorporated; Westinghouse Electric Company; Amana Refrigeration, Inc.; Waage Manufacturing Company; Montgomery Ward Company; Cristensen Manufacturing Company; *Family Handyman* Magazine; General Motors, Inc.; Proctor Electric Company; Matsushita Electric Corporation of America; and Black & Decker Manufacturing Company. In addition, a thank you should be given to McGraw-Hill Book Company for illustrations and material from their books—*Small Appliance Servicing* by P. T. Brockwell, Jr., and *Servicing Electrical Appliances,* Volumes I and II, by the Staff of the National Radio Institute.

In addition, we want to thank Janet Just, who did the typing of the manuscript; Ronald L. Graffius of G. J. Bear Company, who reproduced a portion of the illustrations; and Mary Puschak, who coordinated all the art for this book.

Small Appliance Servicing Guide

Basic appliance servicing techniques

Before delving into the service procedures of specific appliances—both small and large—let us first look at some of the servicing techniques that hold good for most—if not all—appliances. This includes making wiring connections, checking and replacing line cords, basics of appliance disassembly and reassembly, repairing the finish of appliances, and preventive maintenance.

MAKING WIRING CONNECTIONS

On many appliance jobs there will be a need for splicing and terminating wires. A good electric connection has several requirements. Wires must be free of insulating materials, connections should be secure, and splices must always be covered with tape, so that the wire is as well insulated as it was before insulation was removed. To avoid nicking the wire, always cut the insulation at a slant with a knife, as when sharpening a pencil. Move the knife blade at a small angle with the wire; this produces a taper on the cut insulation, as shown in Fig. 1-1. The insulation may also be removed by using a plierlike hand-operated wire stripper. In either case, after removing the insulation, scrape the bare wire ends bright and clean with the back of a knife blade or rub them clean with fine sandpaper.

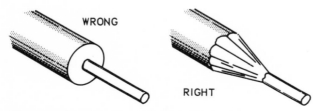

Figure 1-1. Removing insulation from a wire.

Western Union splice. Solid appliance wires may be joined by a simple connection known as the *Western Union splice*. In most instances the wires may be twisted together with the fingers and the ends clamped into position with a pair of pliers.

Figure 1-2 shows the steps in making a Western Union splice. First, prepare the wires for splicing by removing sufficient insulation and cleaning the conductor. Next, bring the wires to a crossed position and make a long twist or bend in each wire. Then wrap one of the wire ends four or five times around the straight portion of the wire. Wrap the other end in a similar manner. Finally, press the ends of the wires down as close as

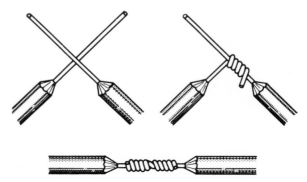

Figure 1-2. Western Union splice.

possible to the straight portion of the wire to prevent the sharp ends from puncturing the tape covering that is wrapped over the splice.

Staggered splice. Joining multiconductor cables presents somewhat of a problem. Each conductor must be spliced and taped; if the splices are directly opposite each other, the overall size of the joint becomes large and bulky. A smoother and less bulky joint may be made by staggering the splices.

Figure 1-3 shows how a two-conductor cable is joined to a similar cable by means of the staggered or lamp-cord splice. Care should be exercised to ensure that a short wire is connected to a long wire, and that the sharp ends are clamped firmly down on the conductor.

Figure 1-3. Staggered splice.

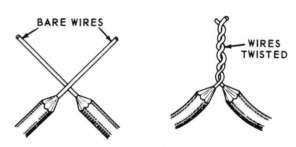

Figure 1-4. Pigtail or rattail joint.

Pigtail splice. To make this joint, strip the insulation from the ends of the conductors to be joined. Then twist the wires to form the pigtail or rattail effect (Fig. 1-4).

Fixture joint. While this joint is used primarily to connect a light fixture to the branch circuit of an electric system, it can be used in appliance work anytime one wire is smaller in diameter than the other. Like the pigtail joint, it will not stand much mechanical strain.

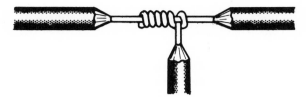

Figure 1-6. Knotted tap joint.

such a junction is called a *tap joint*.

Remove about 1 in of insulation from the main wire to which the branch wire is to be tapped. Strip the branch wire of about 3 in of insulation. The steps in making the tap are shown in Fig. 1-6.

Cross the branch wire over the main wire, as shown in the figure, with about three-fourths of the bare portion of the branch wire extending above the main wire. Bend the end of the branch wire over the main wire, bring it under the main wire, around the branch wire, and then over the main wire to form a knot. Then wrap it around the main conductor in short, tight turns, and trim off the end. Incidentally, the knotted tap is used where the splice is subject to strain or slip. When there is no mechanical strain, the knot may be eliminated.

When working with braided wire, be careful not to cut any of the fine wire strands when stripping the insulation from the individual conductors because this will reduce the current-carrying capacity of the wire. To be on the safe side, therefore, if you inadvertently sever more than one or two strands, trim the wire off flush and begin again.

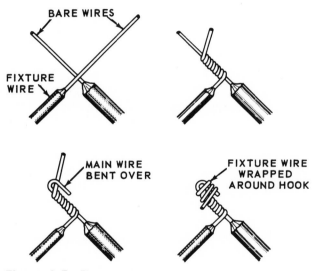

Figure 1-5. Fixture joint.

The drawing in Fig. 1-5 shows the steps in making a fixture joint. The first step is to remove the insulation from the wires to be joined.

After preparing the wires, wrap the fixture wire a few times around the branch wire, as shown in Fig. 1-5. The wires are not twisted, as in the pigtail joint. Then bend the end of the branch wire over the completed turns and wrap the remainder of the bare fixture wire over the bent branch wire. Soldering and taping completes the job.

Knotted tap joint. All the splices considered up to this point are known as *butted splices*. Each was made by joining the free ends of the conductors. Sometimes, however, it is necessary to join a conductor to a continuous wire, and

Soldering

Soldering is a vital part of electrical appliance servicing procedures. It is a manual skill in which practice is required to develop proficiency; however, practice serves no useful purpose unless it is founded on a thorough understanding of basic soldering principles.

Both the solder and the material to be soldered must be heated to a temperature which allows the solder to flow. If either is heated inadequately, "cold" solder joints result. Such joints do not provide either the physical strength or the electrical conductivity required. Appreciably exceeding the flow point temperature, however, is likely to cause damage to the parts being soldered. Various types of solder flow at different temperatures. In soldering operations it is necessary to select a solder that will flow at a temperature low enough to avoid damage to the part being soldered, or to any other part or material in the immediate vicinity.

The duration of high-heat conditions is almost as important as the temperature. Insulation and many other materials in appliance equipment are susceptible to damage from heat. They are damaged if exposed to excessively high temperatures even briefly, or deteriorate if exposed to less drastically elevated temperatures for prolonged periods. The time and temperature limitations depend on many factors—the kind and amount of metal involved, the degree of cleanliness, the ability of the material to withstand heat, and the heat transfer and dissipation characteristics of the surroundings.

Full details on the selection of soldering equipment can be found on p. 167 of *Basics of Electric Appliance Servicing.*

Solder. The three grades of solder generally used for electric-appliance work are 40–60, 50–50, and 60–40 solder. The first figure is the percentage of tin, the second is the percentage of lead. The higher the percentage of tin content, the lower the temperature required for melting. Also, the higher the tin content, the easier the flow, the less time required to harden, and generally the easier to do a good soldering job.

In addition to the solder, flux is needed to remove any oxide film on the metals being joined; otherwise they cannot fuse. The flux enables the molten solder to wet the metals so the solder can stick. The two types are acid flux and rosin flux. Acid flux is more active in cleaning metals but is corrosive. Rosin is always used for light soldering work in making wire connections. Generally, the rosin is in the hollow core of solder intended for appliance work, so that a separate flux is unnecessary. Such rosin-core solder is the type generally used. It should be not-

ed, though, that the use of flux is not a substitute for cleaning the metals to be soldered. The metal must be shiny clean for the solder to stick.

Soldering process. Cleanliness is a prime prerequisite for efficient, effective soldering. Solder will not adhere to dirty, greasy, or oxidized surfaces. Heated metals tend to oxidize rapidly, and the oxide must be removed prior to soldering. Oxides, scale, and dirt can be removed by mechanical means (such as scraping or cutting with an abrasive) or by chemical means. Grease should be removed immediately prior to the actual soldering operation.

Items to be soldered should normally be tinned before making mechanical connections. When the surface has been properly cleaned, a thin, even coating of flux may be placed over the surface to be tinned to prevent oxidation while the part is being heated to soldering temperature. Rosin-core solder is usually preferred in appliance work, but a separate rosin flux may be used instead. Separate rosin flux is frequently used when tinning wires in cable fabrication. (*Tinning* is the coating of the material to be soldered with a light coat of solder.)

The tinning on a wire should extend only far enough to take advantage of the depth of the terminal or receptacle. Tinning or solder on wires subject to flexing causes stiffness and may result in breakage. In fact, the tinned surfaces to be joined should be shaped and fitted, then mechanically joined to make a good mechanical and electric contact. They must be held still with no relative movement of the parts. Any motion between parts will likely result in a poor solder connection. Remember, too, that when forming the loop in a tinned wire under a binding screw, it is not good practice to cross the free end over the starting point, for such an uneven doubling of tinned wire will not yield when the screw is tightened, with the result that the wire will not be uniformly compressed under the screwhead.

The technique for silver soldering is given in *Refrigeration, Air Conditioning, Range and Oven Servicing* of this series.

Solder connections. Frequent arguments occur in appliance shops concerning the proper method of making soldered connections to terminals and binding posts. For many years it was considered necessary to wrap the lead tightly around the terminal to provide maximum mechanical support and strength. But excessive wrappings of leads results in increased heat requirements, more strain on parts, greater difficulty of inspection, greater difficulty of assembly and disassembly of the joints, and increased danger of breaking the parts or terminals during desoldering operations. Insufficient wrapping may result in poor solder joints due to movement of the lead during the soldering operation. After a great deal of research, most appliance manufacturers recommend the joints illustrated in Fig. 1-7. Wrappings of three-eighths to three-fourths turn are usually recommended so that the joint need not be held during the application and cooling of the solder.

Heat the areas to be joined just up to or slightly above the flow temperature of the solder. Control the application of heat carefully to prevent damage to components of the assembly, insulation, or nearby materials. Then apply solder to the heated area. Use only enough solder to make a satisfactory joint. Avoid heavy fillets or beads.

Do not melt the solder with the soldering tip and allow it to flow onto the joint. Instead,

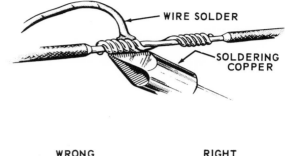

Figure 1-8. Application of solder (top) and right and wrong solder joint (bottom).

heat the joint and apply the solder to the joint. When the joint is adequately heated, the solder will flow evenly. Excessive temperature tends to carbonize flux, thus hindering the soldering operation.

Do not apply liquid to cool a solder joint. By using the proper tools and soldering technique, a joint will not become so hot that rapid cooling is needed. If, for any reason, a satisfactory joint is not obtained initially, take the joint apart, clean the surfaces, remove excess solder, and repeat the entire soldering operation (except tinning).

After the joint has cooled, remove all flux residues. Any flux residue remaining on the surface of electric contacts may collect dirt and promote arcing at a later time. This cleaning is necessary even when rosin-core solder is used. Never solder or desolder connections while equipment power is on or while the circuit is under test. Always discharge any capacitors in the circuit prior to any soldering operation.

Solder splicers. The solder-type splicer is essentially a short piece of metal tube. Its inside diameter is just large enough to allow the tip of a stranded conductor to be inserted in either end, after the conductor tip has been stripped of insulation. This type of splicer is shown in Fig. 1-9.

First heat and fill the splicer with solder. While

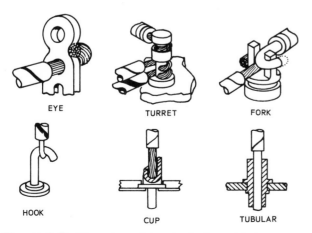

Figure 1-7. Wrapping of terminals for soldering.

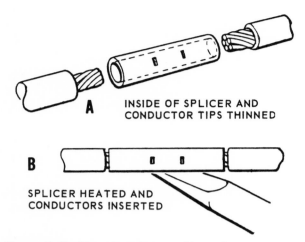

Figure 1-9. Steps in using a solder splicer.

the solder is still molten, pour it out, leaving the inner surfaces tinned. When the conductor tips are stripped, the length of exposed strands should be long enough so that the insulation butts against the splicer when the conductors are tinned and fully inserted. When heat is applied to the connection and the solder melts, excess solder will be squeezed out through the vents. Clean this away. After the splice has cooled, wrap or tie insulating material over the joint.

Solder terminal lugs. In addition to being joined or spliced to one another, conductors are often connected to other objects, such as motors and switches. Because this is the spot where a length of conductor ends (terminates), such connections are referred to as *terminal points.* In some cases, it is allowable to bend the

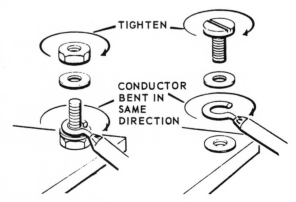

Figure 1-10. Conductor terminal connection.

end of the conductor into a small "eye" and put it around a terminal binding post. Where a mounting screw is used, the screw is passed through the eye. The conductor tip which forms the eye should be bent as shown in Fig. 1-10. Note that when the screw or binding nut is tightened, it also tends to tighten the conductor eye.

This method of connection is sometimes not desirable. When design requirements are more rigid, terminal connections are made by using special hardware devices called *terminal lugs.* They come in many sizes and shapes, but all are essentially the same as the type shown in Fig. 1-11.

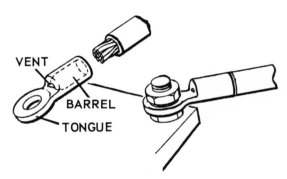

Figure 1-11. Solder-type terminal lug.

Each type of lug has a barrel (sleeve) which is wedged, crimped, or soldered to its conductor. There is also a tongue with a hold or slot in it to receive the terminal post or screw. When mounting a solder-type terminal lug to a conductor, first tin the inside of the barrel. Also strip and tin the conductor tip, then insert it in the preheated lug. When mounted, the conductor insulation should butt against the lug barrel, so that there is no exposed conductor.

Solderless Connectors

Where there will be no strain on the wire, a quick and satisfactory method for joining wires is simply to use convenient solderless connectors (or wire nuts). These connectors—as the name implies—do not require solder. Solderless con-

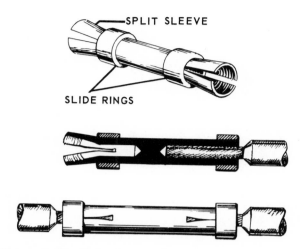

Figure 1-12. Split-sleeve splicer.

nectors are attached to their conductors by means of several different devices. Four of the most common types of solderless connectors, classified according to their methods of mounting, are the *split-sleeve*, *split-tapered-sleeve*, and *crimp-on* splicers and the *wire nut*.

To connect a *split-sleeve splicer* (Fig. 1-12) to its conductor, first insert the stripped wire tip between the split-sleeve jaws. Using a tool designed for that purpose, force the slide ring toward the end of the sleeve. Close the sleeve jaws tightly on the conductor, and the slide ring holds them securely.

A cross-sectional view of a *split-tapered-sleeve splicer* is shown in Fig. 1-13. To mount this type of splicer, strip the conductor and insert it

in the split-tapered sleeve. Turn or screw the threaded sleeve into the tapered bore of the body. As the sleeve is turned in, the split segments are squeezed tightly around the conductor by the narrowing bore. The finished splice must be covered with insulation.

The *crimp-on splicer* is the simplest of the ones discussed. The type shown in Fig. 1-14 is pre-insulated, though uninsulated types are manufactured. These splicers are mounted with a special plierlike hand-crimping tool (see p. 9) designed for that purpose. The stripped conductor tips are inserted in the splicer, which is then squeezed tightly closed. The insulating sleeve grips the outer insulated conductor, and the metallic internal splicer grips the bare conductor strands.

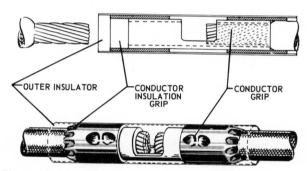

Figure 1-14. Crimp-on splicer.

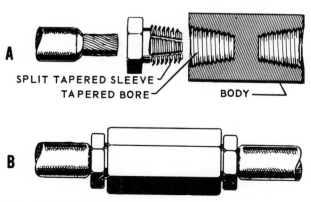

Figure 1-13. Split-tapered-sleeve splice.

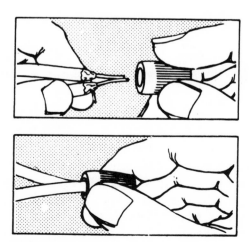

Figure 1-15. Method of installing a wire nut.

If there will be no strain on the wire, *wire nuts* may be used. Figure 1-15 shows how the wires are inserted into the plastic cap, which is screwed tight. Be sure no bare wire is left exposed.

Solderless terminal lugs. Solderless terminal lugs are used more widely than solder terminal lugs. They afford adequate electric contact, plus great mechanical strength. In addition, they are easier to attach correctly because they are free of the most common problems of solder terminal lugs, such as cold solder joints and burned insulation. These lugs come in many sizes and shapes, each intended for service with wires of different size. Only a few are discussed here. Classified according to their method of mounting, they are the split-tapered-sleeve (wedge), split-tapered-sleeve (threaded), and crimp-on types.

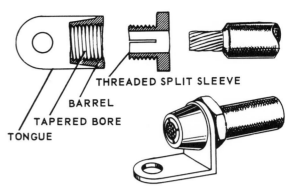

Figure 1-17. Split-tapered-sleeve terminal lug (threaded).

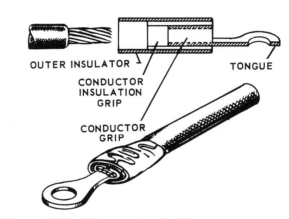

Figure 1-18. Crimp-on terminal lug.

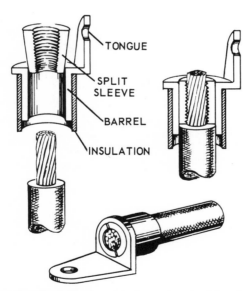

Figure 1-16. Split-tapered-sleeve terminal lug (wedge type).

The *split-tapered-sleeve terminal lug* (Fig. 1-16) is commonly referred to as a "wedge-on" because of the manner in which it is secured to a conductor. The stripped conductor is inserted through the hole in the split sleeve. When the sleeve is forced or "wedged" down into the barrel, its tapered segments are squeezed tightly around the conductor.

The *threaded split-tapered terminal lug* (Fig. 1-17) is attached to a wire in exactly the same manner as a split-sleeve splicer. The segments of the threaded split sleeve squeeze tightly around the conductor as it is turned into the tapered bore of the barrel. For this reason, the lug is commonly referred to as a "screw wedge."

The *crimp-on lug* is simply squeezed or "crimped" tightly onto a conductor by means of the same tool used with the crimp-on splicer. The lug shown in Fig. 1-18 is preinsulated, but uninsulated types are manufactured. When mounted, both the conductor and its insulation are gripped by the lug.

Taping a Splice

The final step in completing a splice or joint is the placing of insulation over the bare wire.

RING TYPE HOOK TYPE FORK TYPE

Figure 1-19. Crimping tool and other solderless types of terminals.

The insulation should be of the same basic substance as the original insulation. In the past a rubber splicing compound was usually employed. However, plastic electrical tape has come into wide use in recent years. It has certain advantages over rubber and friction tape. For example, it will withstand higher voltages for a given thickness. Single thin layers of certain commercially available plastic tape will stand several thousand volts without breaking down. However, to provide an extra margin of safety, several layers are usually wound over the splice. Because the tape is very thin, the extra layers add only a very small amount of bulk; at the same time, the added protection, normally furnished by friction tape, is provided by the additional layers of plastic tape. In the choice of plastic tape, expense must be balanced against the other factors involved.

Figure 1-20 shows the correct way to cover a splice with tape. Apply the splicing tape smoothly and under tension so that there will be no air spaces between the layers. In putting on the first layer, start near the middle of the joint instead of the end. The diameter of the completed insulated joint should be somewhat greater than the overall diameter of the original cable, including the insulation.

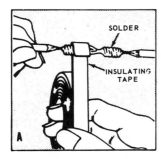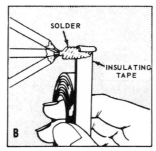

Figure 1-20. (A) Taping a soldered joint. (B) Taping a bunch splice.

CHECKING AND REPLACING LINE CORDS

Line cords of major appliances and cordsets of small appliances are major causes of problems with appliances. This is especially true of the latter; the cordset suffers more abuse than any other part of a small appliance. In either case, however, if no power is getting to an appliance, it is not going to operate.

In Chap. 7 of *Basics of Electric Appliance Servicing*, we covered current-carrying capacities of wire and the various types of insulation employed to protect wire. When it is neccessary to replace an appliance cordset or line cord, it is very important to select a replacement of proper size and insulation for its intended application. If the service manual gives such information or if the original cord has an identifying letter code, be so guided when installing a new cord. Otherwise use the chart below to determine suitable insulation.

Small kitchen appliances	
Knife sharpener, juicer, can opener	SPT-2
Food mixer	SJ, SV
Blender	SJ, SV, SVI

Large appliances	
Washing machine	SJ
Range, clothes dryer	SRDT
Vacuum cleaner, floor polisher	SV, SVT
Air conditioner	S, STP-3

Shop and garden tools	
Heavy-duty power tools	S
Lawn mower, hedge clipper	SJT, ST
Trouble light	SJT, SJO
Miscellaneous shop and garage tools	SJ, SV, SJT, SVT, SJO, SO

(continued)

Heating appliances	
Toaster, coffee maker, fry pan, waffle iron, grill, broiler, portable heater, soldering iron	HPN
Pressing iron	HPD

Miscellaneous appliances	
Hair dryer, portable humidifier, toy transformer	SPT-2
Radio, miscellaneous small appliances	SP-1, SPT-1
Lamp, portable fan	SP-1, SPT-1, SPT-2
Movie projector	SJ, SJT, SVT
Tape recorder	SJT, SVT
Rubber jackets	S, SJ, SP-1, SV
Plastic jackets	SJT, SPT-1, SPT-2, SPT-3, SRDT, ST, SVT
Neoprene jackets	SJO, SO, HPN
Asbestos cotton braid	HPD

There is a great deal of confusion among service technicians as to the code designations used to identify wire insulation. The chart below will help to clarify this.

Most small-motor appliances will operate on either 2-16 or 2-18 cord. These numbers mean that there are two wires, each being size No. 16 or No. 18. (A three-wire cord of the same size would be identified as either a 3-16 or 3-18 cord.) To find the wattage of appliances drawing heavier current such as irons, portable heaters, toasters, and frying pans, check the rating plate fastened to the unit or its motor. It usually reads like this: "120 V ac, 550 W" or "1,375 W, 120 V ac." The table at the top of p. 11 gives the abilities of different cords to carry current (two- or three-wire cord).

The scourge of every service technician is the extension cord. An overloaded cord will overheat, waste current, and often fail to throw

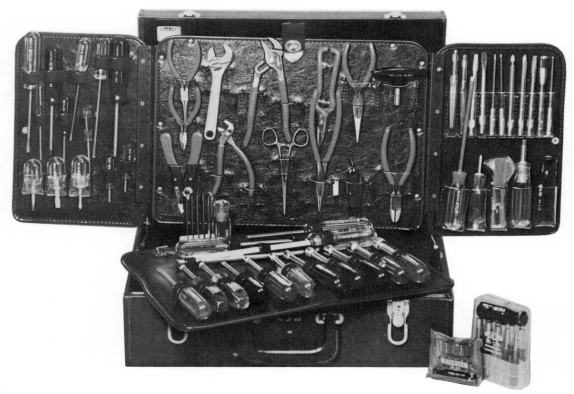

Figure 1-21. Typical tools used by an appliance technician.

Wire size	Type	Normal load	Capacity load
No. 18	S, SJ, SJT, or POSJ	5.0 A (600 W)	7 A (840 W)
No. 16	S, SJ, SJT, or POSJ	8.3 A (1,000 W)	10 A (1,200 W)
No. 14	S	12.5 A (1,500 W)	15 A (1,800 W)
No. 12	S	16.6 A (1,900 W)	20 A (2,400 W)
No. 10	S	27.5 A (2,400 W)	30 A (3,000 W)

Note: As a safety precaution for your customer, be sure to use only replacement cords which are listed by Underwriters' Laboratories, Inc.

Code Designations for Insulation and Jacket Materials

R, Rubber; N, Neoprene; P, Plastic; A, Asbestos.

Cord types*	Conductor insulation	Jacket
General purpose		
SP, SRD	R	...
SPT, SRDT	P	...
S, SV, SJ	R	...
SO, SJO, SVO	R	N
SO, SJO, SVO, SJTO, SVT, SVTO, SVHT	P	P
Heater		
HPN	N	...
HPD	R & A	Yarn braid
HS, HSJ	R & A	R
HSO, HSJO	R & A	N
Tinsel		
TP	R	...
TPT	P	...
TS	R	R
TST	P	P

* SP and SPT have appended numerical codes designating the insulation thicknesses. For example, SP-1 has the thinnest insulation, SP-2 is intermediate, SP-3 thickest.

Note: A code *ending* with the letter T indicates a thermoplastic material, usually vinyl. If the T is absent, the covering is of rubber, or rubber topped with neoprene (or other approved alternate material such as chlorosulfonated polyethylene). A code *beginning* with the letter T identifies so-called "tinsel" construction; these cords have spirally wound copper ribbons instead of wire strands in order to impart extreme flexibility for such applications as with razors. The letter O designates resistance to oil. For example, SJ is rubber, SJO is oil-resistant rubber, and SJTO is oil-resistant plastic.

enough power for the appliance to operate at top efficiency. For this reason, always warn your customers against misuse of extension cords. Manufacturers of some major appliances specify that extension cords should not be used with their products. But when extension cords are needed, use the following table in selecting the length and type.

Light Load (to 7 A)	Medium Load (7–10 A)
To 25 ft, use No. 18	To 25 ft, use No. 16
To 50 ft, use No. 16	To 50 ft, use No. 14
To 100 ft, use No. 14	To 100 ft, use No. 12

Heavy Load (10–15 A)
To   25 ft, use No. 14
To   50 ft, use No. 12
To 100 ft, use No. 10

Continuity Testing

All appliance circuits must have a complete "electrical path" or continuity in order to operate. For example, in a typical small appliance, the current comes in at one side of the plug, goes through the wiring, switch, motor and/or heating element, and then back to the other side of the plug. If there is a break in the circuit or the switch is turned off, the appliance will not operate. Therefore, when an appliance is completely dead, the first check to make is for continuity.

As was stated in Chap. 8, *Basics of Electric Appliance Servicing,* several devices or instruments are available for checking continuity.

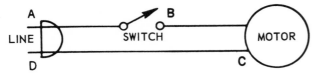

Figure 1-22. Checking for continuity of a line cord with an ohmmeter.

While all will do a good job, the fastest and most accurate is an ohmmeter. To use an ohmmeter to check an appliance such as the fan shown in the simple schematic diagram, Fig. 1-22, first adjust the needle for zero reading when the test probes are brought together. Under normal conditions and with the switch turned on, there should be continuity (or a reading of very low resistance) when the test probes touch the two prongs of the line plug (points *A* and *D*). If there is no continuity, there is an open circuit.

To identify the faulty component, proceed with the continuity test as follows. Leave one probe on the prong of the plug (point *A*) and touch the other probe to the far side of the switch (point *B*). Then turn the switch on. If the meter goes to zero reading, the switch is good. Now move the test from point *B* to point *C*, which is the other side of the fan motor. A low-resistance reading here means that the windings of the motor are all right. Next, move the probe from point *C* to point *D*. If you get no reading, the trouble is that one side of the line cord is broken. Once the line cord is replaced, the fan should operate.

Continuity checking is a simple task. But when using an ohmmeter or any other continuity tester with its own current supply, remember that the test must be performed with the appliance completely disconnected from the power source.

Line cords and cordsets should also be given a "hi-pot" continuity and short-circuit check. Full details of how to conduct these tests are given in Chap. 8, *Basics of Electric Appliance Servicing.* But when undertaking any hi-pot test, be sure that this procedure is recommended by the maker in its service manual. Some small-appliance manufacturers do not suggest the hi-pot test, while others recommend that their authorized service technician "hi-pot" every appliance after repairing it, regardless of the trouble and kind of repair job that was done. Also keep in mind when using a hi-pot tester that extreme caution should be exercised since some of these testers are capable of delivering lethal currents.

Line Plug Replacement

Among any service technician's customers there are undoubtedly some "cord jerkers." Yanking a cord out of a wall plug by force, rather than grasping the plug itself, can loosen connections or even break the wires. Rough handling is probably the greatest single cause of extension cord mortality. Of course, damaged cords are fire hazards. However, from the standpoint of good workmanship and economy, and to ensure long, uninterrupted service from a repaired appliance, replacement of the entire cordset is preferred to repairing—except possibly for the replacement of the attachment plug or some such minor service. But inasmuch as you will be called upon to disconnect and reconnect many cordsets in the course of your work, this text would hardly be complete without instructions for the professional handling of this vital subassembly. And you must not conclude that, because many do-it-yourselfers repair and install their own appliance cords, cord servicing is something anyone can do and is a subject to be passed over lightly. True, many do-it-yourselfers do repair their own—some with disastrous consequences. Suffice it to say, therefore, that an amateurish cord repair or installation can render an appliance just as inoperative—and just as deadly—as any other incompetent servicing.

For major appliances, round conventional attachment plugs are usually best since they provide enough space to allow the use of the "Underwriters' knot" to help secure the wires. Here is the step-by-step procedure of installing

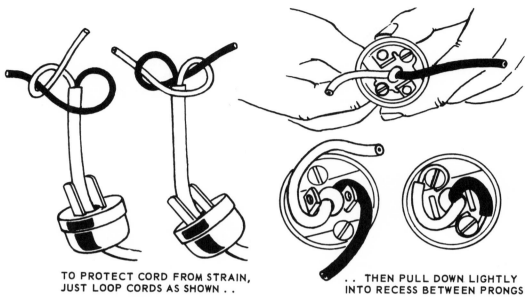

TO PROTECT CORD FROM STRAIN,
JUST LOOP CORDS AS SHOWN . .

. . THEN PULL DOWN LIGHTLY
INTO RECESS BETWEEN PRONGS

Figure 1-23. How to make the Underwriters' knot.

this type of plug (Fig. 1-23):

1. Cut off the old plug or cut off the end of cord straight above the worn or frayed part. Thread the cord through the hole in the plug.
2. Split the outside covering $2\frac{1}{2}$ in down from the end of the cord.
3. Cut off the outside covering at the end of the split. This will expose two separate wires that are insulated.
4. Tie an Underwriters' knot. To do this, hold the wire in the left hand with the two separate ends up and across in the front of the left-hand wire. Loop the left-hand wire toward you, down in front and under the right-hand wire, then back of both wires, and through the loop formed in the right-hand wire. Place the two ends together and pull to a firm knot at the end of the separated wires. The more insulation on each wire, the better. Pull the knot down into the well inside the plug. The purpose of this knot is to fit into the depression in the cap so that if the cord is pulled, the strain is on the knot and on the plug instead of on the wire

around the screws. When the strain is on the wires around the screws, the wires often break or pull away from the screw posts, causing short circuits.

5. Scrape insulation from each set of wires back 1 in from the end, being careful not to cut or break any of the fine wires. Twist the strands of each set firmly. Do not uncover the loose ends entirely, only enough to wrap around the screw or contact posts. Usually 1 in is enough.
6. Lead the wires around the flat metal prongs to form an S. Loop the bare copper ends around screw posts in a clockwise direction.
7. Hold the wires firmly in place and tighten the screws. Be sure that all fine wires are smooth and together, with no ends protruding that could cause unwanted contact.
8. Place a paper disk insulator over the prongs and press into place.

When the three-wire type of cable is used, the black and white wires go to the regular flat prongs, while the green ground goes to the round prong. Full details on this type of plug are found on p. 138, *Basics of Electric Appliance Servicing.*

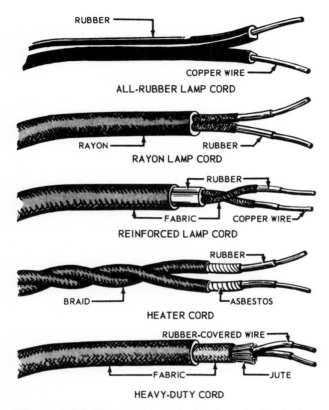

Figure 1-24. Types and methods of method stripping various cords.

For small appliances, the snap-on plugs may be suitable. These plugs are equipped with sharp prongs which, when closed on the cord, bite right through the covering and insulation and make a firm connection with the wires.

While they are seldom used today, fuse plugs are found on some of the older heating-type small appliances. These "fusible" plugs give added protection, of course, to the appliance; the size of the fuse is determined by the operating wattage of the unit. The major problem that a service technician has with this type of plug is a blown fuse. To correct this trouble, just replace the blown cartridge fuse with a good one of the same size. Incidentally, the fuse size can usually be determined by dividing the wattage shown on the rating plate by 100. For example, a 600-W appliance uses a 6-A fuse, a 1,000-W unit a 10-A fuse. Fusible plugs are usually referred to as "3AG-type"; a 10-A fuse would be designated 3AG-10.

Connecting the line cord to the appliance. The electric connections at the appliance end of the line cord may be made to terminal screws on an insulating strip, to solder lugs on a terminal strip, under nuts on existing bolts, etc. Some heating-type appliances are connected with special appliance plugs, but every electric connection in an appliance must be tight. A loose connection not only impairs the efficiency of an appliance, but also is the forerunner of more serious trouble. Indeed, a loose connection in the terminal enclosure of an iron, for example, can develop enough excess heat to burn off a terminal in a short time. Make it a habit, therefore, to check all accessible connections when you service an appliance—no matter what your specific task happens to be. In fact, many experienced service technicians loosen screw or bolt connections at electric terminals and then retighten them. This procedure helps to break up any oxidation that may have formed at the connections and reduces resistance at these points. Remember that resistance at electric connections—no matter how small—can cause problems. Resistance causes the connection to heat up—especially in resistance-heating appliances such as toasters, grills, roasters, and irons—and this heat creates further oxidation and more resistance. This vicious cycle will continue until there is a breakdown at the terminals.

In order to ensure that no "wild" strands of wire protrude from a connecting terminal, strip and trim each conductor to fit the connecting device precisely so that every strand is caught in the connecting clamp and the insulation butts right up to the binding screw, for one "wild" strand could cause a ground.

When untinned stranded wires are to be fastened directly under the head of a binding screw, make one complete right-hand turn of the conductor under the screwhead—no more, no less—and cross the free end of the wire over the starting point at a right angle. Less than one full turn would cause the strands to spread and the connection would soon loosen; more than a full turn would create a windlass effect, and you would break more than half the strands before you could get the connection tight. It is a convenience, however, to strip a little more wire than is actually needed so that it is possible to retain the hold on the free end of the wire as the screw is tightened. That is the quickest method of attachment with no risk of splaying the strands. After tightening the connection, nip off the surplus wire close to the binding screw with a diagonal pliers.

Where a binding-screw terminal is provided with special circular washers which are intended to hold the strands together, make one full turn of the wire, but in this type do not cross the free end over the starting point. Instead, form the bared tip of the wire into a U and then bring the free end as close to the starting point as is necessary to fit the cavity in the terminal washer.

Do not attempt to improve on a straight-tip pressure connection by making a circular turn of the wire under either the screwhead or the special washer. Rather, strip exactly the right length of wire to fill the channel in the terminal or its washer and tighten the screw securely. If the channeled washer or its terminal is distorted and will not grip the wire firmly, replace the necessary parts to ensure a tight connection.

It is most important to make sure that there is an anchorage for the cord so that no strain will be put on the electric connections. That is, the cord must be made mechanically secure at some point ahead of the electric connections. The most common cord-holding devices are the so-called "strain reliefers." With these, the cord goes through the two halves of the plastic or metal device. By squeezing the halves together with a pair of pliers, you can slip the smaller end through the hole in the appliance case. A notch in the strain relief catches on the case itself and holds the cord firmly in place.

In some appliances, the cord is protected against strain by being held in place against the case on the inside with a small metal clamp or clip. Whenever a cord goes through a sharp-edged hole in an appliance with a sheet metal case, a plastic or rubber grommet should be used. When replacing a cord, always be sure to replace the strain-relief devices and any grommets, if employed.

Appliance plugs. These heavy-duty plugs are used with some heat-type appliances such as table ovens, electric skillets, and waffle irons. Some plugs of this type have built-in switches and thermostats; in other words, the plug contains the control device for the appliance.

Because most units that use appliance plugs operate on fairly high wattage (1,000 W or more), the terminals, or "shoes" as they are often called, must be heavy and are designed for a special cord known as a *heater type*. The wires making up this cord usually fit in grooves on the inside of each half of the plug shell. When the two halves are put together, they form channels so that the wires cannot get together. The ends of

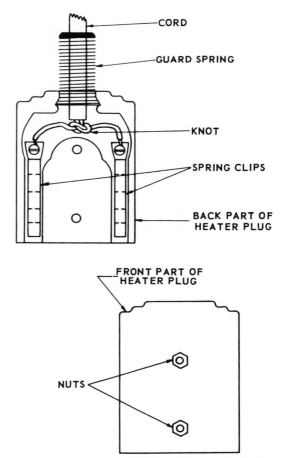

Figure 1-25. Open and closed views of an appliance (heater) plug.

the wires are then fastened to the terminal or shoe screws on the end of each contact of the plug, thermostat, switch, or heat probe. (More on the latter plug can be found on p. 186.)

Because of the various kinds of appliance plugs, it is difficult to describe how they are assembled—but most are very simple. To replace a defective appliance plug, remove the wires from the contact terminals or shoes and cut off all bared wire. Open the two halves of the new plug, remove the insulation from the wires, and attach these wires to the new contact shoes. Be careful to remove just enough of the asbestos insulation to attach the wires around the screws on the contact terminals or shoes. When doing this, remember that the fabric outer sheath of heater cord is readily removed by unraveling with an awl. Begin unraveling at the end and work back a little at a time as far as is required, then trim the ragged ends of the outer braid with a pair of scissors. Now strip the tips of the individual conductors and attach eyelets if required, or, if straight wire tips are to be used, twist the strands tightly so that they will hold together well. If you are preparing the cord for connection in a heating-appliance terminal enclosure, bear in mind that because of the extreme heat at this point no friction tape should be used for insulation on this end of the cord. (One manufacturer applies a small band of friction tape to the outer braid of the cord as padding under a metal strain-relief clamp, but drying out of the tape in this instance, of course, can do no harm.) Instead of tape, asbestos string should be used to bind the fuzzy asbestos insulation of the separate

conductors as well as the ragged ends of the outer braid of the cord.

Here is an easy way to lace the end of the cord neatly. After the outer braid has been cut back the required length and the tip ends of the wires have been properly stripped, form this prepared end of the cord into a Y. Begin with the asbestos string or a fine wire at the bottom of the Y by whipping down the ragged ends of the outer braid; then, using a coarse spiral overlay, bind the fuzzy asbestos insulation of one separate conductor almost to the end of the insulation. Make two or three turns at this end and return, making a similar overlay in the opposite direction to the bottom of the Y. Repeat the process on the second individual conductor, finishing at the bottom of the Y where, after making a turn or two, you can tie the ends of the asbestos string or fine wire together. When properly done, the individual asbestos insulated wires will be whipped with a crisscross effect.

DISASSEMBLY AND REASSEMBLY

Disassembly of an appliance sometimes becomes a match of wits between the service technician and manufacturer. This is especially true in the case of small appliances. In fact, small-appliance technicians working on appliances with which they are not totally familiar sometimes must be part detective and part genius to locate the secret latch or clip that holds some units together. As a general rule, manufacturers of major appliances usually provide fairly complete disassembly and reassembly instructions. Unfortunately, however, this is not always true with small appliances.

To disassemble a small appliance, first look for the fasteners used to hold the case together. Today, most units have cases made of two or more pieces of sheet metal or plastic, and they are generally held together by small bolts or screws, usually the self-tapping type. While some standard slotted screws are employed, most are of the Phillips or hex-head type, and most are

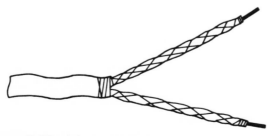

Figure 1-26. Asbestos insulation wrapped with heavy thread or thin wire.

located where only the right tool will get to them. In addition, there has seemingly been a growing trend by manufacturers in recent years to hide the appliance's disassembly screws under nameplates or decorative trim. But since most nameplates and escutcheons are of the "glued-on" type, it is wise to make sure that there really are screws beneath them before tearing them off. One way of doing this is to use a sharp-pointed instrument such as an ice pick to make a small hole in the escutcheon so that you can see underneath. If you must pull off a plate or escutcheon, obtain a new one from the manufacturer or parts supplier and fasten it on after you make the repair.

Some chromium-plated escutcheon plates are still in use. To protect their finish, slip a piece of soft cardboard between the surface and the shank of the tool. And since it does not matter what type of wrench is used—box or open end—on cap screws or nuts which are tightened against a finished surface, slide a cardboard washer over the screwhead or nut so that the jaw or the face of the wrench will not scar the finish as it is turned.

A few small appliances are riveted together or use rivets to hold control plates in place. When small hollow brass rivets are used, remove them by carefully filing one side flush. Then push the rivet out on the other side. When solid rivets are employed, use a drill and punch to remove them. Select a drill the same size as the rivet and then drill off the rivet head. To complete the task, use a pin punch with a flat end exactly the size of the rivet shank to drive the rivet through the metal or plastic. By the way, if the case is sheet metal, never use a center or pointed punch; it will expand the shank of the rivet and thus enlarge the hole. This will make it necessary to replace the old rivet with a larger one, which can be very unsightly.

When reassembling some small appliances that were riveted together, it is sometimes possible to use small screws and nuts rather than rivets. This method is usually easier. But when it is necessary to employ replacement rivets, use only a flat-headed punch or a smooth-faced hammer.

To peen the end of a rivet properly, use a series of light taps rather than a few heavy wallops for a much better looking and more durable job. Employ the same method when it is necessary to resort to a flat-end punch to reach a rivet: light taps until the rivet shank is peened over and held securely in place.

Some assembly methods may test your ingenuity the first time you encounter them, but after that there will be no trouble at all. For example, when a part appears to have no more screws holding it in place yet will not come free, gentle upward, downward, or sideward pressure may turn up a hidden catch of some type. A few appliances have spring catches that make it necessary to push a pointed tool through a small hole or crevice to release some parts. Another assembly method that gives service technicians problems is the so-called "hook and rivet" arrangement. With this, and all new and strange disassembly situations, always "take it easy." All too frequently when the disassembly is rushed, the result is a pile of unidentified parts. For this reason, when disassembling an unfamiliar appliance unit, always make rough sketches of the part locations. Also mark the internal parts with a light scratch and a mating scratch on another part of the assembly. Many parts can be reassembled in two ways (one right and one wrong), and the light mark will quickly lead to the correct one. When disassembling an appliance, keep all the parts in one place, or perhaps in a small box. In this way, you will not have to hunt for the parts when you reassemble the item.

When disassembling an unfamiliar appliance, do not use an excessive amount of force. If you apply too much force, the usual result is the bending of some part out of shape or even breaking it. On the few occasions when it is absolutely necessary to apply force, be sure to do it in just the right manner and at the right place to prevent damage to the other parts. A plastic hammer is a good tool for any service technician to have available. It can frequently be employed to decouple parts that would be severely damaged by the use of a steel hammer.

There are times during a disassembly when you may run into an excessively tight or a damaged screw. Of course, the best way to avoid difficulty with tight screws is to use the proper screwdriver for the occasion. For example, for slotted screws, which are more often a source of trouble, use the largest screwdriver that will fill the slot. If the angle is too short or if ground too sharp, the blade may come out of the slot and round off the edges, making the task even more difficult. Sometimes it may be necessary to grind the screwdriver blade to an exact size to remove the screw.

If the screw slot is damaged, it often can be reshaped enough to allow for the removal of the screw. Use a proper-sized pattern file to do the job. Once the screw has been removed, throw it away. Never reuse a badly damaged screw or bolt.

Here are three aids that may be used to remove a tight screw; select one or more of them as necessary.

1. Apply penetrating oil and rust solvent to affected screw or bolt. Allow the chemical solvent sufficient time to work its way in and then loosen the screw.
2. Try tightening the screw before attempting to loosen it.
3. If there is no risk of damaging other parts, strike the screw a sharp blow squarely on its head; if it is a slotted-type screw, hold a solid shank screwdriver in the slot and hit the screwdriver with a plastic hammer. If the head of a fillister-head screw breaks in half or the slot is damaged, the screw frequently can be removed with a small pipe wrench. Hexagonal and square-headed screws seldom present a problem, since good leverage can usually be obtained on these with a box-type or socket wrench without damaging the head.

When a screwhead is broken completely off, or when it is so badly damaged that normal removal is impossible, the best procedure is to drill a pilot hole in the damaged screw and attempt to remove it with a screw extractor. This tool—resembling a left-hand-thread tap but with coarse threads like a drive screw—may be tightened counterclockwise into a previously drilled pilot hole in a broken-off screw stump to facilitate its removal. Drill a hole through the top of the screw or bolt, and screw the extractor into it. As you turn the extractor, its left-handed thread forces it down into the hole, so that the screw can be turned with it and thus be extracted.

If there is not enough space to use an extractor, or one is not available, select a drill that is slightly smaller in diameter than the screw. Carefully center punch the top of the broken screw and drill straight down into the screw. When properly and carefully done, this will cut out all the defective screw except the thread portion, leaving the threads in the hole almost undamaged. The hole can be rethreaded if damaged by the drill. A damaged nut and bolt in an unthreaded hole can usually be removed by chiseling or grinding off the head of the bolt and driving the shank through the hole with a flat-ended punch.

Breaks in plastic appliance cases frequently can be repaired with epoxy resin cement. Apply this material as its container directs. Small cracks in the thermoplastic cases can be repaired with a hot soldering iron. Hold the cracked portions of the case tightly together and run the tip of the iron along the crack on the inside of the case. (This type of plastic will soften.) Then push the edges back together so that they will harden again when cool.

When a bad break occurs to a plastic case and a replacement is not available or feasible, cut a strip of thin fiberglass cloth and spread the epoxy resin the length of the crack and about an inch wide. Lay the cloth over the crack and press it down firmly. Brush on more cement, smooth it out, and let it set at least overnight before checking the repair for strength.

When reassembling appliances with plastic cases, make sure not to run them up too tightly. Since self-tapping screws are usually used in plastic cases and they cut their own threads, it is easy to strip them if you use too much force. Once they are stripped, it is difficult to make them hold again. When this occurs, frequently it is best to use a larger screw, or to fill the hole with plastic cement and run the screw in lightly. Let this set overnight and then tighten the screw.

Most reassemblies are performed in the reverse of the disassemblies. But when making a reassembly, it is important to check that all electric connections are tight. Also make certain that no electrical wires or cords are pinched. Frequently the wiring must be placed in specific channels or otherwise the case cannot be put back together. As in disassembly, force is seldom required to reassemble an appliance.

At the beginning of this section, we stated that some manufacturers publish excellent small-appliance manuals. When instructions are given, follow them carefully. Who knows better what the appliance is supposed to do than the manufacturer who made it? To illustrate how to follow a "good" set of disassembly and reassembly advice, steps for disassembly and reassembly of a typical electric can opener follow (Fig. 1-27).

Disassembly

To disassemble the case

1. Withdraw the cutter screw; remove the cutter and the cutter spring.
2. Remove the handle screw and the handle.
3. Withdraw the three backplate screws and remove the strain relief with a pair of needle-nose pliers. Remove the backplate and the spacers.
4. Withdraw the case screws and the motor-mounting screws.
5. Lift the motor enough that those parts clear the holes in the case and then move the clamping lever down to the bottom of its slot. Remove the motor assembly from the case. *Note*: The mounting bracket may be stuck to the case because of the fire-retardant point. It will be necessary to pry up on the motor with a little force to break the bond.
6. Withdraw the switch screws. Remove the switch blade, the spring, and the switch plate.

To disassemble the motor assembly

1. Grip the driver with pliers, using paper or cloth to cushion the jaws, and back it off from its shaft. Remove the shim washers and back off the thrust nut and washer.
2. Withdraw the two front bearing screws; lift off the cordset guard and the front bearing strap. Remove the driver shaft assembly.
3. Withdraw the two rear bearing screws; lift off the rear bearing strap. Remove the brush-holder assemblies, the two brushes, and the two brush springs.
4. Spring the field strap and remove the armature, field assembly, front bearing, and actuator.
5. Withdraw the two governor support screws and remove the governor support assembly and the mounting bracket.
6. Lift off the cutter plate. This may be somewhat difficult because the stud which protrudes through the cutter plate may have a rough shoulder edge. It may be necessary to sand or file this edge so the stud will clear the plate hole.
7. Disengage the clamping-lever spring and remove the clamping-lever assembly.

Reassembly

1. The limits on armature end play are 0.001 to 0.004 in. If it requires adjustment, be sure the thrust bearing is turned so that the screw bites a new place in the bearing.
2. After replacing the carbon brushes, run the can opener for approximately 20 min to wear in the new brushes.
3. When mounting the motor assembly into the case, make certain the slide plate on the clamping lever is above the switch blade. When the clamping lever is depressed, the slide plate moves down with the lever. When the clamping lever is approximately one-third depressed, the phenolic insulator on the slide plate strikes the switch blade and forces it down until the contacts close. (The clamping lever should be approximately two-thirds depressed at the time the contacts close.) As the clamping lever is further depressed toward the bottom, the slide plate is forced to slide up on the lever. The slide plate is automatically "cocked" to its original position when the clamping

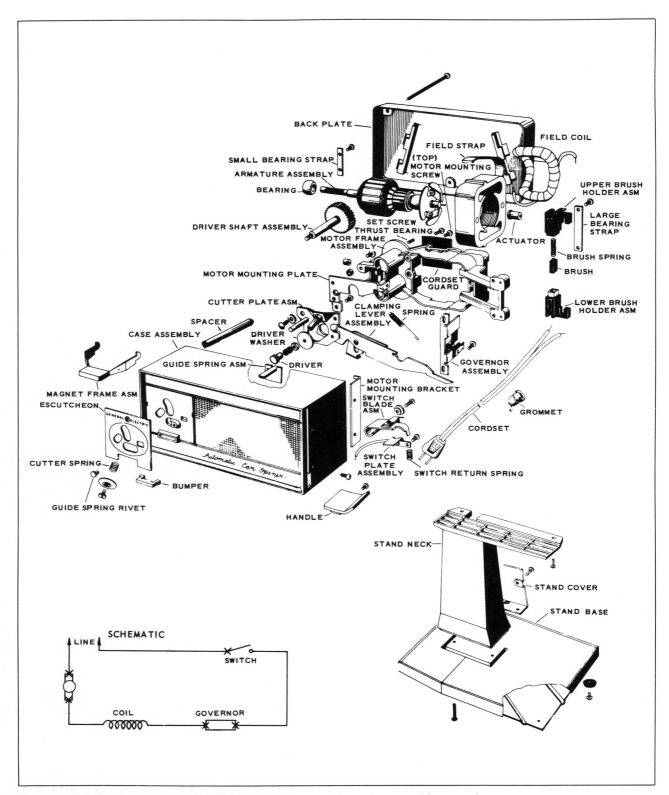

Figure 1-27. The can opener described in the disassembly and reassembly procedures.

lever is returned to its uppermost position.

4. To install a new escutcheon, clean the surface of the case with isopropyl alcohol (rubbing alcohol). Peel off the protective film and press the escutcheon firmly in place, paying particular attention to the edges and corners. This escutcheon does not require soaking or activating.

5. The cutter must be assembled with the curved face toward the escutcheon.

6. The gear on supply-part driver-shaft assemblies must be correctly positioned on the shaft after the assembly is installed in the unit. When one requires replacement, follow these steps:

   a. Install the driver-shaft assembly as received. Place the driver washer and driver nut on the shaft and tighten firmly.

   b. Hold the motor assembly in a horizontal position with a firm support under the threaded end of the driver shaft. Place a bushing with a minimum $\frac{1}{2}$-in hole on top of the gear around the knurled section of the driver shaft and with a drill press or light arbor press push the gear onto the shaft so there is no end play. If neither a drill nor an arbor press is available, the gear may be tapped on, but care must be taken to drive it on evenly.

   c. Holding the motor in the hand, tap the end of the drive shaft where it protrudes through the gear until there is perceptible end play, not to exceed 0.004 in.

TROUBLE DIAGNOSIS

In this chapter, we are not going to discuss the various trouble diagnosis techniques. Those for small appliances are covered in the remaining chapters of this book and those for major appliances are fully covered in *Major Appliance Servicing* and *Refrigeration, Air Conditioning, Range and Oven Servicing*. But when servicing any appliance, the most important part of trouble diagnosis is to first determine exactly what the appliance is designed to do and how it is designed to perform. Then trouble diagnosis becomes a routine elimination of those components, parts, or problems that prevent the appliance from operating in the manner it was designed to do.

As mentioned in the other books of this series, almost all appliance manufacturers publish service manuals or bulletins. The more you use and read the manuals, the more familiar you will become with the appliance's operation. Another good source of service information is asking the customer about the behavior of the appliance. Some manufacturers even key their service information around customer complaints. The following is an example of an "analysis of complaints" for a typical electric grill taken from a manufacturer's service manual:

Analysis of complaints

1. *Waffles do not brown on top.* Check the cold resistance of both heating elements individually. Each element should measure between 5.00 and 5.20 Ω. If the elements are within tolerance, examine lead crimp connectors carefully for bad contact. The problem may be caused by the customer's technique in cooking waffles. Insufficient or too-thin batter or keeping the top open too long after batter is poured will cause the bottom to cook longer than the top.

2. *Waffles brown unevenly.* This may be due to grids that have been allowed to accumulate too much browned grease in some areas or to batter that is too thick to spread evenly over the entire grid.

3. *Waffles stick.* Check the grill for operation within the temperature limits specified as follows:

   a. The wattage will be 0 at OFF.

   b. The wattage in operating range will be 1,275 W ± 5 percent at 120 V.

   c. The temperature at the third or fourth cut-in will be 445°F ± 15°F with control set at DARK.

If the grill does not conform to these limits, replace the thermostat. Among other possible causes of this complaint are the following:

a. Improperly seasoned grids
b. Grids scoured and not reseasoned
c. Grill being opened before waffles are baked
d. Waffle mixes prepared without enough shortening
e. Special recipes with too much sugar
f. Insufficient preheat (batter poured before light goes out)

REPAIRING THE FINISH OF APPLIANCES

You can give nearly every appliance you recondition a more professional touch by washing it thoroughly, polishing its glossy trim, and touching up its minor blemishes. The modern paint finishes used on appliances are most durable. But should a surface become gouged, scuffed, or scratched, it can be restored to its original factory finish by the use of aerosol spray paints. Most appliance manufacturers and their jobbers carry in stock cans of aerosol paint for their own appliances. For small blemishes, they carry containers of touchup paint.

It must be remembered that a can of aerosol spray paint is, after all, a tool, and the proper use of any tool requires a little guidance and practice. Fortunately, the important elements of good touchup technique are few, and the skill needed to do a professional-looking job is easily gained.

Surface preparation. Proper preparation of the damaged area and the surrounding surface is of the greatest importance. Nothing contributes more to a first-rate touchup job.

Sand lightly with No. 400 or 10/0 wet-or-dry sandpaper and water or oil-free naphtha. Do not use a glue-sized sandpaper, as any traces of glue left behind will harm the finish. Sand along the scratch until it is gone, taking off no more paint than is necessary. Then feather outward for $\frac{1}{2}$ in on each side of the scratch, using a light rotary motion and tapering out to the full thickness of the paint. Keep the sanded area as small as possible.

For round spots, use a rotary sanding motion, blending out gradually to the full paint thickness as described above. Be careful to avoid cutting through to the bare metal, if at all possible. On edges and corners sand very lightly, as there is greater danger of exposing bare metal at these points.

When the area is thoroughly blended in and feels smooth, wipe it clean with a lint-free cloth moistened with oil-free naphtha or "leveler." All the surface that was sanded must be painted with the finish coat. Mask the area with paper and masking tape to avoid painting more than necessary.

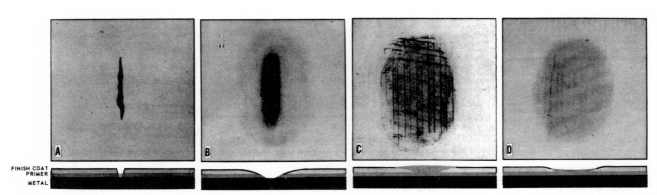

Figure 1-28. A typical scratch. (A) Lower view is a cross section, enlarged. (B) The scratch after sanding. Note feathering of the paint. (C) The scratch after applying the aerosol primer. (D) The primer has been sanded, the primer overspray removed.

When to use primer. If you happen to expose the bare metal in sanding out a deep scratch, and if the original finish has a prime coat, use an aerosol primer before proceeding with the finish coat. Spray on several light coats of primer (see spraying instructions given below under Applying the Finish). Allow 5 min drying time between coats. Wet-sand lightly with No. 400 or 10/0 sandpaper and water or oil-free naphtha. Remove any primer "overspray" outside the prepared area with rubbing compound. Wipe clean with a lint-free cloth moistened with oil-free naphtha and proceed with the touchup finish. Do not use leveler for this; it will soften the primer.

When to use putty. For deep dents and gouges where primer fails to build up the damaged area to the level of the surrounding surface, fill in the depression with pyroxylin putty. If the putty is not pliable enough, add lacquer thinner or leveler. Allow plenty of time for putty to dry—1 to 2 h air-drying or $\frac{1}{2}$ h with a heat lamp. Sand smooth with No. 400 or 10/0 wet-or-dry sandpaper. Apply at least one coat of primer over the putty. Do not use leveler to clean the puttied area before painting. It will soften both primer and putty.

For wrinkle finishes. Surface preparation for touching up wrinkle finishes is almost the same as for smooth. Scuff marks need not be prepared, except for thorough cleaning of the area with oil-free naphtha or leveler. Clean out gouges with No. 400 or 10/0 sandpaper and slightly bevel the edges. After thorough cleaning, fill in the gouge with pyroxylin putty. Stipple the soft putty with the fingertips or a stiff brush to imitate the look of the surrounding wrinkle finish. Allow 1 to 2 h for putty to air-dry, or $\frac{1}{2}$ h with a heat lamp. Then proceed with the finish coat. No primer is needed.

Applying the finish. The following three rules are the most important in aerosol touchup. Disregarding them is the most common cause of an unsatisfactory job.

1. For best results, the temperature of the spray paint should be about 70°F. If it is too cold, the pressure will be low and the

Figure 1-29. Method of shaking aerosol can.

paint will not atomize completely. If the can is chilled, soak it in hot tap water (no more than 120°F) for 5 min before using.

2. Shake the can well before using (Fig. 1-29). Perfect color match in aerosol touchup depends on very thorough mixing of the contents of the spray can. Shake the can back and forth until the agitator ball inside rattles. Then hold the can at the top and swirl the ball around the bottom V of the can for at least $\frac{1}{2}$ min to be sure of mixing in all the pigment.

3. Do not "freeze" on the valve. In other words, do not keep the valve open continuously as if you were spraying a room with insecticide. Aerosol painting calls for a series of short bursts. For touching up a small spot, open the valve for the shortest possible time. For larger areas, release your finger at the end of each stroke.

In addition to these three rules, some others are good to keep in mind, too. Do not work in a drafty area. Air currents can deflect the spray and carry it a considerable distance. Where possible, have the work surface horizontal to help prevent runs and sags. Open the valve completely, pressing it down all the way when spraying. Opening it part way reduces the pressure, prevents proper atomization of the paint, and causes droplets to form. Keep your fingertip out of the spray or droplets will form

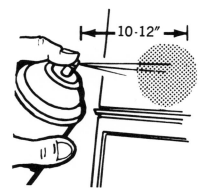

Figure 1-30. Proper distance between aerosol paint can and the work surface.

The secret of a perfect color match lies in two things: thorough agitation and mixing of the paint before use, and the application of a sufficient number of light mist coats to build up the correct color value. The most frequent problem encountered by a beginner with aerosol paints is the occurrence of runs, sags, or "curtains" (Fig. 1-32). This is caused by one or more of the following errors:

1. "Freezing" on the valve
2. Holding the can too close to the surface
3. Holding the can still or moving it too slowly

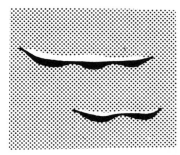

Figure 1-32. One of the major problems: sags.

and be blown onto the work surface. Hold the can 10 to 12 in from the work surface (Fig. 1-30). If you hold it too close, the propellant gas cannot completely escape and the paint will pile up and bubble. If you hold it too far away, the paint will not "hide" or cover properly. The spray will be too thin, and the results will be spotty.

Use short, dusting strokes, moving the can constantly at a steady rate parallel to the surface (Fig. 1-31). Shut off the valve at the end of each stroke. Do not try to cover with one coat, but apply repeated light coats until the color and finish blend into the surrounding area. Wait long enough between coats for the paint to become tacky. Lacquers will do this in a few minutes; enamels take a little longer.

Any of the above will cause too much paint to build up on the surface and create runs. It is very important always to practice painting on a piece of scrap material before starting the actual job. Get the "feel" of the spray can. See in what manner the color builds up. Make any mistakes where they will not matter.

Finishing touches. After an area has been touched up with lacquer, it should be over-sprayed with clear leveler to blend in the dry overspray ring. Be sure the finish coat has had time to set up (2 to 5 min) before spraying on the leveler. Apply it very lightly. Then allow 1 h for drying (or 30 min with a heat lamp). If any lacquer dust still remains on the area around the touchup, remove with rubbing compound. Finally, polish the entire area to a uniform luster with polishing wax. *Note:* The rubbing compound and the polishing wax must be white for both white and pastel finishes.

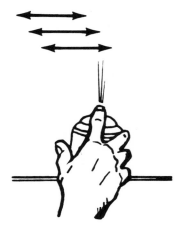

Figure 1-31. Proper method of painting stroke.

Figure 1-33. (Left) Typical dry overspray ring after touching up with lacquer. Blend with clear "leveler." (Right) After polishing wax—a perfect touch-up job.

Baked enamel finishes. When new, most white or pastel-colored appliances have a slight cream tint to their color. This is caused by the high heat at which the enamel finishes are baked at the factory. Normal exposure to ordinary light causes this creamy tone to disappear in a few days to a month. The finish then becomes a bright white or its true pastel shade. Because of this bleaching and the other color changes due to aging, it is very important that all new appliances be touched up only with the lacquer specified by the maker of the appliance. The patch at first will seem a shade or two whiter than the rest of the finish, but it will match perfectly in a short time as the rest of the finish bleaches.

When working with a baked enamel finish, it is desirable in many cases to keep the "orange peel" texture to the finish that gives most appliances today their porcelain-like appearance. In this case, do not use rubbing compound to clean up the overspray ring around the touched-up area—it may result in too smooth a surface. Instead, spray lightly with an aerosol leveler to blend in the overspray without applying so much that the touchup will run.

Care and storage of spray cans. The opening in the plastic spray head must be clear and free from all residue to prevent irregular spray and clogging. To clean it, scrape the face with a fingernail or some other dull edge (a knife

would cut the plastic face). Twist the spray head one-half turn to clear the interior passages. In severe cases the spray head can be pulled off the can for more thorough cleaning. When replacing it, hold a finger over the opening to prevent accidental spraying. Leveler can be used to soften stubborn residue.

When the can appears to be empty, you can often get enough more spray to finish the job by rotating the spray head about one-half turn. This positions the interior tube to get the last drop out of the can.

After finishing a job, clean the valve passages before storing by turning the can upside down and opening the valve briefly. You will get a short spray of paint followed by the plain gas that cleans the valve.

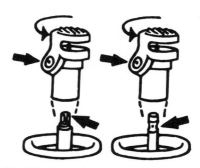

Figure 1-34. Method of cleaning a spray head.

Do not store aerosol paint cans where they will be exposed to temperatures over 120°F. Be sure to avoid places where the tools will be exposed to sunlight—store windows, the rear-window shelves of automobiles, and the like. Do not use a spray paint near fire or open flame, as the solvents used in aerosol paints and lacquers are inflammable. When touching up gas-fired appliances, be sure to put out the pilot light first. Never puncture or incinerate an aerosol can.

Repairing Porcelain Finishes

At one time porcelain enamel was the traditional appliance finish. Today only a few manufacturers still employ porcelain as an interior finish for their "top of the line" refrigerator models. About the only other use of porcelain in the making of appliances is for dishwasher tubs and washer and dryer baskets.

Porcelain chippage and crazing can be repaired with epoxy patching kits. These kits are usually available from the appliance manufacturer in the proper color. Here are the steps necessary to repair a crazed or chipped porcelain surface.

1. Roughen the metal surface inside the crazed or chipped porcelain area and feather the edges of porcelain using a fine-grade sandpaper or aluminum oxide paper (either 2/0 or 3/0).
2. Clean the metal surface and the surrounding porcelain area with a household cleaner and wipe dry.
3. Squeeze beads of epoxy resin and epoxy hardener of equal lengths on a cardboard pad and mix thoroughly as directed by the manufacturer.
4. Apply a sufficient amount of epoxy to cover the crazed porcelain surface.
5. Smooth the epoxy to conform to the porcelain surface and wipe off excess epoxy. But do not wipe the epoxy on the crazed area.
6. Instruct the customer not to use the appliance for 24 hours to allow the epoxy to cure.

If the manufacturer's instructions are properly followed, the epoxy patch will have a lifetime equivalent to that of the porcelain.

Plastic Finishes

Appliance cases and parts that do not need great structural strength or are not subject to high heat frequently have plastic surface finishes. These are attractive and give service; but never use any type of abrasives or wax cleaners on these surfaces. Most plastic finishes clean easiest with warm water and soap.

Nameplates, escutcheons, decorative trim, control panels, and other similar appliance components are usually made of plated metals or anodized aluminum. (A few are made of plastic or paper.) Most of these can be cleaned

with a damp cloth. When working on an appliance, take every precaution to protect its finish. For example, when servicing an iron, clear a space of any loose parts and tools about 18 in square near the front of your workbench and pad that area with several thicknesses of soft cloth or a piece of a blanket. Such an arrangement will give you sufficient room to work in and will protect the appliance's finish. A carefully planned work area should be provided for all small appliances.

Various chemical products are available for cleaning terminals, switch contacts, insulating wiring, cemented parts, etc. But when using them around very hot parts, make certain that they are not flammable. Two of the most common cleaners used around an appliance shop—acetates and acrylates—are flammable.

FINAL TEST

To do a good servicing job, you must test the appliance before returning it to the customer. Repair it as if it were your own. Test the completed repaired appliance by subjecting it to the "final test" recommended in the manufacturer's service manual. For example, the manufacturer of the can opener illustrated earlier in this chapter (Fig. 1-27) recommends the following final test for the appliance.

Final Test

Check the cutter gap, the driver speed, and the operation of the clamping lever.
1. *Cutter gap*
   a. *Specification*: The clearance between the cutter and the driver must be more than 0.010 in and less than 0.020 in.
   b. *Method of measurement*
      (1) There is a certain amount of "slop" in the fit of the cutter on the cutter stud. The cutter must be pushed up against the stud as

would be the case if a can were being punctured.

(2) The driver must be pushed in toward the case to eliminate any shaft end play from the gap measurement.

(3) The clamping lever must be latched down.

(4) When these conditions are established, the gap between the rounded rear surface of the cutter and the flat front surface of the driver must accept a 0.010-in feeler and reject a 0.020-in feeler.

  c. *Correction*: If the gap is too wide (accepts a 0.020-in feeler), increase the total thickness of driver washers by withdrawing the driver and adding driver washers as required. If the gap is too small (accepts a 0.010-in feeler), decrease the total thickness of driver washers by withdrawing the driver and removing driver washers as required.

2. *Driver speed*

  a. *Specification*: The governor must be adjusted so that driver rotates at 125 r/min ± 25 r/min.

  b. *Method of measurement*

(1) Remove the cutter.

(2) Depress the clamping lever and hold it down to keep the motor running.

(3) Measure the driver speed with a tachometer.

  c. *Correction*: If driver speed is not within specification, bend the contact arm of the governor against which the actuator rides as follows:

Bend the arm in against the actuator to increase the contact pressure and speed.

Bend the arm away from the actuator to decrease the contact pressure and speed.

3. *Clamping lever* (Power Piercing)

  a. *Specification*: When the clamping lever is depressed, the motor must start when the lever reaches a point measuring $1\frac{1}{4}$ to $1\frac{1}{2}$ in from the underside of the lever to the bottom of the case (see diagram).

  b. *Method of measurement*

(1) Plug the can opener into a 120-V 60-c ac outlet.

(2) Depress the clamping lever very slowly until the motor starts.

(3) Holding the clamping lever at this position, measure the distance from the bottom of the case to the underside of the lever.

  c. *Correction*: Remove the backplate and adjust the switch as follows:

If the motor starts too soon, bend the lower contact plate *down and away* from the upper contact.

If the motor starts too late, bend the lower contact up toward the upper contact.

4. Hi-pot at 1,100 V for 1 min between each of the two cordset terminals and the cutter or driver.

While small appliances need little or no preventive maintenance, many major appliances such as washers and dryers do. In fact, any belt-driven appliance, because its parts generally operate faster and under greater load than most other appliances, must have regular attention, at least once a year. Other appliances have "special" preventive maintenance needs. For instance, some washer water pumps are equipped with an oil wick for lubrication. As a rule, this wick should be removed and soaked in oil as recommended by the manufacturer once or twice a year.

Full preventive maintenance and lubrication instructions are usually given in both the user's booklet and the service manual. These instructions must be carried out very carefully by either the customer or the service technician. Preventive maintenance techniques are also described for major appliances in *Major Appliance Servicing* and *Refrigeration, Air Conditioning, Range and Oven Servicing* of this series, while general lubrication techniques for small appliances are given in this chapter. But in all cases, it must be remembered that lubricating properly—the single most important item of preventive maintenance—is not simply a matter of a few drops of oil. Here are three important points to keep in mind when lubricating any appliance.

1. *Do not use too much.* When manufacturers supply specific instructions as to the amount and frequency of lubrication, follow them to the letter. They know their products best. It is a grave mistake to think that if a little lubricant is good, a lot will be wonderful; that can be a wonderful way to foul up an appliance. If instructions are not provided, follow the general information given in the books of this series.

2. *Use the correct lubricant.* Manufacturers have taken great pains to create lubricants ideally suited to specific jobs. Use the right one for maximum benefits.

3. *Clean what is being lubricated.* This is not always possible, but it is always desirable.

As repeated many times in this series, your responsibility to a customer does not end with a good repair job. It also includes customer education. That includes complete instructions in the proper use and care of the appliance whenever you find evidence of abuse, but it must be done most tactfully. Tell customers how to avoid recurrence of trouble without capitalizing on their ignorance of mechanical items. Make them feel as though their lack of care was a natural oversight or a misunderstanding of the operating instructions which might easily occur to anyone. Remember: good customer education makes your own job a great deal simpler.

Motor-driven small appliances

Small appliances are generally divided into two broad basic categories: motor-driven and resistance-heating appliances. A few appliances such as heater fans and some of the personal-care items combine the two basic principles into one item. But for now, let us take a look at the motors that drive small appliances.

Motors, of course, play a very big part in the operation of any motorized small appliance. In fact, most troubles with these units can be traced to faults in the motor and its controls. Furthermore, most experienced service technicians are of the opinion that the major proportion of these troubles are caused by the lack of care or abuse on the part of the customer rather than normal wear and tear or poor appliance design on the part of the manufacturer. Therefore, it behooves service technicians to be sure that their customers are well "educated" on the care and use of motorized small appliances, and the only way to do this is to know how small motors work.

TYPES OF MOTORS

Motorized small appliances use all sizes of motors ranging from the tiny ones in electric clocks and shavers to powerful jobs that drive vacuum cleaners and portable power tools. As was stated in Chap. 5 of *Basics of Electric Appliance Servicing,* motors found in most modern small appliances are of three major types: the shaded-pole induction motor, the series-wound universal motor, and the permanent-magnet dc motor. The latter is found in most cordless small appliances.

There are several variations of the basic types, including some that combine features of the shaded-pole induction motor and the universal motor. But all motors used in small appliances operate in much the same way and are subject to similar faults. For this reason, we consider it best to discuss motor troubles for all appliances in this chapter and cover the special features of particular appliance motors when necessary with the information on that appliance in the appropriate chapters. But before going into the actual servicing technique, we would like to quickly review some of the more important basic motor principles described in detail in Chap. 5, *Basics of Electric Appliance Servicing.*

All motors operate on the same basic principle of magnetism, which states that like magnetic poles repel one another with a physical force, whereas opposite magnetic poles attract. In small-appliance motors there are usually two magnets; one is fixed and does not move in any direction; the other is free to rotate inside the fixed-magnet's field. The fixed magnet is called either the *stator* or *field,* while the free-moving magnet is known as either a *rotor* or *armature.*

The actual motor movement occurs whenever the magnetic poles of the armature and stator are out of alignment with each other. The armature reacts to this unbalanced magnetic condition by rotating in a direction which attempts to locate its poles with the opposite one of the stator. But, because of the polarity-switching arrangement in every motor, the poles of the armature and stator are never permitted to completely align themselves. In fact, owing to this continuous switching of polarity, the armature is continuously "chasing" the magnetic poles of the stator, thus developing the rotary motion and torque necessary to drive an appliance. Of course, different types of electric motors, as described in *Basics of Electric Appliance Servicing,* employ various techniques of develop-

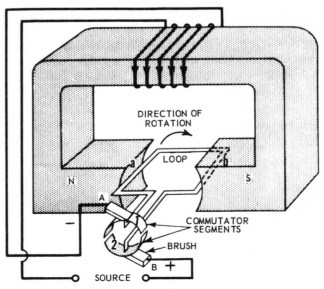

Figure 2-1. Basic motor action.

ing these magnetic pole and switching actions that are required to provide smooth motor operation.

Shaded-Pole Induction Motors

This type of motor, which is used in some small appliances including electric clocks, small fans, some shavers, can openers, hair dryers, and phonograph turntables, is very simple and has only two major parts (Fig. 2-2): a field coil (stator) and a rotor (armature) which includes the bearings. The field coil consists of many turns of fine wire on a laminated iron frame. The rotor is suspended inside the frame so that it is in the magnetic field of the core. When the shaded-pole motor is plugged into a 60-Hz ac line, the coil builds up an alternating magnetic field, which changes direction every $\frac{1}{120}$ s. Thus this magnetic field created by the field coil gives the rotor a tiny "push" every $\frac{1}{120}$ s. Once the motor starts, the rotor keeps receiving these tiny pushes and keeps operating. In other words, its speed is tied in directly, or synchronized, to the frequency of the power, which is, of course, 60 c/s. For example, if each cycle causes the rotor to turn one-half revolution, the 60-c current will cause the rotor to turn exactly 30 revolutions in 1 s, which is equal to 30 times 60 or 1,800 r/min.

Since the speed of shaded-pole motors depends completely on frequency, the applied voltage does not affect their operation, unless it drops so low as not to provide enough "push" to keep the rotor moving. You will find this type of motor used where a constant speed is important and relatively little power is required. It has the advantages of simple construction, low cost, no sliding electric contacts, reliability in operation, and being self-starting. By the way, the shaded-pole motor receives its name from the fact that one winding in its stator and a portion of each pole face is usually "shaded" by a copper band or strap to provide the rotating magnetic field necessary for starting.

To check the stator or field windings, use any of the continuity and ground tests described in Chap. 8, *Basics of Electric Appliance Servicing.*

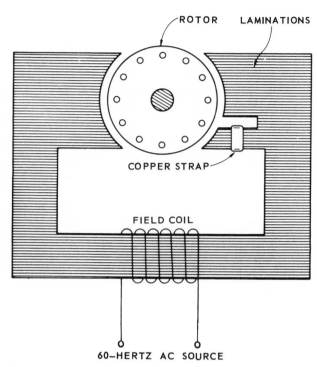

Figure 2-2. Diagrammatic representation of a shaded-pole induction motor.

But when performing the series lamp test, since it uses the 120-V current, remember to touch only the insulated parts of the clips. For the ground test, clean a portion of the stator and attach the clip to it. Then touch the other test probe to each of the motor leads, one after the other. If the bulb lights dimly, a partial ground is indicated. A bright light means a complete ground. If the resistance is high enough, there could be a slight ground even though the bulb does not light at all. If a shaded-pole motor is grounded, or open, it is best to replace the entire motor, since the cost of replacement is usually less than having it rewound. Actually, this is true of most small-appliance motors.

Series-Wound Universal Motors

To provide the required starting torque, which is the ability to start even though under considerable load, most small appliances use series-wound universal motors. The speed of this type of motor is determined by the load; the greater the load or the more work the motor has to do, the slower the speed will be; that is, universal motors have enough torque to work under heavy loads at slow operating speeds. Motor speed also varies with voltage. However, universal motors will maintain a fairly constant speed under a given constant load, but they are not as reliable as the shaded-pole types. On the other hand, their power output is considerably higher. Thus, where constant speed is not as important as power, the universal motor is used. Incidentally, these motors operate equally well on a dc or ac power source. It is this ac/dc feature that gives universal motors their name.

The design of a universal motor is similar to that of the shaded-pole motor. The major difference is that everything is bigger: Larger wire is used in the field coils and more current is drawn through them. This creates a more powerful magnetic field which in turn permits the motor to deliver more power. The rotor or armature also has its coils wound in such a way as to get a bigger "push" of power from the fields. Since this armature turns, a split-ring commutator is employed to make an electrical contact with it while it is rotating. (The word "commutator" means "switch.") The segments of commutator are usually made of copper or brass and are fastened to the armature shaft. Mica or a hard fiber electrically insulates the commutator segments from each other and from the metal armature shaft. The soft carbon brushes are mounted in fixed insulated holders so that they slide across the commutator segments as the armature rotates.

Permanent-Magnet dc Motors

As mentioned earlier in the chapter, most cordless small appliances employ permanent-magnet dc motors. The commutator and brush assemblies in a dc motor operate in the same fashion as do those of the universal motor. Their function in both is to switch relative magnetic polarities in such a manner that the armature is continually chasing the opposite poles formed by the stator. The major differences between the two types of motor are as follows.

1. The universal motor can operate on both alternating and direct current, while the permanent type works on direct current only, usually supplied from batteries.
2. The dc motor usually employs a permanent magnet to generate the stator's magnetic field, while the universal type always uses an electromagnet. (A few dc motors also develop the stator's field electromagnetically.)
3. The armature windings in a universal motor are placed on a laminated soft iron core to reduce the heating effects of eddy currents generated by the ac power to the windings. Since direct current does not generate eddy currents, the dc motor may use either a laminated or solid core. The latter is the one most commonly used.

Because dc motors operate on much the same basic principles as universal motors, they share most of the operating characteristics mentioned earlier for the ac/dc type. But it must be remembered that most dc motors used in small appliances operate from a battery power source. Many cordless appliances have a charging diode built right into the unit, while others have an external charging device. But in the vast majority of cordless appliances the power source is the batteries whether or not the unit is connected to the ac outlet. When the unit operates when connected to a power source and will not run when not connected, the problem is generally due to the fact that the batteries are discharged and that the charging current from the ac outlet is furnishing the necessary power to the motor. More information on cordless small appliances and their source of power is given in the next chapter.

TESTING AND SERVICING MOTORS

The testing and servicing information given in this portion of the chapter applies primarily to the universal type of motor. True, most of the testing procedures as well as the servicing techniques apply to all three types of small appliance motors. But, since most shaded-pole induction and permanent-magnet dc motors are so inexpensive, it is usually more economical to replace them than to repair them. In fact, certain repairs on universal motors are such that necessary labor and charges will exceed the replacement cost of the complete motor. For this reason, the servicing information given in this chapter is limited to those things that a service technician should do. Such servicing procedures as how to rewind a faulty armature or grind a pitted commutator have been omitted; the only feasible servicing procedure for correcting such problems is to replace the defective part or the complete motor.

Brush Troubles

The carbon brushes used in universal motors and most dc motors are a major cause of problems. There are usually two brushes, and these are generally spring-loaded carbon blocks that press against the segments (bars) of the motor armature (commutator). If either brush does not touch the copper or brass segments of the armature, the appliance will be dead. This is generally the result of the brushes wearing down—when the brushes get too short, the unit stops.

While several different types of brush fasteners are employed, the majority of them have screw caps of plastic or an insulating material, which holds the brush spring down in the holder (Fig. 2-3). To change the brush and/or check its condition, remove the brush cap. The brush spring should pop out a little. If it does not, this means that the brush has worn down too much and needs to be replaced or that the spring no longer has tension and brush and spring should be replaced. Most commonly, the wire, or "pigtail" as it is often called, that goes from the brush itself to a small metal cap, soldered to the end of the spring, is twisted. This compresses the spring, stopping it from pressing the brush against the armature. The cause of the twisting is the turning of the brush cap as it is screwed onto the brush holder. To correct this

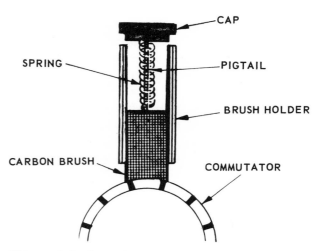

Figure 2-3. An enlarged view of a typical brush-holder assembly.

last condition, twist the brush pigtail wire (not the spring) counterclockwise about five turns. Reinstalling the brush and turning the cap clockwise will untwist the pigtail wire, leaving it its full and proper length.

A few brush arrangements are held in position by flat pieces of plastic, while others are kept in place by a springy strip of brass fastened at one end. To remove the brush in the latter setup, lift the end of the strip and turn it to one side; you can then pull out the brush by the spring. But most of these heavier brushes have internal pigtails, which are fine wires inside the spring. They aid in the carrying of electric current and prevent the spring overheating. For this reason, check the pigtail visually for continuity. If it is broken, the spring will show certain signs of overheating—generally it will turn a light reddish color and lose most of its tension. This is because, if the pigtail is broken, the spring will have to carry the current and as a result will overheat, lose its temper and tension, and the brush will bounce. In a few brush arrangements, however, the spring is designed to carry the current.

Dirt between the brush and the brush holder will also prevent the brush from touching the segments of the armature. You will know immediately if this is the case because as the brush is removed from its holder, it will not come out easily, as it should. If the spring stretches or if you feel any kind of drag, something is holding the brush back. Chances are it is dirt. Remove the brush, clean it with a rag, and push the rag into the brush holder with a wood probe to clean the brush holder. Try the brush again in the holder. It should go into place easily. By the way, when removing the brushes, it is wise to mark them L and R as well as for their position. In this way, proper seating of the brushes is ensured when they are returned to their holders.

While the brush is out, look at the end that presses on the commutator. It should be curved (concave), very shiny, and smooth; also, its curved bottom should be 90° to its sides. If it is angled, the brush holder has turned. As previously mentioned, most brush holders are kept in place by a small screw in the tool casting. These screws should be tight.

If the brush holder is twisted, the holding screw probably is loose. Set the brush holder straight and tighten the screw. Loose brush holders can cause other problems besides turning. If the brush holder is accidentally pressed in too far, its inside end will touch the commutator.

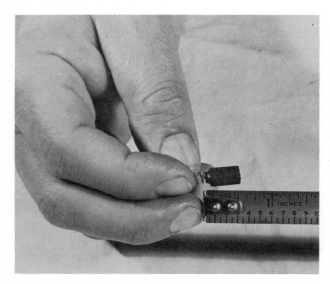

Figure 2-4. The length of the brush from the spring should be at least $\frac{1}{4}$ in; if it is less than that, both brushes should be replaced.

The inside end of the brush holder should be about $\frac{1}{16}$ in away from the commutator. When the brush is pushed into its holder and the cap secured, you should be able to see the brush protrude this $\frac{1}{16}$ in and touch the commutator. If it does not come through, this means that the end of the brush holder was pushed against, or worked its way against, the commutator. In this case, the revolving commutator will form a burr on the end of the brass liner of the brush holder, and the brush will get hung up on this burr. The burr can be removed with a small, flat file without removing the brush holder.

Whenever you take apart a small-appliance motor, be sure to check the brushes for length. If they are worn down to approximately $\frac{1}{4}$ in, replace them. When replacing a set of worn brushes, make every effort to obtain those recommended by the appliance manufacturer. In this way, you are certain that brushes fit properly and that they are of the correct chemical composition. Most manufacturer's replacement brushes are already ground to fit properly against the commutator.

If the manufacturer's replacements are not available, standard-size brushes can be used and are available at appliance supply shops. When selecting a set of replacement brushes, be sure to get the proper length and, above all, the proper size. A brush that is too long can be cut off, but one that is too big must be painstakingly sanded down to fit, by rubbing it on a piece of sandpaper on a flat surface. Also make certain that the brushes fit snugly in the holders, but are still free enough to move back and forth. If necessary, the old springs can be used on the new brushes, provided that they have not lost their tension from being overheated. More on the running in or seating of new brushes can be found below under Running-in New Brushes.

Some small appliances are of what we call *clamshell construction*. This type of appliance is cast in two halves, like a clamshell. In most of them, to get to the brushes, one of the halves must be removed (the half with the screwheads available). Keep the appliance flat with the half to be removed facing up. When all the screws are loose, lift the top half off slowly. Do not turn the other half over, for all the parts can fall out. In most cases, the brush holders are not brass lined and are not held in place by setscrews. The brush has no wire, and the spring itself conducts the current. One can easily see whether the brushes are worn too short.

Armature and Commutator Troubles

Most problems in a commutator are indicated by either excessive sparking around the brushes or a motor that does not run or runs with insufficient power. If there is very heavy sparking at the brushes and the commutator, be sure that both brushes are long enough and free to slide in their holders. If the cause of the sparking or arcing is not in the brushes, check the commutator surface very carefully for scratches, pits, or bits of metal in the insulating space between the segments. Also be sure that the commutator is not fouled with excessive dirt and grease. Not only will this cause sparking, it also can make a motor run slowly.

The grime which frequently accumulates on a commutator can be cleaned with a good grease solvent such as perchloroethylene, trichloroethane, or trichloroethylene. (Many older service manuals recommend carbon tetrachloride for this job, but this material is difficult to obtain since it has been banned by the government for household use.) Make a pad of cloth about the width of the commutator and long enough to encircle it, saturate the cloth with the cleaning solvent, then squeeze the pad tightly around the commutator with one hand while revolving the armature with the other. If this scouring is not effective, cut a strip of very fine sandpaper (6/0 to 10/0) or crocus cloth, no wider than the commutator but long enough to encircle it, and repeat the process described above using the sandpaper instead of the rag. Be careful that the edge of the sandpaper does not touch the windings and that no grit gets into the windings. Never use emery cloth or steel wool to clean a commutator since it will leave bits of conductive material that can short-circuit the commutator. Also do not use any of the silicone contact

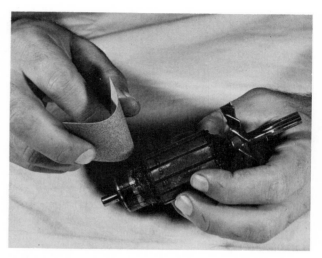

Figure 2-5. Remove dirt and minor nicks from the commutator by rubbing lightly with a very fine sandpaper.

cleaners that are available since they seem to combine with the protective copper-carbon film that coats the commutator to create an undesirable chemical. Incidentally, this coating on the commutator is a tough, highly conductive film that enhances the electrical properties of the commutator/brush assembly and acts as a dry lubricant to reduce friction. It is created by the normal friction between the copper or brass commutator segments and the carbon brushes, and should not be cleaned off unless it becomes excessive.

Sometimes, because of sparking and wear, tiny particles of commutator metal lodge between the insulating material of adjacent segments. To remove these metal bits or any other similar foreign materials, scrape out the spaces between the segments with a knife blade, hacksaw blade, or a sharpened wooden dowel. To complete the job, scrub the commutator and insulating material with a toothbrush saturated in isopropyl alcohol.

Minor scratches and pits on the commutator segments can be removed by rubbing with fine sandpaper or crocus cloth. However, when the commutator is badly worn or pitted, the armature assembly or the complete motor should be replaced. If your examination reveals that the segments have worn away, leaving the mica or hard fiber material flush with the surface of the commutator, this situation is all right, provided that the mica or fiber is not "high" enough to make the brushes bounce as they rotate over a surface. (In a new armature, the mica or hard fiber material is undercut between the segments, leaving a small space.) To undercut the mica or fiber, scrape out the slot between the segments with a very fine hacksaw blade or the tip of a knife until enough insulating material has been raked out so that the slots can be seen. Be careful when undercutting not to gouge the surface of segments.

If the appliance motor has been in operation for an extended period, the commutator may have a groove worn in it. This is not necessarily an indication of trouble if the surface is a smooth, even color all the way around.

An unbalanced armature, easily detected by violent vibration of the motor at full speed, is an extremely rare fault, but every household has an amateur mechanic, and when one of these hammer-happy do-it-yourselfers sails into a small appliance motor, anything can happen. Armature unbalance may be caused by a bent shaft, lost or improperly placed balancing or insulating wedges, improperly positioned or broken governor member and/or cooling fan, or any other damage which would alter perfect distribution of the weight of the armature. It is possible, of course, to replace or repair some damaged appurtenances on the armature, but if the armature proper is out of balance, it should be replaced. In fact, the armature assembly or the complete motor should be replaced if your inspection of the unit reveals loose commutator leads, severely damaged commutator, defective shaft, open or burned windings, or grounds. But, since a motor is usually the most expensive replacement item in a motor-driven small appliance, you, as a service technician, must be perfectly sure of your diagnosis before replacing it. This is especially true when diagnosing armature- and field-winding problems.

Shorted armature windings. A motor that has a shorted armature winding will usually run slower than normal, run hot, and use more power than usual. Excessive sparking at the brushes is also another indication of shorted armature windings. As a rule, most winding problems in an armature will show up on the commutator. If the winding is shorted or open, the brushes will make a small spark every time they pass over the faulty commutator segment. This will show up as a very bright segment, or, more likely, as a darkened, "spark-marked" segment. Either means trouble.

Several tests can be performed to indicate a shorted armature winding. Here are the more popular ones.

1. *Growler test.* The testing device known as a *growler*, available from most electrical supply houses, is used in place of the motor's field. The external growler creates a strong alternating magnetic field for the armature. To test for a short, place the armature assembly into the growler, plug in the growler, and turn the armature slowly by hand while holding a hacksaw blade or similar piece of metal loosely on top of the commutator segments (Fig. 2-6). If the armature has no shorted windings, there will be no magnetic field and the blade will not

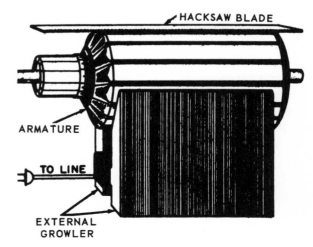

Figure 2-6. A method of locating a shorted armature, using an external growler and holding a hacksaw blade on the top slot of an armature as it is slowly rotated.

react except for a small amount of sparking that will normally occur as the blade shorts the segments. If there is a short-circuit, the blade will vibrate with a distinctive "growling" sound at one or more positions as the armature is turned a full revolution.

2. *Continuity test.* Check the resistance from segment to segment of the commutator with an ohmmeter; in order to do this, measure the resistance between consecutive pairs of segments—between the first and second, then the second and third, the third and fourth, and so on. The manufacturer's service manual will give the proper amount of resistance. If such a manual is not available, a good armature will show pretty much the same amount of resistance between each adjacent pair of segments, but if a pair of commutator segments is attached to a shorted winding, the resistance will be either zero or very low when compared with the resistance between all other pairs of segments.

3. *Power-on test.* To conduct this test, remove the brushes and connect a jumper wire between the brush connections in the motor. Then apply power to the motor's field windings, while slowly rotating the armature by hand. A normal armature—one with no shorted windings—will rotate freely through the complete turn, but if the armature has a shorted winding, it will tend to bind at one or more points during its complete revolution.

Open armature windings. A motor that has an open armature winding will either run more slowly than normal, producing insufficient power, or not operate at all. In either case, the motor will draw an above-normal supply of current; but this higher current drain—about 15 percent above the motor's nameplate rating—is not enough to seriously overload the windings, blow a fuse, or trip a circuit breaker.

The simplest tests to determine an open armature winding are as follows.

1. *Growler test.* With the armature in place on the growler, take a hacksaw blade or similar thin piece of metal and drag it slowly and gently between the two adjacent segments on the commutator. If the armature is normal, the voltage induced into the windings will produce a small electric spark when the blade shorts them out. But if the armature winding is open, no spark will appear. Be sure to check between each two segments as you rotate the armature in the growler.

2. *Continuity test.* The continuity test for checking an open winding is the same as for a shorted one. Check the circuit continuity between the commutator segments—between the first and second, then between the second and third, between the third and fourth, and so on. An armature operating normally will indicate good continuity between all commutator segments. A motor with an open armature winding will indicate no continuity between the pair of adjacent segments connected to the defective portion of the circuit. A test lamp may be used in place of an ohmmeter to find an open winding. Simply check between each two segments all around the commutator and see whether the test lamp lights in each position. If you find two segments where the lamp does not light, the winding is open there.

In a few instances, it is possible to observe an obviously open wire on the end of a bar and to resolder it successfully. Most of the time, however, any resoldering done on the commutator will throw the armature off balance and cause further difficulties. Therefore, it is usually best to replace the armature. A new one for a small appliance motor is not prohibitively expensive, especially since it will save you time and money.

Testing an armature for grounds. The final test on an armature before determining that it is normal is to make sure that there is no short between any of the windings and the metal shaft and frame. The armature need not be turned during this test, since if any coil is shorted it will show up at every commutator segment. Frequently, when intermittent grounding is suspected, the armature can be placed in a jig or vise, tapped with plastic mallet, and rotated as the test for ground is conducted. If the test lamp lights even intermittently while the armature is tapped, the assembly must be replaced. Often, grounds of this kind will appear only when the armature is rotating at high speed, and on occasion this sort of trouble will respond to no stationary test.

Shorted and open field windings. A motor that has shorted field windings will usually run a great deal more slowly than normal, draw rather heavy current, and run at higher than normal temperatures—over 120°F. The symptom of a motor with an open field winding is quite simple: it does not run at all.

To check the field for open or shorted coils, measure the resistance of each coil. As a rule, field coils are arranged in pairs, although a few motors, usually those employed in some mixers, may have a field coil on only one side. Since the coils are generally identical pairs, they should have the same resistance (be careful not to be fooled by taps on the windings). This field coil resistance for most small-appliance motors will be less than 20 Ω. If either of the coil readings is significantly different, it is a good indication that one of the coils is faulty.

If, for instance, the ohmmeter test indicates that one of the coils has infinite resistance, it is safe to assume that the winding is open. On the other hand, if one coil has a much lower resistance than the other, it is a good indication that this winding is shorted. In either case, the fields or the entire motor should be replaced.

If the field windings are cloth-wrapped, the condition of this wrapping is a good indication of whether the motor has been overheating. With enameled wire coils, any cracking or flaking of the enamel is usually a fairly good sign that the fields should be replaced.

Running-in new brushes. As was stated earlier in this chapter, if the manufacturer's replacement brushes are not available, substitute

brushes must be "run-in." This means running the motor until the ends of the new brushes are worn off sufficiently that they fit the commutator snugly. True, almost any set of poorly fitting brushes will eventually wear down until they fit the surface of the commutator smoothly. In the meantime, however, the excessive sparking caused by ill-fitting brushes could cause severe damage to the commutator.

To prevent this, sand the ends of the brushes to a slight concave contour before installation. (Remember that a good brush should be slightly concave on the end and very smooth and shiny.) Wrap fine sandpaper or crocus cloth—never, as previously mentioned, use emery cloth—around a cylindrical object (a dowel stick or pencil will do), and run the brushes back and forth several times. Then install the brushes, being sure that the concave surface is set correctly on the commutator and not crosswise, and permit the motor to run for approximately a minute. Then remove the brushes and note the shiny spots on the surfaces where they touch the rotating commutator. Using a sharp knife and fine sandpaper, scrape and sand away some of the carbon along the shiny areas.

Install the brushes again and operate the motor for about a minute. Again remove the brushes, and scrape and sand the shiny areas as before. Continue this procedure of sanding and running until the complete contact surface is very smooth and shiny. This, of course, indicates a proper fit.

Another method of seating a new set of brushes is to grind them down on a piece of fine sandpaper or crocus cloth fastened around the commutator. With the brushes installed in their holders, exert a little pressure on them with the fingers of one hand and turn the armature approximately one-half turn in the proper direction of rotation with the other. Then release the pressure on the brushes and back off the armature one-half turn to its original starting position. Reapply finger pressure on the brushes and turn the armature one-half turn in its proper rotating direction. Continue this procedure until the brushes are ground to fit the commutator snugly. When using this method of running-in new brushes, it is usually necessary to clean the commutator segments, after removing the sandpaper, with a cleaning solvent to remove any abrasive or carbon particles.

Motor Bearings

Bearings are the number-one source of *mechanical* trouble in small motors. But unlike most electric motor problems, bearing problems can usually be traced to user abuse. The most common bearing problems are due to lack of or improper lubrication. Another common cause is the accidental damage from a sharp blow or a fall which knocks the bearings or armature out of alignment. Operating a motor with the bearings or armature out of alignment wears down the bearings rather rapidly. Worn bearings allow the armature to bounce from side to side as it rotates, which may cause sparking, fast brush wear, and noisy operation.

Every appliance, as stated in Chap. 1, should be lubricated as directed in the user's booklet or the service manual. Unless the manufacturer's instructions state otherwise, use a pure automotive oil such as SAE-20 or SAE-30 (without detergent additives) to lubricate the motor bearings. Do not use the so-called "household machine" oil sold at variety stores, since such oil frequently breaks down into a gummy substance when subjected to the heat generated in a high-speed motor. Also, because of the high speed at which small appliances operate, motor bearings should never be packed with any type of cup or bearing grease.

The frequency of lubrication is determined primarily by how often the appliance is used, the operating time period, the size of the motor, the type of bearings, and the capacity of the lubrication system. Many small subfractional horsepower motors such as those employed in personal-care appliances, seldom need any lubrication. On the other hand, larger horsepower motors used in fans, vacuum cleaners, and some portable power tools require fairly frequent application of lubricant.

Several types and designs of bearings and lubrication systems are used in conjunction with

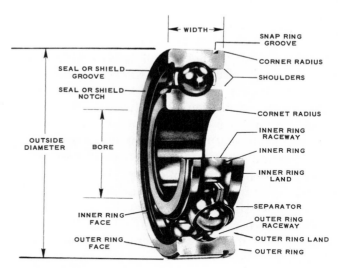

Figure 2-7. Typical ball bearing.

small-appliance motors. The most common of these are ball (roller) bearings and sleeve bearings. In addition to these metal bearings, plastic types—made of such materials as Delrin, Teflon, and nylon—are found in increasing numbers in small appliances. These materials are resistant to corrosion, water, and detergent, and are excellent as light-duty, self-lubricating bearings.

Some bearing lubrication systems are very simple. For instance, many cordless appliances count upon a drop or two of oil or a dry

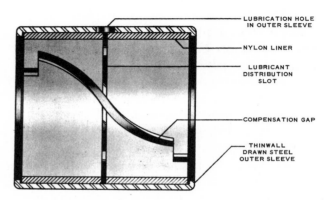

Figure 2-8. Nylon bearing units such as shown here are increasing in popularity with small-appliance manufacturers. They are a great deal easier to replace than metal sleeve bearings.

lubricant applied at the time of manufacture to do the job throughout the expected life of the unit. The great majority of today's appliances use a simple sleeve bearing with a tiny oil hole, mounted in a shaft or cage with a felt wick around it which should be saturated with a light oil. This oil then works its way into the bearing. While some ball bearings also use a wick-type lubrication system, most are usually lubricated with a lifetime supply of grease when assembled. Overlubrication of ball bearings can cause problems since it may result in excessive churning, friction, and heat, which will break down the lubricant and eventually damage the bearings. Sleeve-type bearings are not generally damaged by overlubrication. The problem here is one of oil leaks onto clothes or food.

Ball bearings are the easiest of all bearings to replace. As we know, ball or roller bearings ride between an inner and outer race, and the only motion in the device should be that of the bearings within their raceway. To replace a ball bearing, pull off the defective bearings and replace with a new set by press-fitting them over the shaft and into the bearing retainer. When doing this, be sure to apply the heaviest amount of pressure to the inner race and not to the outer race. If too much pressure is applied to the outer race, it could cause misalignment of the bearings. Also be sure the bearings are tight in their position; a loose fit will allow the bearing raceways to move and wear their mating parts.

Sleeve bearings are used a great deal more today in small appliances than the ball type. The reason is simply one of cost; in addition, sleeve bearings also perform well in small appliances. But once a sleeve bearing becomes defective, it is most difficult for a service technician to replace. As a rule, the inside surface of the sleeve bearings is reamed and then press-fitted into the motor's end plates at the time of manufacture. To replace these sleeves, special tools are generally required. It is not difficult to press out a bad sleeve bearing, but when you press in the new one, very often the bearing constricts, changing its inside diameter. This must be reamed out to a precision fit, or the appliance will bind or overheat at the

new bearing. Since very little metal is removed in reaming, it can easily be done by hand. The reamer should be held in a reamer wrench to ensure that the reamer cuts true in the hole. If machine reaming is utilized, it must be done at slow speeds (below 300 r/min) with the housing carefully supported at right angles to the reamer. Under no circumstances should the holes be reamed with a *hand*-held electric drill or similar tool. The reamer should be cleaned and oiled before reaming each hole. The holes must also be *very* carefully cleaned after the reaming. Examine the holes to be sure that there are no small metal chips remaining.

It is wise to always follow the manufacturer's recommended servicing procedure regarding the replacement of bearing to the letter. For example, one manufacturer may suggest sending the entire motor to the factory for bearing replacement; another may supply the bearings separately and also have available at nominal cost the aligning reamer or burnishing tool which, if required, you must have to fit the bearings correctly; still another may furnish the bearings as a subassembly with the motor end caps ready for use without reaming. In any case, be sure to evaluate the relative costs of new assembly or bearing parts, the time required to do the job, the general condition of the complete appliance, and the tools needed to do the work. Once these points have been carefully considered, make your recommendation to your customer. Many times it is cheaper to replace the complete appliance. While such advice may lose a service call, it may gain you a customer for life.

The symptoms of bad bearings are various. The moving parts may be frozen, very stiff, or hard to rotate. There may be a screeching or grinding sound. A bad bearing at the commutator end of the armature can cause excessive sparking. But as a rule, when a bearing starts to fail, the first indications are noisy operation and excessive "slop" in the motor shaft. When these warnings occur, it is usually too late to attempt to lubricate, clean, or otherwise salvage the bearing because the situation will continue to get worse. Complete failure of the bearing will result in a seized or frozen shaft. When the motor shaft becomes frozen into a bearing, try to get some SAE-20 or SAE-30 oil to soak into the bearing for an hour or so until it frees enough to be rotated by hand. Continue to flush it with oil until the shaft runs freely. If the motor will then start, keep applying oil on the seized bearings until the motor comes up to its normal speed and holds it.

If oiling does not free a seized bearing, disassemble the motor. Pull the bearing from the shaft; if it is a ball-bearing type, replace it; if it is a sleeve-type bearing, the freezing may have been caused by a lack of oil, or in some cases by the owner's use of a household motor oil, as noted above. To clean a sleeve bearing, tear off a strip of clean, dry cloth about 1 by 5 in, saturate it with a grease solvent (see Armature and Commutator Troubles above), and twist the cloth into a long cylindrical swab, thick enough to fit the bearing snugly. Twirl and move the swab through the bearing with a seesawing motion. The portion of the motor shaft that fits into the bearings should be cleaned in the same manner. Washer-type oil wicks can be cleaned by soaking and rinsing them in a cleaning solvent. Wicks which are extremely dirty should be replaced.

After the sleeve bearings have been cleaned thoroughly, apply a liberal coat of motor oil (SAE-20 or SAE-30). Before assembling the motor, remove all excess oil. If the service manual does not give the oil capacity of the wicks, a good general procedure is to slowly inject the oil into the wicks with a syringe until they look full. Unsealed ball bearings that might become dirty may be cleaned by removing them from the motor and then soaking them in solvent. Once the ball bearings are cleaned, coat them with light oil. But be sure to wipe off all excess oil before fastening them back on the motor shaft.
End play. Every motor must have some end play to prevent binding which will slow down the motor and cause overheating after being in use a short time. (*End play* is defined as the motor shaft in-and-out travel from end to end.) But too great an end play will permit the motor

shaft to move back and forth in its bearing as it spins and may allow the brushes to ride off the commutator or the armature to move partially out from under the fields. Nearly all motors use thrust washers, of either metal or fiber, on the motor shaft to control or limit the amount of end play.

The adding of thrust washers to the motor shaft will reduce the amount of end play, while removal of washers or replacing them with thinner ones increases the end play. The end play can be checked with a feeler gauge. Do not force the components together, but take sufficient time to get this part of the job as perfect as possible.

The final adjustment of end play in some small appliances is made from the outside of the unit by means of a thrust-regulating screw and hence can be achieved after the motor is completely assembled. Care must be taken, though, even in this type, that the correct thrust washer, if required, is in place on the opposite end of the armature shaft before closing the motor unit. To adjust the end play in such an appliance, tighten the thrust-regulating screw just to the point where the armature starts to bind, then back it off about a quarter of a turn and tighten the check nut. When making end-play adjustments, remember that the commutator must be located so that the brushes will center upon its running surface.

MOTOR TROUBLES

The following are some problems that occur with small-appliance motors and some of the possible causes of troubles. A defective motor can be caused by one or more of the basic problems, but there is generally only one cause.

Motor will not run. Assuming that power is reaching the motor, the following are the most frequently encountered reasons for a "dead" motor.

1. Brushes too short or brushes defective

2. Brushes right length but not touching commutator or sticking to the commutator
3. Shorted armature winding
4. Open armature winding
5. Shorted field winding
6. Open field winding
7. Broken wire from switch to field, field to brush holders, or line cord to field
8. Armature shaft bent or rubbing against the field windings
9. Bearing seizure

Motor hums but fails to start. This problem may be caused by any of the following.

1. Shorted armature winding
2. Shorted field winding
3. Field winding grounded
4. Armature shaft bent or rubbing against the field windings
5. Bearing seizure
6. Armature or bearings dirty and gummy

Motor starts but heats up rapidly. The maximum temperature of a motor housing should be about 120°F under normal load conditions. If the customer complains of the motor running hot, check for binds, shorted armature winding, or shorted or grounded field winding.

Brushes spark excessively. Brushes that spark are a good indicator of one or more of the following problems.

1. Armature winding shorted
2. Partially burned-out armature
3. Commutator dirty or out of round
4. Commutator segment(s) shorted
5. Commutator pitted or badly worn
6. High mica between commutator segments
7. Worn brushes and/or annealed brush springs
8. Oil-soaked brushes

Motor runs slow. If the motor runs slow but at a fairly steady rate, the cause may be insufficient lubrication or foul lubrication that tends to bind the motor shaft.

Motor slows down and runs with insufficient power. The following are the major reasons for

insufficient power or the slowing down of a motor when put under working conditions.

1. Armature winding shorted
2. Armature winding open
3. Field winding shorted
4. Field winding open
5. Armature shaft bent or rubbing against the field winds
6. Worn or defective brushes
7. Commutator oily or dirty
8. Commutator pitted or badly worn

Motor runs rough. If the motor on any appliance runs in "spurts" or if it will not start every time it is turned on, the trouble is either defective brushes or brush holder, an intermittently open line cord, or a defective armature.

Motor is noisy or vibrates. Because of the power developed by the motor, a reasonable amount of motor noise can be expected with any small appliance. Here are the common causes of a motor being more noisy than normal.

1. Worn bearing
2. Excessive end play
3. Bent armature shaft
4. Binding armature shaft
5. Loose parts on or near the motor
6. Foreign material in the air gap between the armature and field winding

Motor shocks user. Grounding to the case sometimes accompanies windings failure whether it occurs in the fields or in the armature, and careless placement of the wiring during motor assembly also will cause a ground. Any power ground to the case will give the user a shock.

MOTOR SPEED CONTROLS

A good number of small appliances such as fans, mixers, blenders, floor polishers, vacuum cleaners, and small power tools require variation in the speed of their motors. While there are several ways to vary the speed of a motor, the five most commonly used methods are as follows.

1. Tapped field
2. Adjustable brush
3. Governor
4. Rectifier
5. Solid state

Tapped-Field Speed Control

When variable-speed appliances first made their appearance in the late 1930s, the early designs contained a variable resistor or rheostat which was connected in series with the motor, and when this resistance was varied, the amount of voltage applied to the motor was also changed. A variation in voltage changed the speed of the motor. But because the resistor dissipated a great amount of the line power, this method was most wasteful in terms of overall power consumption and operating cost. The variable-resistance method of speed control is not used in any modern small appliances.

A few years after introduction of the resistance speed controls, the series-induction or tapped-field motor control came into use. This operates in much the same manner as the resistance type, except that the series-connected inductor does not dissipate its absorbed energy as heat. Rather, it absorbs electric energy during one portion of an ac cycle and returns it during other portions of the cycle. In this way, inductors change the voltage applied to the motor in the same manner as resistors, but they do not have any of the wasteful characteristics of resistors.

While the early series-induction appliances used a reactance choke to obtain control, today those that employ this type of control achieve it by using the inherent inductance of the motor field windings. The schematic drawing in Fig. 2-9 shows a standard design series: wound universal motor having a tapped field which, being switched in section by section, provides the three operating speeds. Since changing the amount of current through the field varies the magnetic force, the power and hence the speed are varied in proportion. With the speed control switch in the LOW position, the field windings

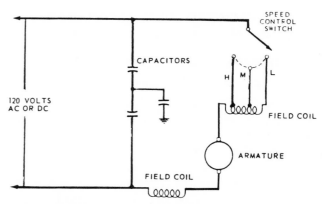

Figure 2-9. Schematic drawing of a typical tapped-field speed control.

are in series, thus giving the lowest possible speed. With the speed-control switch in the MEDIUM position, a portion of the field winding is disconnected from the circuit, and an intermediate speed is achieved. Finally, when the speed-control switch is placed on the HIGH position, an additional portion of the field winding is cut out, with a resultant increase in the current and motor speed. In this way, any desirable number of speeds may be had from the universal-type motor by simply adding the desired number of field-winding taps. Incidentally, inductor motor controls cannot operate in dc circuits.

Troubles in tapped-field speed controls usually occur in only two areas, the speed-control switch and the tapped-field windings. In most speed controls of this type, each of the speed-field coil windings has about the same resistance. A quick test of the condition of each winding can be made by checking with a test lamp between adjacent taps on the motor field coil. The lamp should glow with about the same intensity across each winding. Of course, a more exact diagnosis can be achieved by using an ohmmeter and checking its readings against the service manual.

In this chapter we shall not cover in detail the servicing of motor controls since their operations vary somewhat with the type of appliances that use them. For this reason, complete troubleshooting procedures for these controls accompany descriptions of the specific small appliances.

Adjustable-Brush Speed Controls

The adjustable-brush speed control is seldom used today except in some "economy" food mixer models. This method, also called the *movable-brush control*, utilizes the principle whereby the speed of the motor can be varied by slightly rotating the brushes concentrically with the commutator. This operation is accomplished by means of an externally located brush shift control lever or knob. Since there is only one position of the brushes, with respect to the motor field, at which the motor will develop its optimum speed and power, any movement of the brushes from this optimum position will result in a continued decrease in speed. That is, as the user moves the control shift lever away from the highest speed setting, the brushes leave their optimum operating point, and speed begins to decrease.

Most problems with the adjustable-brush speed controls occur with the mechanical linkage between the control shift lever and the brushes. The electrical problems with this system generally are simple brush and commutator troubles which are the same as those described earlier in this chapter.

Governor Speed Control

One type of centrifugal governor speed control comprises a set of spring-loaded arms, attached to a hub or disk on the armature shaft, which are gradually impelled outward by centrifugal force as the motor gathers speed. The springs in this type are so designed that the arms do not reach their outer extreme until the motor has attained its full speed. Then, as the motor slows down, the springs pull the arms toward the starting position. Thus, the arms of the governor assume a definite position for every speed variation. For example, the arms are fully extended at full speed; half, at half speed; one-quarter, at quarter speed, and so on, with proportionate changes in between. This gradual movement of the governor arms is transmitted by

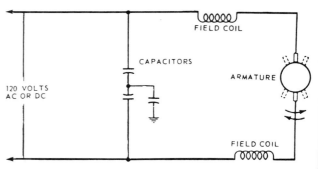

Figure 2-10. Schematic drawing of a typical adjustable-brush speed control.

mechanical means to the governor switch which is opened and closed intermittently to maintain the speed selected on the control dial. Setting the dial at a lower speed shortens the required travel between the governor-actuating member and the switch; moving the dial to a higher speed increases this travel. There is not, however, a complete cessation of power with these rapid openings of the governor switch. To prevent "bumping" and to reduce arcing at the switch points, a resistor and a condenser are connected in parallel across the governor-switch terminals. Because of the resistor, this circuit does not pass enough current to cause the motor to run fast enough to meet minimum requirements, but it does serve to maintain torque when the governor switch opens and thus eliminates any possible backlash.

Though all centrifugal governors utilize similar principles, there is another somewhat different method of operation by which the same purpose is carried out. To state it simply, the governor arms in this type must move to their outer extremes by centrifugal force to open the governor switch at any speed, even the lowest. Speed control is accomplished by varying the pressure on the governor arm spring (or springs) through the control dial. Increasing the pressure obviously forces the machine to run faster in order to throw out the governor arms against this opposing tension, while a very light pressure causes the motor to creep. Hence, numerous speeds are possible between these two extremes.

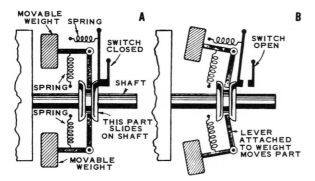

Figure 2-11. The centrifugal speed-control mechanism.

As in the preceding type, this governor control also requires a resistor and a condenser to prevent pulsating operation which would otherwise be present in all but the higher speed.

Armature speed is usually controlled by the governor switch in the regulator assembly. Within the limits of speed and load for which this motor has been designed, the armature accelerates when the governor switch is closed and decelerates when the governor switch is open. The governor switch responds to two opposed forces acting on the regulator leaf on which one of the contacts is mounted. The spring pulls against the leaf tending to close the switch and thus accelerate the motor, while the actuator pushes against the leaf tending to open the switch and thus decelerate the motor. The governor switch is open or closed depending upon which of these two forces is the greater; both of these forces are variable. The force exerted by the spring varies with the position of the control knob. The cam is keyed to the knob and control arm so that, when the knob is turned away from OFF, tension is placed on the control spring which, in turn, pulls on the leaf. The control rod and the governor spring form a flexible connection between the cam and the leaf. Thus, turning the control knob away from OFF progressively increases the force of the spring against the leaf. As is true with any centrifugal type of governor, the force exerted by the actuator varies directly with the speed of the armature. When the armature attains a speed at which the

force of the actuator against the leaf approximately equals the opposing force of the spring, any further increase in speed opens the governor switch, and any other subsequent decrease in speed closes it. At this point, the governor switch oscillates between the open and closed positions with great rapidity, and armature speed is stabilized within very narrow limits. To increase the armature shaft speed, the control knob is turned to a higher number. This puts additional force on the spring and thereby requires additional force from the actuator to open the governor switch. Additional actuator force can be obtained only from greater armature speed which, of course, results in greater output speed.

Rectifier Speed Controls

Rectifiers, also known as *diodes*, allow current to pass in one direction only; thus they have the effect of allowing only half of the alternating current voltage to reach the other side. Operating as a speed control, a rectifier or diode connected in series with the motor will permit current flow in only one direction, and the motor will run at only about half its rated speed.

When the control switch is set in the HIGH position, the motor will receive full ac power since the diode is shorted out of the circuit. But when the control switch is in the LOW position, the diode is placed in series with the motor, and the motor will operate at about one-half its rated speed.

The rectifier speed control contains only two components: the control switch and silicon rectifier or diode. The switch can be checked by giving the standard continuity test, while the diode can be checked with an ohmmeter. A low resistance reading (under 100 Ω) should be obtained when the test probes are placed across the rectifier in one direction, but when the test probes are reversed, a high reading (over 1,000 Ω) should be obtained. If the meter reads the same in both directions, the rectifier or diode is defective.

Unless you are completely familiar with the circuit, it is a good idea to disconnect one lead of the rectifier from the circuit before making the test; otherwise, you may damage other parts, or other parts in shunt with the diode could give an erroneous reading. When disconnecting one end of the diode, apply heat with a soldering iron (never use an iron of more than 125 W) to the connections for as short a period of time (5 to 10 s) as possible. The same precautionary steps should be employed when resoldering the diode. When replacing a diode, make certain that it is installed in exactly the same manner as its predecessor. Be sure to follow all reference marks. If an exact replacement is not available, a rectifier rated at 1,000 V and no less than 2 A will usually serve most small appliance rectifier speed control circuits.

Solid-State Speed Controls

Solid-state speed controls are becoming very popular in some blender and food mixer circuits. Some of these controls will enable the user to select any one of 14 speeds.

A detailed presentation of the operation of solid-state speed controls would require a complete understanding of solid-state electronic circuits, and the troubleshooting procedures demand special electronic test equipment that few appliance service technicians have in their shops. For this reason, theory and trouble-shooting procedures for solid-state controls have been omitted. The basics of solid-state circuits, however, are given in some detail in *Major Appliance Servicing* of this series. Much of the theory holds true for solid-state speed controls.

The electronic components that make up a solid-state control circuit for small-appliance motors are always mounted on an electronic printed-circuit board. Actually, the servicing of a solid-state control circuit means the replacing of the complete printed-circuit assembly if the control is determined faulty. That is, solid-state speed control units are in the printed-circuit "package" so that replacement of individual parts is generally impractical or even more expensive than the purchase of the entire unit. When replacing the printed circuit of a speed control unit, be sure to follow the manufacturer's directions to the letter.

PROCURING OF PARTS

Before going into the actual servicing of small appliances, we would like to answer one question that troubles many a beginning service technician—where to buy parts? The answer is a simple one—from the appliance manufacturer's nearest authorized service center. In other words, make it a rule to use genuine parts exclusively on all the appliances you service. This holds true for procurement of major appliance parts as well.

In some communities, certain items—such as resistors and condensers of similar specifications to the original—may be available as bulk electrical supplies for a few cents less, and perhaps you can adapt some of these to several makes, but this practice will often lead to butchering and unnecessary additional labor expense. Do not let fancy price lists lure you away from a genuine-parts-exclusively policy. Bear in mind, too, that no small armature is worth rewinding if a new one is obtainable—whatever the apparent saving. Armature burnouts in appliance motors are rare, but when you do need a replacement, install a new one with complete assurance that it is all new and that it is in perfect balance. The best investment you can make, therefore, in your customer's behalf, and in your own future, is the replacing of inoperative parts with new, genuine parts.

One reason procurement of replacement parts for appliances is a problem is that, unlike many other devices such as radio and television sets, most appliance parts will fit only a few makes, and sometimes only one. However, the situation is not completely impossible.

In most large cities several appliance parts supply houses are within easy reach. Often, in medium-sized cities, major manufacturers or independent suppliers maintain branch offices where you can buy locally. In small towns,

GENERAL ⊕ ELECTRIC

SUPPLY PARTS – 23R30 ROTISSERIE-OVEN

Screw, Foot	XQ 1X 20	Bracket, Heater	XQ 2X135	
Screw, Liner Attachment	XQ 1X 20	Lead Assembly, Switch Timer	XQ 2X136	
Thumb Screw	XQ 1X 28	Lead, Cordset	XQ 2X137	
Screw, Liner Attachment	XQ 1X 42	Handle Support	XQ 2X138	
Screw, Switch Attachment	XQ 1X 43	Spring, Temperature Knob	XQ 2X139	
Screw, Window Frame	XQ 1X 47	Clip, Thermostat Bulb	XQ 2X140	
Screw, Button Frame	XQ 1X 48	Tab Receptacle	XQ 2X141	
Screw, Motor Mounting	XQ 1X 51	Spit Handle	XQ 3X 10	
Screw, Timer Mounting	XQ 1X 52	Handle, Door	XQ 3X 17	
Screw, Thermostat Mounting	XQ 1X 57	Knob, Timer	XQ 3X 19	
Screw, Handle Support	XQ 1X 58	Knob Assembly, Temperature	XQ 3X 20	
Rivet, Bracket	XQ 2X 11	Thermostat	XQ 4X 3	
Bracket, Thermostat Bulb	XQ 2X 15	Switch	XQ 7X 4	
Skewer Assembly	XQ 2X 39	Cordset Assembly	XQ 8X 8	
Strain Relief	XQ 2X 66	Heater, Broil	XQ11X 13	
Foot	XQ 2X 81	Heater Assembly, Bake	XQ11X 14	
Button, Frame	XQ 2X 90	Shell, Top	XQ14X 6	
Spit Slot Cover	XQ 2X 94	Bottom Shell	XQ14X 7	
Spring Clip	XQ 2X 99	Liner, Upper	XQ15X 15	
Tab Receptacle	XQ 2X101	Center Liner Assembly	XQ15X 19	
Lead, Timer	XQ 2X102	Liner Assembly, Right	XQ15X 20	
Lead Asm., Broil Heater Switch	XQ 2X105	Liner Assembly, Left	XQ15X 21	
Tab Receptacle	XQ 2X108	Escutcheon	XQ17X 5	
Button, Rotiss	XQ 2X125	Timer	XQ19X 3	
Button, Broil	XQ 2X126	Motor & Collet Assembly	XQ20X 6	
Button, Bake	XQ 2X127	Window Glass	XQ21X 6	
Lamp	XQ 2X129	Door Assembly	XQ21X 7	
Lead Assembly, Timer-Lamp	XQ 2X130	Spit Assembly	XQ22X 7	
Lead Asm., Bake Heater Switch	XQ 2X131	Tray	XQ24X 8	
Lead Assembly	XQ 2X132	Broil Rack	XQ24X 9	
Retaining Ring	XQ 2X133	Shelf	XQ24X 12	
Lead, Bake Heater	XQ 2X134	Reflector	XQ28X 10	
		Shelf Support	XQ28X 11	

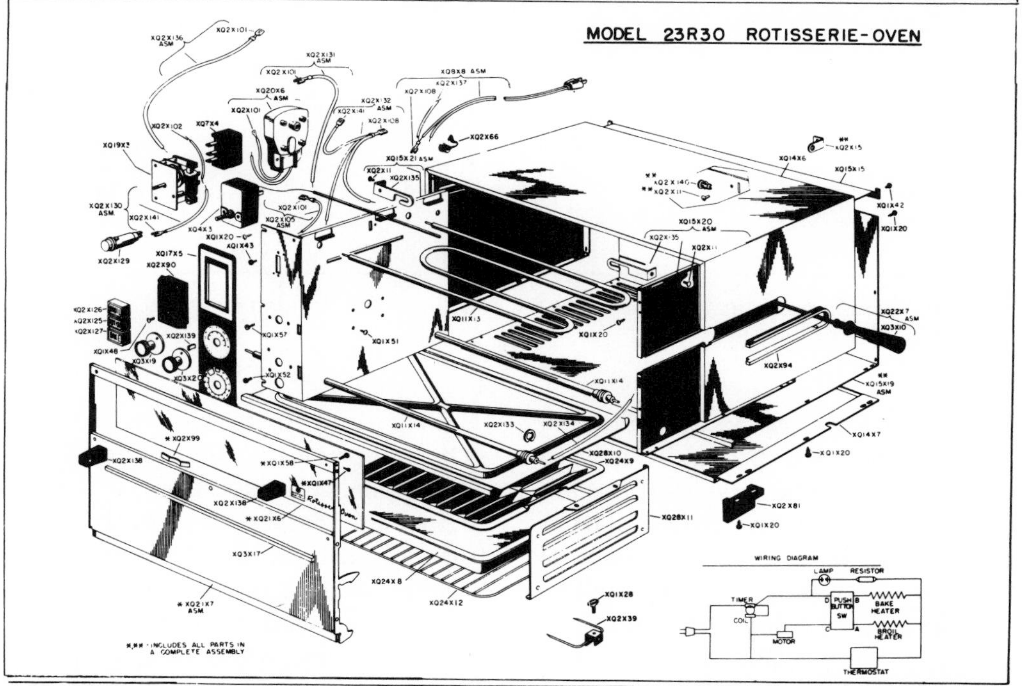

MODEL 23R30 ROTISSERIE-OVEN

*,** INCLUDES ALL PARTS IN A COMPLETE ASSEMBLY

WIRING DIAGRAM

Page: R30-2   Form PS6-15   November 1974

Figure 2-12. Parts callouts generally found in an instruction sheet.

HOUSEWARES DIVISION			GENERAL ELECTRIC COMPANY			PARTS PRICE LIST
Cat. No.	List Price or Supersedure	Description	Cat. No.	List Price or Supersedure	Description
XUT23X1	S .40	Arm	XUW4X1	S .70	Cover Handle
XUT23X3	.40	Arm	XUW4X2	1.30	Handle – Left
XUT23X4	.10	Terminal	XUW4X3	1.30	Handle – Right
XUT23X5	.10	Sleeve	XUW4X4	.30	Dial
			XUW4X5	.50	Cover Handle Asm.
XUT23X6	.40	Shield			
XUT23X7	.20	Connector	XUW4X6	.90	Body Handle
XUT23X8	.40	Connector	XUW4X7	.40	Adj. Knob
XUT23X9	.40	Connector	XUW4X8	1.10	Cover Handle Asm.
XUT23X10	.40	Connector	XUW4X9	.70	Body Handle
			XUW4X10	.50	Adj. Knob
XUT23X11	.20	Cord Grip			
XUT23X12	.10	Cord Grip			
XUT23X13	.50	Spring	XUW7X1	2.15	Thermostat
XUT23X15	.50	Insert	XUW7X2	4.40	Thermostat Asm.
			XUW7X3	2.10	Thermostat Asm.
XUT23X16	.50	Insert	XUW7X4	XUW7X5	Thermostat
XUT23X17	.10	Clip	XUW7X5	5.00	Thermostat Asm.
XUT23X18	.40	Slide Rod	XUW7X6	4.40	Thermostat Asm.
XUT23X19	.40	Slide Rod			
XUT23X20	.20	Spacer			
XUT23X21	.10	Cam	XUW8X1	.30	Lead Asm.
XUT23X22	.30	Connector Asm.	XUW8X2	.30	Lead Asm.
XUT23X23	.40	Latch Keeper	XUW8X3	.30	Lead Asm.
XUT23X24	1.70	Latch Release Asm.	XUW8X4	.20	Lead Asm.
XUT23X25	.20	Latch	XUW8X5	.30	Lead Wire
			XUW8X6	.30	Lead Wire
XUT23X26	.20	Spring	XUW8X7	.30	Unit Lead Wire
XUT23X27	.10	Rivet	XUW8X8	.40	Lead Wire – Upper
XUT23X28	.20	Insulator			
XUT23X29	.10	Insulator	XUW10X1	1.90	Cordset
XUT23X31	.60	Lifter Arm	XUW10X2	1.90	Cordset
XUT23X32	.50	Piston Asm.	XUW10X3	2.30	Cordset
XUT23X33	.20	Bearing			

Figure 2-13. Typical parts number identification and price list in a manufacturer's catalog.

however, because usually no local source of supply is available, you will probably have to depend on mail-order service. Few locations are more than 100 miles from an appliance parts distributor; in fact, there is usually a choice of four or five within this range. So the problem boils down to one of availability and transportation. Because of road conditions and parcel schedules, sometimes the city nearest you geographically is the farthest away in terms of time required for deliveries. You will learn the fastest means of transportation in your area by experience.

In general, there are two sources for parts. The first is the state distributor for the maker of the appliance in question. It is usually located in the state's largest city, but there may be branch offices in other cities if demand warrants. Your second source is the independent appliance parts wholesaler, who handles parts for all major appliances and whose offices also are usually in the largest or most centrally located city in the state.

The independent wholesaler is generally a much faster and more reliable source of replacement parts than the state distributor for that brand, possibly because the excessive paper work necessary in the manufacturer's operations slows down delivery. Another problem is confusion about part numbers: manufacturers often turn down an order for a part, stating "there is no such number," even though the number is stamped right on the part. Exhaustive catalogs, issued by many appliance parts wholesalers, may help to solve this problem, since they not only list parts for all appliances but also cross reference between models and manufacturers.

Sometimes a manufacturer will market the identical appliance under two brand names. Moreover, even though the units are identical, they may carry different part numbers. As a result, parts ordered from one are often refused with the suggestion that the order be sent to the alternate manufacturer. To add to the confusion, the latter sometimes refers you to the one you wrote to originally.

If a part is unobtainable from any nearby sources, you will have to order directly from the factory, and you can count on a delay of several days, or even weeks, for delivery. For emergency repairs where cost is a minor item, telephoning the factory service manager and requesting shipment by air express will usually get you quicker service than ordering by letter. Often a telegram will work. For the same rate as a day letter, night letters allow you to use more words and hence longer descriptions of the part, yet with only a few hours' lag in delivery time.

The number of parts to stock is determined by your volume of appliance business and the nearness of suppliers. Of course, you should stock as many as possible, in order not to waste time running after parts—but within limits. Information on how to run a small business as well as how to stock necessary parts may be obtained from the associations listed in Appendix F of *Basics of Electric Appliance Servicing*.

One last word about parts procurement: Customers are often impatient to have an appliance repaired. For some reason, they do not mind waiting a reasonable length of time, as long as you deliver when promised. But if they have to wait one more day, you are in deep trouble. Customers in smaller towns do not seem to be as demanding as those in larger cities; they are accustomed to waiting and as a rule are quite patient. For the "cannot wait" customers who must have the job done immediately, tell them that you will be very happy to oblige, but there will be a small extra charge for long-distance telephone calls, bus charges, etc. When you tell them how much the charge will be, they may change their minds and give you all the time you need.

Servicing motor-type small appliances

CHAPTER

3

As was stated previously, there are a great many motorized appliances on the market and this means that they must be serviced. In this chapter we give the information necessary to service most small appliances that are motor-driven. Remember, however, that this information is of a general nature and more specific information is given in the manufacturer's service manual.

FANS

Electrical fans are one of the simplest types of motor-driven appliances. They usually consist of only three components: a propeller assembly with two to five blades, a motor, and a selector control switch. The function of these components is to circulate air for cooling purposes or to move air about the home. The capacity of a fan is usually expressed in cubic feet per minute (ft^3/min) and is determined largely by the length, pitch, and speed of the blades. Naturally, the more air the blades are capable of moving, the more power is required from the motor—or rather, the more powerful a motor is needed.

Today most small- and medium-sized fans employ shaded-pole induction-type motors to drive their propeller assemblies. There are two reasons why the shaded-pole motor has replaced the universal type in this size range.

1. Because of continuous operation that most fans are subjected to, brushes of universal motors tend to wear out fairly rapidly. Shaded-pole motors, of course, have no brushes and offer no problems on this score.
2. Since shaded-pole motors have a slow starting torque, the blades stop easily when a foreign object is placed into them. A universal motor, which has a high starting torque, does not have this safety feature.

Larger fans usually have split-phase or capacitor-start motors; some motors, such as those used in large attic fans, have a belt connection between the motor and the blade assembly. The only fans that may be operated on either alternating or direct current are those driven by a universal-type motor.

Fans are made in many different designs. Some have only two speeds, some have three or more, and others are even reversible, with three speeds in each direction. This reversible feature permits the user to set the fan in a window, for example, and blow out hot air during the day and pull in cool air during the night. All these operations are controlled by the selector control switch.

A few fans have a built-in thermostat which will turn on the unit automatically when a preset temperature is reached. A complete explanation of thermostats is given in Chap. 5.

Some fans have a mechanism that permits them to oscillate, to move back and forth, as the motor and fan rotate. In this way they can move a larger volume of air. The oscillating mechanism usually consists of a small worm gear and arm on the rear of the motor housing, which frequently has a metal covering. There is usually a thumb nut on a rod coming down out of the motor; this nut can be tightened to hold the arm in such a position that the fan will not oscillate. It can be made to oscillate by loosening the nut, thus letting the arm drive the fan back and forth again. In other words, this design feature permits the unit to be employed either as an oscillating model or as a stationary type of fan.

Modern household fans are usually divided, depending on their operation and method of mounting, into these five general classes.

1. *Portable or table fans.* Available in either oscillating or stationary (nonoscillating) designs, this class of fan is generally mounted on a heavy base or pedestal. The blades (two or three) are usually connected directly to the motor shaft and are protected by a suitable wire guard. An ON-OFF switch is provided, and sometimes a two- or three-speed control.
2. *Floor fans.* These fans, as their name implies, are usually placed on the floor and are designed for either horizontal or vertical operation. The horizontal-type fan moves air in a horizontal or straight plane, while the vertical type moves the air from the floor outward in a circular motion. They are available in either oscillating or stationary models.
3. *Window fans.* This type of fan may be either built into a window or portable. The built-in type is fastened in the window for permanent installation, while the portable one may be

moved about as desired. Most modern window fans have the reversible feature of expelling the air from a room or drawing it in from the outside.

4. *Exhaust fans.* Usually built into the ceilings and walls of kitchens, baths, and laundry rooms, exhaust fans come in various shapes and sizes, but they all operate in basically the same way. They usually use a shaded-pole motor, and the blades—usually four —are of fairly high pitch so that they draw out of a room a good amount of air for their size. The exhaust from these fans may go directly outdoors or may be directed by ducts to the outside.

5. *Attic fans.* These fans are usually just big versions of the exhaust fans described above. While they are mounted in various ways, the service technician is usually concerned only with the control switch, motor, and blade assembly; the latter is usually belt-driven rather than having a direct drive. The motors used on larger attic fans are generally of either the split-phase or capacitor-start types. Full information on these motors can be found in *Basics of Electric Appliance Servicing* and *Major Appliance Servicing*.

Servicing of Nonoscillating Fans

As was mentioned earlier, a fan consists of a motor, blade assembly, and selector control switch. Most fans use a series-inductor or tapped-field motor control system (see Motor Speed Controls in Chap. 2). The majority of three-speed fans employ the tapped-field control, while the two-speed fans use a reactance choke. As shown in Fig. 3-1, in a three-speed fan, not all the field windings are used except in the LOW position. Then, if the fan motor will run only on high or medium speed but not on low speed, the problem is usually in the portion of the field winding between the MEDIUM and LOW terminals. Or, if the fan motor runs only in the HIGH position, the trouble is in the field windings between the HIGH and MEDIUM terminals. Of course, also check for a dirty switch and a loose contact or wire. Frequently, a dirty switch contact will cause one or more speeds to go dead. If a fan will run on any one speed, the chances are that the motor is in good operating condition. For example, if it will run on MEDIUM but not on HIGH or LOW, the switch is almost sure to be defective. Most of the time, cleaning the control switch contacts with a little contact cleaner will clear up the problem; if it does not, replacement of the switch is the only solution. While many fans that use tapped field windings for speed control have replaceable windings, it is usually best to replace the entire motor. If a reactance is found by continuity test to be either open or shorted, it should be replaced.

A fan that runs slow may have one of the motor troubles outlined in the previous chapter, or it may have a defective speed control. Lack of proper lubrication can also cause a fan to operate slower than normally. For specific lubricating information, check the user's guide or the service manual.

If the fan motor refuses to run at all, check the cordset. This is bent and pulled continually, and many troubles are found in the line cord itself, near the plug. The wires may break inside the insulation. Since most of the plugs are of the molded-on type, the only repair usually is cutting off about 3 or 4 in of the cord and installing a replacement-type line plug. If the jacket of the cord itself is worn, frayed, or broken in places, replace it. If the problem is not discovered after checking the selector control switch,

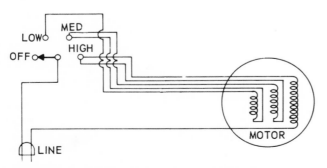

Figure 3-1. Schematic diagrams of three-speed fan.

the cordset, and all connections in between, then testing the motor in the various ways mentioned in Chap. 2 will be necessary.

The blade assembly can be the source of problems, particularly if the complaint is that the fan is noisy. Actually, the commonest cause of noise problems is loose parts, especially in the safety guard over the blades or the blades themselves. In the latter case, if the hub loosens on the motor shaft or one of the blades is loose, this will create a great deal of vibration and rattle.

The hub is usually fastened to the motor shaft with one or more setscrews, and any looseness here can be corrected by tightening these screws. As a rule the blades are riveted to the hub assembly. To retighten a loose rivet, take the hub and blade assembly off, hold the rivet over a very solid vise or a good-sized piece of flat steel, and tap the other end of the rivet lightly with a ballpeen hammer. If this does not do the job, a new rivet will have to be installed in place of the old one. If you have any difficulty in removing the blade hub from the motor shaft, sometimes lubricating it with isopropyl alcohol will help.

While working on the blades, be very careful not to bend them out of position. This can cause much more trouble than the insignificance of the problem would indicate. If the blades are bent through handling or being hit accidentally, their symmetry is destroyed. This lack of symmetry will cause the whole body of the fan to vibrate and rattle and may also affect the speed of the fan. To check this, select some point of reference on the cage or guard and measure with a ruler to some easily identifiable point on a blade. Then turn the blade slowly by hand, and check each blade to see that it has the same angle with the shaft as the others.

A more exact way of checking the blade angles is to cut a template of cardboard in the shape of a triangle at the proper angle, which will be about 15 to 20° in most fans. (Too high a pitch on the blades will increase the load, making the fan run slower than normal, and the fan will actually

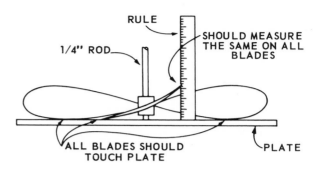

Figure 3-2. Another method of checking a fan blade for pitch and tracking.

move less air than it should.) By placing this triangular gauge under a blade, it is possible to see if the blade is properly set. If a blade is out of alignment, bend it in the direction required to correct the condition. It is usually wise to run and check the reaction anytime a blade is bent, even slightly. Frequently it requires several "slight" adjustments to remove all vibrations from blades that are out of alignment. Remember that it is most important for all the blades to have the same angle of attack. Being a few degrees off angle will not matter if the fan is otherwise in good condition, but one blade at one angle and the remaining at another surely will create vibration and rattle.

Fan blades, especially in kitchen exhaust fans, can become unbalanced by accumulations of dirt on one or more blades. To correct, remove the dirt with a good cleaning solvent such as perchloroethylene, trichloroethane, or trichloroethylene. Bearings that have worn in excess of two or three thousandths of an inch will also cause noisy operation. In the case of excess bearing looseness, it is best to install new bearings.

Here are a summary and analysis of complaints usually received about nonoscillating (stationary) fans.

Motor will not run.

1. Check the cordset, selector switch, field winding, and all connections for continuity. Replace any faulty components.
2. Check for binding rotor. Remove bind.

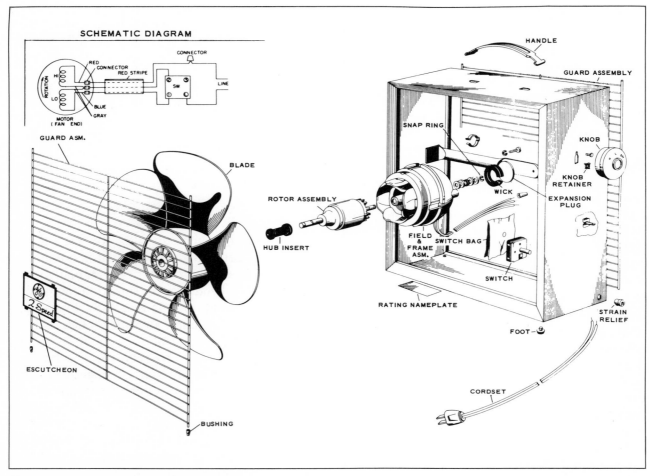

Figure 3-3. Parts and schematic drawing of a typical portable fan.

3. Check the thermostat (if used) to see that the contacts are not stuck in the OPEN position. Correct the situation if necessary.

Motor does not respond properly to changes in speed of selector switch.

1. Check the speed control switch. Replace if defective.
2. Check field winding taps or reactance choke. Replace any faulty part.

Fan with reversible feature does not reverse.

1. Check the selector switch and switch connections. Repair or replace as necessary.
2. Check continuity of field windings. If component is faulty, replace entire motor.

Motor runs hot, slowly, or intermittently; fan consumes above-normal power.

1. Check for shorted field winding. If defective, replace complete motor.
2. Check for rotor bind. Remove bind.
3. Check for bound or frozen bearings. Clean and lubricate the bearings. Replace any faulty components.

Fan is noisy or vibrates.

1. Check the blade for distortion, breakage, warpage, balance, and alignment. Repair or replace the blade as necessary.
2. Check the blade for loose hub or elements. Repair or replace the blade, if faulty.

3. See if bearings need cleaning or lubrication. Clean and lubricate if necessary.
4. Check the rotor shaft. If loose or bent, replace the rotor.
5. Check for steel chips in the field. Remove the rotor and blow out the field with compressed air, if any chips are found.
6. Check for loose guards. Repair.
7. Rotor may be striking the field. Repair or replace.
8. Check for loose or missing screws. Replace as needed.
9. In some two-blade models, the upper blade may be out of diametric balance with the lower blade. Replace the upper blade; if the problem is not corrected, replace the lower blade also.

Fan has magnetic hum.

1. Check air gap for unevenness.
2. If the field cannot be adjusted to eliminate the hum, the replacement of a rotor or a complete motor may be necessary.

Servicing of Oscillating Fans

The typical oscillating-fan mechanism, as previously mentioned, consists basically of a worm gear on the motor shaft that engages a gear on a short rotor shaft. This shaft has a worm on the other end and is enmeshed with a spur gear on a vertical shaft. A pinion or stud attached to the lower end of this vertical shaft rotates at a very slow speed, and, by means of a short lever fastened to the pinion or stud at one end and the motor at the other end, the fan is made to turn back and forth.

In addition to those problems already mentioned for the stationary fan, the oscillating type poses some additional ones. Here is an analysis of complaints that are received regarding the oscillating-fan mechanism and what should be done to remedy them.

Fan will not run.

1. Check all connections, cordset, field, and switch for continuity. Replace any faulty components.

2. Check for binding oscillating mechanism. Repair as needed.

Fan will not oscillate.

1. Check compression stud, worm gear, and pinion. If worn, replace.
2. Check spur gear for broken teeth. Replace if necessary.
3. Check for bent rotor shaft. Repair or replace as necessary.
4. Check spur gear pin for proper setting. If loose, knurl end slightly and press into place or replace the complete gear assembly.

Fan has magnetic hum.

1. Check the air gap for unevenness. If incorrect, loosen field screws and correct the position of the field.
2. Check the armature for a bent shaft. Repair or replace as needed.
3. Check for worn or loose bearing fit. If defective, replace. When replacing, clean the gear case of all old grease. The bearing swivel stud washers and rotor shaft should be lubricated with a light film of SAE-30 motor oil.

Oscillating mechanism's bearings rattle.

1. Check for worn bearings, particularly the

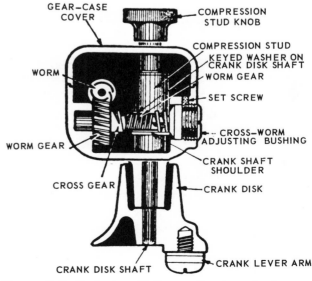

Figure 3-4. Typical oscillating mechanism of a fan.

motor end. Replace gear case assembly if necessary.

2. Check rotor for excessive wear. Replace if defective.

3. Check for proper grease. Clean out gear case and replace with the grease recommended by the service manual.

Oscillating mechanism is noisy. Excessive end play between the compression stud and the cover may be the cause of this noise. To prevent excessive end play, use spacers between the compression stud and worm gear.

Before returning a fan to a customer, check the fan's operation against the manufacturer's specifications for current or power rating. (This holds true in the servicing of all small appliances.) The power consumption should not exceed the nameplate rating by more than 10 percent at any speed. A test of the power consumption of a fan should always be made in a draft-free room.

If a hi-pot test is recommended as a final test (see Continuity Testing in Chap. 1), be sure that the fan has run for about 5 min before conducting it. A warm motor will usually indicate high-voltage leakage sooner than a cold one. Always follow the service manual instructions for conducting a hi-pot test.

VACUUM CLEANERS

All vacuum cleaners, irrespective of type, operate on the same basic principles and contain the same basic parts: a cordset to supply the necessary electric power, a switch of some type to control the operation of the unit, a motor-driven fan to provide the needed air suction, a duct system to channel the air flow, a nozzle to pick up the dust and dirt, a container to catch the dust and dirt, and a case to keep the components together. While there are many different vacuum cleaners on the market today, all can be divided into three basic categories.

1. *Canister/tank.* The cleaning ability of the canister/tank model comes from the suction

produced by the motor-driven fan. This suction pulls at the surface, and the air flow carries the litter and dust into the dust bag.

2. *Upright.* The cleaning ability of the upright comes from both agitation and suction. The suction lifts the rug against the nozzle; the brushes and/or beater bars vibrate the dirt and litter loose from the rug; the air flow carries the dirt into the dust bag.

3. *Combination.* Some manufacturers combine the strong suction of the canister/tank and

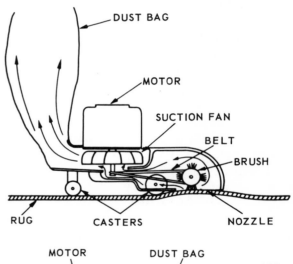

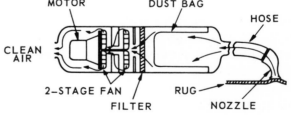

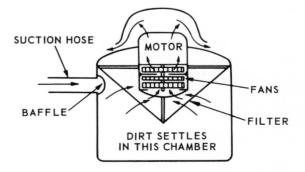

Figure 3-5. Working operation of canister type (top), cylinder (center), and upright (bottom).

the powered agitator head of the upright. They do this by adding a powered rug nozzle to the canister/tank models.

There are several variations of these three basic models. For instance, some vacuum cleaners feature a water filter system rather than using the bag as a filtering device. In these, an air stream passes through the water and deposits the dust and dirt on the water surface. There are also built-in systems (see Chap. 8, *Refrigeration, Air Conditioning, Range and Oven Servicing*) in which the dust and dirt goes into a collector tank.

Let us take a closer look at the three basic categories of household vacuum cleaners.

Canister/tank models. This type of cleaner, as previously stated, is strictly "suction" operated; it relies on suction alone without the motor-driven brush. The area of the nozzle in contact with the rug can be smaller, and therefore the effective suction is greater. Brushes and combs have been built into these nozzles to aid in picking up threads, hairs, etc.

In most canister/tank cleaners, the collection bag is located at the intake end of the machine. The dirt-laden air sucked up through the nozzle and the hose passes through the bag, where most of the dirt is separated from the air. The air then passes through the motor and out the other end, where there is usually an additional dust filter to clean the air before it reenters the room. When a cleaner, such as the typical one shown here, is plugged into a 120-V ac power supply and the switch is pushed on, the motor circuit is energized. The armature with fans mounted on the shaft rotates in a counterclockwise direction (looking at the motor from the top). The rotation of the fans creates the air flow that starts at the swivel cap and goes through the disposable bag and inner bag, which filters out the dirt and dust. The air, after being filtered, continues to flow through the motor housing and fans and out the exhaust hole in the casing.

Upright models. As previously mentioned, the upright cleaner agitates the nap of a rug or carpet with a motor-driven brush to help to loosen dirt embedded in the pile and allow it to be sucked up into the bag. This brush also helps in picking up lint, threads, hairs, etc. The motor, usually of the universal type, is connected directly to the suction-fan assembly and is connected by means of a belt-and-pulley arrangement to the revolving rotary-type brush. The fork handle is attached to the cleaner and serves to guide it over the carpet surface. Wheels attached to the casting incorporate an adjustment screw by means of which they may be raised or lowered for cleaning rugs of different thicknesses.

In the typical upright unit shown here, when the switch is turned on, it completes the circuit and energizes the motor. The pulley mounted on the end of the armature drives the belt which turns the brush. The rotating brush loosens the dirt particles and tends to pick them up with centrifugal force. Air flow created by the fan of the armature pulls the dirt through the opening, around the brush, and through the motor base to the bag. The air passes through the bag, depositing the dirt in it. The air then enters the motor through a filter, passes through the motor, and is exhausted through a grill in the motor frame. Many units have a HIGH-LOW speed switch which regulates the motor speed and also mechanically closes off the intake from the bottom of the unit when in the HIGH position. This switch also lifts the brush free from the floor surface when in the HIGH position.

Combination models. Double duty is the feature of a combination unit. The canister, with attachments, offers the powerful suction that is important for general floor care and above-the-floor care and the versatility that is unique to the canister. The powered nozzle attachment with its separate motor-driven brush agitates the rug and provides rug cleaning efficiency. In this way, the combination unit is similar to the upright but differs in its ease of operation.

Washer-dryer vacuum cleaners. When the switch is turned on in the typical unit shown here, the motor circuit is energized. This causes the fan on the armature to turn, creating air flow. The air, with dirt particles, enters the nozzle

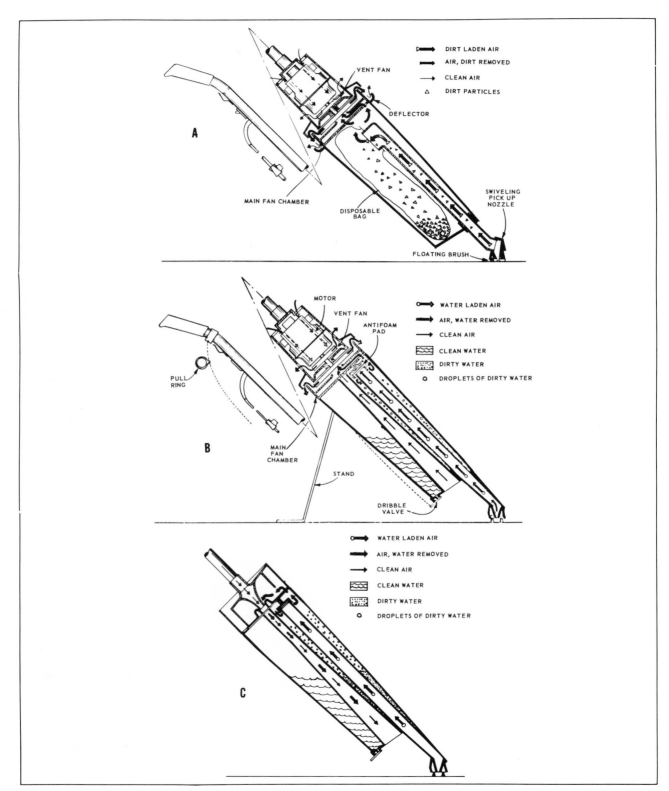

Figure 3-6. Washer-dryer vacuum cleaner operation.

and goes through the inner tube, and the dirt is deposited in the bag. The air then goes through the bag, the gasket, the lid, and the orifice and is expelled by the fan through the sides. Cooling air is pulled in the top vent through the motor and is expelled by the vent fan through the side vents in the motor housing (Fig. 3-6A). Incidentally, most *electric brooms* operate in much the same manner.

To operate as a washer, the water and detergent (most cleaners hold about $1\frac{1}{2}$ qt) are usually poured into the lower portion of the tank and assembly unit. A chain or similar device is pulled to release the detergent and water onto the floor. When the switch is turned on, as previously stated, the motor is energized, causing the fan and the armature to turn; the fan causes the air to flow through both the center duct in the base and the nozzle. The dirty water is picked up with the air at the nozzle, pulled through the inner pipe, and dropped into the dirty-water reservoir. As shown in Fig. 3-6B, the air then passes through an opening in the gasket and lid and is expelled through the side vents in the housing along with the air which is picked up through the center duct. Cooling air is pulled down from the top vents through the motor and is expelled through the side vents by the vent fan.

Washer-dryer vacuum cleaners are available which attach to a standard vacuum hose and which in turn are connected to the exhaust port of a standard vacuum cleaner. An adapter is usually provided if the exhaust port is in the bottom of the cleaner. The water and detergent are added through the filler hole in the top and released from the tank to the floor by means of a pull chain and valve. With the washer-dryer attached to the cleaner and the switch turned on, the air flow from the exhaust of the cleaner is forced through the wand assembly, top assembly, lid assembly, and center duct of the tank. The shape of the top assembly and lid assembly forms a venture which creates a suction at the nozzle through the inner tube and deposits the dirty water in the reservoir in the tank (Fig. 3-6C).

The analysis of complaints for washer-dryer vacuum cleaners, as well as electric brooms and sweepers, is very much the same as for standard vacuum cleaners given on page 61. Servicing techniques are about the same, too.

Servicing Vacuum Cleaners

The most common trouble with canister/tank cleaners is a lack of suction. Since this type of cleaner depends on a high velocity of air movement, anything which impedes this flow will reduce the suction. Among the causes are a clogged nozzle, obstruction in the hose, a faulty hose, very full bag, improper installation of the bag, slow motor speed, and dirt-laden filter. A poor connection of any of the fittings—such as nozzle to pipe sections and/or to hose and to hose cleaner—will cause a loss of suction at the nozzle, where it is needed. A tiny hole or tear in the hose will also cause lost suction. Thus, when a canister/tank cleaner's fans run to full speed but there is little or no suction at the hose end, disconnect the hose in order to determine whether the suction is more nearly normal at the intake port on the cleaner. If it is, chances are good that the hose is clogged. If no obstruction can be seen in either end and there is no external evidence of collapsing of the hose, attach it to the blower port, put the free end outdoors, and run the machine at full speed in an effort to blow out whatever is blocking the hose. This usually works. If not, use an electrician's fish tape to pull out the obstruction.

The hose employed in some canister/tank models can be a source of assorted suction problems. For instance, plastic hoses are subject to tears and pin holes; in other types of hose, the inner lining collapses and flutters, thereby partially blocking the passage and causing an intermittent lapse of suction. A weak spot, easily kinked and noticeable from the outside, usually reveals such a rupture. Regardless of the type of hose used, when it is discovered to be faulty, the best procedure is to replace it. In other words, it is good servicing practice never to try to repair a cleaner's hose, except in an emergency.

Two or three hairpins or the like sometimes wedge crosswise in an upright model's exhaust orifice, where they will trap a mass of lint, rug wool, hair, and dirt, ultimately blocking the passage completely. This trouble is easy to detect, for the bag will be extremely slow to inflate, if it inflates at all. To clear the passage, merely disconnect the bag and pull out the obstruction with your fingers. If a clogged condition cannot be found, check the motor speed. If too slow, the cause should be determined and corrected.

Most experienced service technicians have developed a "sixth" sense for ascertaining the adequacy of a cleaner's vacuum. One of their tricks is to hold the palm of their hand over the suction end of the hose and then, pulling the palm away, to judge the degree of suction by the strength of the "plunk" sound they hear. Of course, a more accurate way of judging a cleaner's suction pull is to use a vacuum gauge.

A simple shop-type vacuum gauge can be made from a suitable glass bottle or jar, a yardstick, a ruler, two pieces of rubber or plastic tubing, and a 50-in piece of glass tubing. These parts are assembled as shown in Fig. 3-7. To run a

suction test, insert the open end of the tubing into the cleaner's intake and pack a plastic bag around the hose to make as airtight a connection as possible. When the cleaner is turned on, it causes a partial vacuum to form in the glass tube which permits the atmospheric pressure to force the water to rise inside it. This water lift, in inches, depends on the vacuum being pulled; it usually ranges from about 30 in for small portable models to over 50 in for large, deluxe canister/tank cleaners. The recommended water lift in inches is usually given in the manufacturer's service manual, or the suction strength of the cleaner under test may be compared with that of a new cleaner of an identical model or type.

The complaint of not picking up lint and threads is rare with canister/tank models because their rug nozzles are usually fitted with a rather trouble-free stationary brush or "combing" device to loosen stubborn particles of lint and the like which do not always yield to suction alone. These parts are readily accessible and quite simple to adjust or renew. Indeed, in most cases where stationary brushes wear out, users will buy new ones and install them themselves. The combing device is virtually indestructible.

Failure to pick up lint and threads is more common to the revolving-brush upright. Since this machine is noted for its ability to pick up almost anything that will go through its nozzle, most users are quick to cry "it will not pick up!" if the machine misses one particle of lint on the first pass. And this type cleaner gets more than its share of abuse. It is used indiscriminately by some persons to pick up needles, pins, buttons, coins, nails, tacks, and even matches. Bags have been set afire by the latter. Obviously, some of the sharp or pointed objects will either get under the belt and cut it in two or wedge in the nozzle and jam the brush, thus also destroying the belt. So when a revolving-brush upright does not pick up lint, you may suspect some trouble with the nozzle brush or its belt. Remember that a belt with too much tension exerts an additional load on the front bearings, whereas one with too

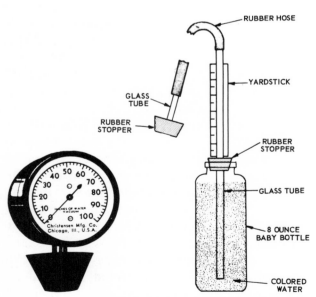

Figure 3-7. A commercial gauge for testing vacuum cleaners is shown at left. A simple vacuum gauge suitable for checking most home vacuum cleaners at right.

little tension will slip when the brush contacts the carpet. Belt tension may be checked by setting the cleaner on the edge of the rug and placing your hand underneath the rug. If you feel vibration, the brush is turning.

Almost invariably when servicing an upright model which fails to pick up lint, you will find that the customer has run the nozzle down to its lowest position in an effort to force the ailing cleaner to pick up clinging particles. In such cases it must be explained tactfully that an abnormally low nozzle position will make the cleaner extremely hard to push and will not improve its cleaning action in any way. Brushes must, of course, be clean to operate at proper efficiency.

To find the ideal nozzle position for a specific rug, follow these three steps: (1) Raise the nozzle to its highest position, (2) turn on the motor, and (3) lower the nozzle gradually only until you hear the sound of compression (at which time the motor accelerates noticeably). This will indicate that the rug has sealed the nozzle's mouth. In other words, the ideal nozzle adjustment is the highest position where compression can be sustained. Thus, wear on both rug and brush is minimized, for it should be quite clear that the rug will be gently swept on a cushion of air. On the other hand, too low an adjustment could mean that the rug will be hammered by the brush with the floor beneath serving as an "anvil." Some hard-to-push troubles point to a binding spindle in a swivel caster. A drop of light oil usually will remedy such a fault.

To eliminate the problem of squeaking, the wheel hubs should be lubricated occasionally. But see to it that not a single drop of oil remains on the outside of the wheel where it could trickle down to the tread. Infrequently you may have to oil the handle-fork pivots to stop a squeak there. Just be sure that you wipe the outside of these parts thoroughly clean, too.

Upright cleaners which have only a detent to lock the handle in any of several positions present no service problems in respect to that mechanism to anyone who can tell one end of a

screwdriver from the other. Satisfactory installation of a handle-balancing spring in certain models, however, is something else. Anyone's patience can be strained almost to the breaking point when undertaking a handle-spring job for the first time, yet it is not a difficult task if a few facts are fully understood. Lacking specific instructions, the following general suggestions are helpful. The handle-spring tension is extremely important. For example, sufficient tension should be provided to pull the handle to its upright position from about a 15° tilt. With more tension than that, the entire cleaner may rock backward when the handle is lowered; with less, the handle will fall. To achieve the happy medium when loading the spring, it may be necessary to "wind it up" about one full turn with the handle fork in its vertical position.

In some models, one end of the spring may be hooked in either of two places 180° apart on the handle-fork shaft, thus affording a two-position adjustment. So with this type it is a good idea to fit the parts together loosely at first, in order that you can select the ideal hooking point before you struggle with loading the spring.

Analysis of complaints. Almost all vacuum cleaner complaints, whether the cleaner is of the upright or canister/tank type, fall into one of the following categories.

Motor will not run. If the lamp lights normally but the motor does not run, make these tests.

1. Check the following for continuity: switch, line cord, field, armature segments, brush to armature, and all lead connections. Make necessary repairs or replace any defective components.
2. Check for worn motor brushes or brushes hanging up in the brush holders. Replace the motor brushes or otherwise correct the situation.
3. Check for a jammed fan. Remove cause of bind. If fan blade is bent or damaged, replace it.
4. Check for frozen motor bearings. Clean and

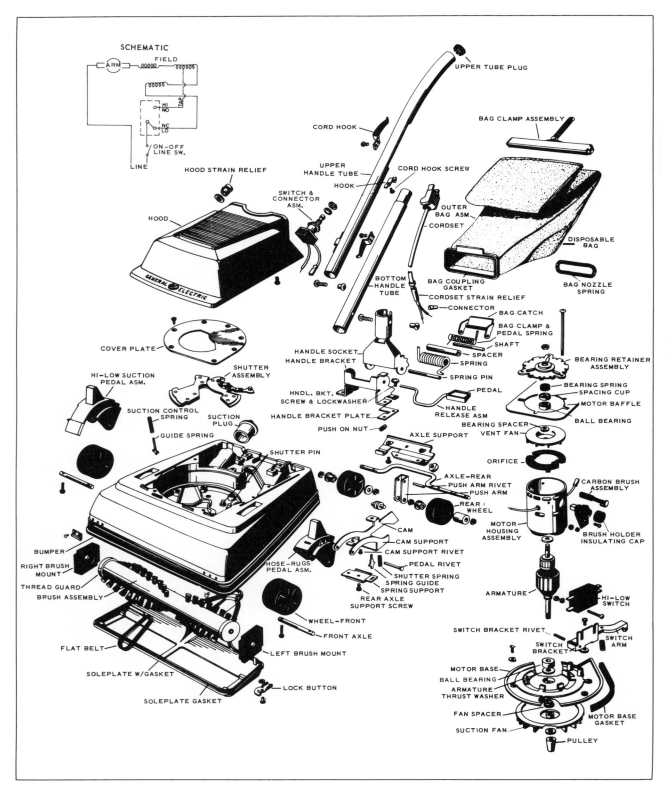

Figure 3-8. Disassembled view of upright-type vacuum cleaner.

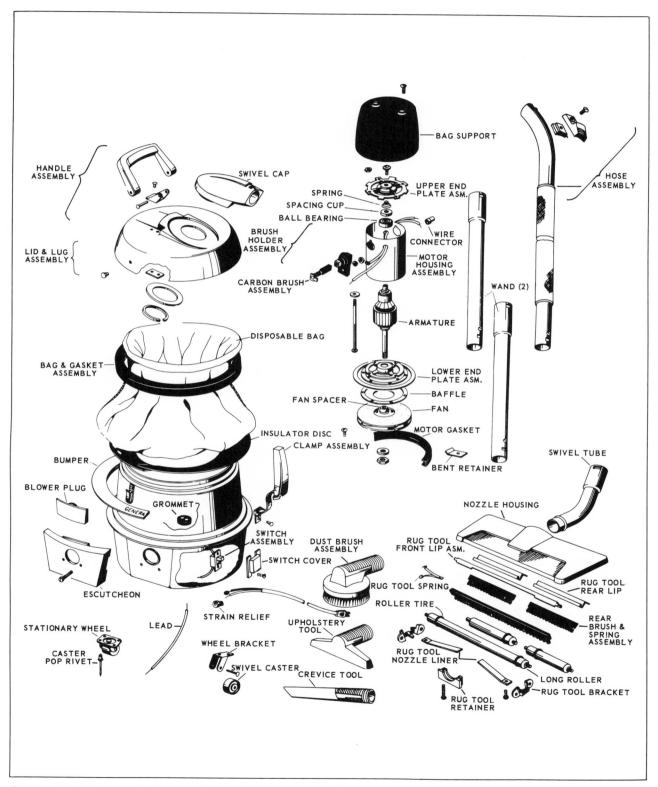

Figure 3-9. Disassembled view of canister-type vacuum cleaner.

lubricate. If too worn, replace bearings.
5. With some models (upright) check brush block contacts for proper contact with rings on the cord reel. Correct the problem.
6. With a model that has a lamp and it does not light, check for only the following:
   a. Open-line cordset.
   b. Faulty ON-OFF switch.
   c. Open connection in the wiring between the cordset and lamp.

Motor overheats.

1. See if the ventilation openings are filled or clogged with dirt. Clean as necessary.
2. Check the armature and motor field windings for continuity. If you find a short, replace the complete motor.
3. See if the field windings are grounded. If they are, correct the problem or replace the motor.
4. Check the armature for contact with the field, in which case worn bearings may be the cause. If badly worn, replace them.
5. Check for frozen motor bearings or a bent armature shaft. Remove the source of bind or replace the armature.
6. In some models (upright), check for foreign material binding the brush assembly. Remove the source of the bind.

Motor running too slow; insufficient power.

1. Check for foreign material caught on the armature or the fan. Remove the source of the bind.
2. Check for burnt-out armature or field windings or poor brush contact in the motor. Repair or replace defective parts or the entire motor.
3. Check for misaligned motor bearings. Realign bearings or replace them if badly worn.

Motor running too fast.

1. Check for shorted field windings. Replace motor.
2. See if fan is loose on the shaft and is not turning with the armature. Tighten the fan.

3. Check the dust bag. If overfilled, replace, if disposable; otherwise thoroughly clean.

Motor starts and stops while vacuum cleaner is being used.

1. Check for intermittent break in cordset. If defective, replace.
2. Check the control switch. If defective, replace.
3. Check for short in the wiring or a bare spot on wiring insulation. Repair as needed.
4. Check for loose connection in wiring. Repair as needed.

Brush will not rotate although motor runs (upright models only).

1. Check for broken or loose belt, or belt off. Replace or repair as needed.
2. Check for tight or locked bearings. Replace or repair as needed.
3. Check for extraneous material binding brush. Remove any foreign material.

Motor runs only on high or low speed (two-speed models only).

1. Check speed (HIGH-LOW) selector switch. Replace if defective.
2. Check mounting on speed selector for looseness or improper adjustment. Take necessary corrective measures.
3. Check for defective field windings. Replace motor if you find faulty tapped field windings.
4. Check mechanical linkage from lever to switch. Repair as needed.

Little or no suction but motor runs.

1. Check for plugged hose. Remove obstruction, if you find one.
2. Check for plugged disposable or cloth bag. Also check the dust level in the bag to see if it is too full. If overfilled, replace, if disposable; otherwise thoroughly clean.
3. Check for loose fan on motor shaft. Tighten as needed.
4. Check the seal at the lid and motor base for leakage. Repair or replace as needed.

5. Check for excessive leakage at the swivel cap. If you find excessive leakage, take necessary steps to correct.
6. Check for clogging of the cleaner with dirt deposits. Clear out obstruction if necessary.
7. With some models, check suction control for proper position. Take proper corrective action.
8. Check for contamination of motor filter. Clean filter, if needed.
9. With some models, check for loose belt or belt slipping. Tighten or replace belt as needed.
10. With some models, check for excessive contamination or plugging of exhaust screen assembly. Clean exhaust screen, if necessary.
11. With some models, check for excessive leakage between top and bottom compartments. Take proper corrective action.
12. Check hose for air leaks. Replace hose, if defective.
13. Check for clogged hose. If necessary, blow or push obstruction out of hose.
14. Check for defective motor and fan assembly. Replace any faulty components.
15. Check motor speed. If low, take necessary corrective action.
16. Check attachments for cracks or damaged portions causing leakage. Replace as needed.

Cleaner does not pick up properly (upright model only).

1. Check the nozzle-adjustment mechanism and handle-tension spring. There are three types of nozzle adjustments: thumbscrew, lever, and automatic. Replace any defective parts. The tension spring must be replaced if broken.
2. Check the floor brush; see that the bristles are flush with the bottom casting lip. If of the adjustable type, the brush may be set to the next lower position; otherwise, the brush should be replaced.
3. Be sure that nozzle adjustment is correct for carpet nap. If not, adjust nozzle for correct contact to carpeting.

4. Check the belt. Make sure it is the proper one for that cleaner. If the belt is broken, replace it.
5. Check the rear caster mechanism; see that it swings freely and that the wheel revolves freely. The height of the caster governs the position of the nozzle above the rug; therefore, if the wheels are worn, they should be replaced.
6. Check for stuck agitator brush. Clean bearings free of dirt if necessary.
7. Check the dust bag. If overfilled, replace, if disposable; otherwise, thoroughly clean.

Noisy operation.

1. Check for loose fan. Repair as necessary.
2. Check for fan or armature striking. Take necessary action to realign parts.
3. Check for loose rivets, screws, or ports. Tighten as necessary.
4. Check for defective bearings; if faulty, replace.
5. Check for foreign objects in sweeper. Clean out, if necessary.
6. Check for broken or bent fan; if defective, replace.

Cord reel will not rewind (with some models).

1. Check for defective cord jamming reel. Remove bind.
2. Check for broken cord reel spring. Replace spring, if defective.
3. Check for loose parts. Tighten as needed.

Dust leaks into room.

1. Check for holes in dust bag. If necessary, replace dust bag. If machine still uses cloth bag, replace dust bag if dirty.
2. Be sure bag is correctly installed. Follow directions in owner's manual.
3. Check sealing gasket. Replace gasket, if defective.

As with all small appliances mentioned in this chapter, the power rating of vacuum cleaners should not exceed the nameplate power rating by more than 10 percent at full load.

FLOOR POLISHERS

Virtually all models of floor polishers intended for home use consist of a high-speed motor that drives a set of buffing wheels or brushes through a worm-gear train. The polisher's motor, which is usually of the universal type, is mounted in a vertical position. A pair of oppositely pitched worm gears on the motor shaft engage with a pair of vertically mounted spindles. These spindles power the buffing wheels or scrubbing brushes. Because of the step-down characteristics of the gear train, the brushes and buffing wheels turn a great deal more slowly than the motor shaft. (In a typical 16,500 r/min high-speed motor, the twin brushes are driven at about 500 r/min through a worm-gear train with a speed reduction of 33 : 1.) This speed reduction, however, effectively multiplies the available torque which is most important in the operation of a floor polisher.

When the switch of a typical floor polisher is turned to the ON position, the motor circuit is energized, causing the armature to turn. The armature shaft protrudes at both ends of the motor with a worm gear cut on each end. The worm gear engages with the spindle gears which drive the right and left spindles in opposite directions. The buffing wheels or scrubbing brushes are placed on the spindles before the floor polisher is turned on.

Most floor polishers feature a two-speed (high or low) selector switch, which achieves its speed control with a set of tapped field windings.

Analysis of complaints. Little can go wrong with a floor polisher, but here is a summary of the major complaints and how to cope with them.

Polisher will not run.

1. Check the cordset, switch, field, and motor lead assembly for continuity. Replace any faulty components.
2. Check the armature for signs of arcing. If the armature is arcing excessively across two segments of the commutator, replace it.
3. Check the brushes for contact with the commutator. Replace worn brushes.
4. Check the unit for binding and alignment by rotating the armature by hand. Repair or replace as needed.

Polisher is noisy.

1. Check for worn floor brush fitting. Replace brushes or fittings as required.
2. Check grease supply in gear box. If necessary, replenish.
3. Check for stripped or worn spindle gears. Replace gears, if defective.
4. Check for worn spindle bearings. Replace frame assembly if necessary.

Floor brushes fall off.

1. Check for missing or worn snap ring on spindle head. Repair or replace as needed.
2. Check for worn spindle head or brush. Replace parts, if necessary.

Polisher runs hot.

1. Check for shorted armature windings. Check commutator for signs of arcing. If defective, replace armature.
2. Check for shorted field windings. Check for signs of overheating. If the windings are defective, replace the motor.
3. Check end play. In most models the adjustment must be less than 0.010 in and the armature must rotate freely. Take whatever corrective measures are needed.
4. Check for loose or high resistance electric connection. Repair as necessary.
5. Check for dirt in motor which may be blocking the flow of cooling air. Clean the motor housing if necessary.

Motor grounded.

1. Check all electric connections. Make necessary repairs.
2. Check for grease at the brush holders. Clean as needed.
3. Check field coil. If faulty, replace motor.

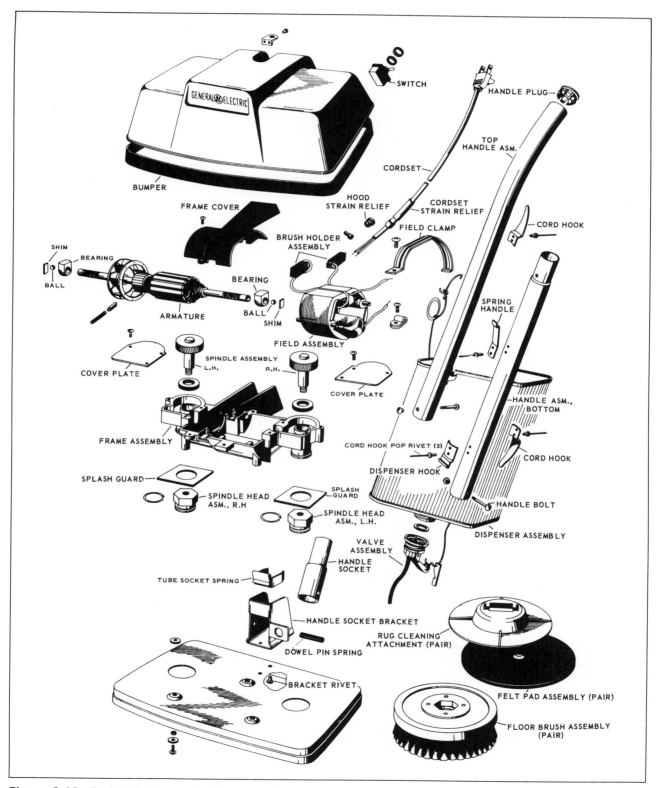

Figure 3-10. Exploded view and major parts of a typical floor polisher.

4. Make certain that all wiring is properly dressed and not pinched or cut by metal parts.

Shampooers are available as an accessory with most floor polishers. They generate foam from a liquid shampoo which is dispensed between the spinning brushes. These two large disks work the foam into the rug and loosen the dirt.

FOOD MIXERS

While there are minor mechanical differences in the various models of food mixers on the market today, fortunately there is a striking similarity in their construction. The heart of the mixer is, of course, a small, high-speed universal motor. But the speed control arrangement can be as simple as a three-step tapped field winding control or as complex as a solid-state electronic control. The motors in the majority of mixers are mounted in a horizontal position, and a worm gear on the motor shaft meshes with a pair of spindle gears that convert the rotation axis to that of vertical action and also reduce the operating speeds to anywhere between 300 and 1,300 r/min. In operation, the food is usually mixed or beaten by a pair of revolving beaters, which are attached to the motor shaft and gear assembly.

Types of Mixers

There are basically two types of food mixers: the stand (upright) type and the portable type. Until a few years ago the stand type dominated the mixer market, but today the portable mixer is the most popular. Some manufacturer's produce a so-called "convertible" mixer which is similar to the stand type except that the head can be detached for portable use or storage. A few food mixers are available that can be built into kitchen counter tops. Techniques for servicing them are the same as for either the stand or portable type.

Portable mixers. The portable mixer is supported by hand during operation. It has a light-

duty universal motor that draws from 100 to 150 W of power; models with relatively high wattage can stir thicker mixtures. While a few portables feature a variable-speed control which permits dialing up to 14 speeds or any speed in between, most mixers of this type have three or five speeds, usually selected by a thumb or finger switch on the handle. Speed control in most three- and five-speed units is attained by various taps on the field windings. In the so-called "infinite-speed models," the beater speed is directly related to the armature speed. The armature speed is controlled by the governor

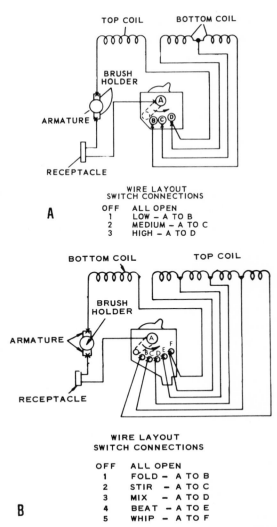

Figure 3-11. Schematic drawing of typical (A) three- and (B) five-speed units using taps on the field windings.

switch. Within the limits of speed and load for which the motor has been designed, the armature accelerates when the governor switch is open. That is, the governor switch responds to two opposed forces acting on the regulator leaf on which one of the contacts is mounted. The spring pulls against the leaf, tending to close the switch and thus accelerate the motor, while the actuator pushes against the leaf, tending to open the switch and thus decelerate the motor. The governor switch is open or closed depending upon which of these two forces is the greater.

Both of these forces which act on the leaf are variable. The force exerted by the spring varies with the position of the control knob. The cam is keyed to the knob and control arm so that, when the knob is turned away from OFF, tension is placed on the control spring which, in turn, pulls on the leaf. The control rod and the governor spring form a flexible connection between the cam and the leaf. Thus turning the control knob away from OFF progressively increases the force of the spring against the leaf.

As in the case with any centrifugal type of governor (see Governor Speed Control in Chap. 2), the force exerted by the actuator varies directly with the speed of the armature. When the armature attains a speed at which the force of the actuator against the leaf approximately equals the opposing force of the spring, any further increase in speed opens the governor switch, and any subsequent decrease in speed closes it. At this point, the governor switch oscillates between the open and closed positions with great rapidity, and armature speed is stabilized within very narrow limits. To increase the beater speed, the control knob is turned to a higher number. This puts additional force on the spring and thereby requires additional force from the actuator to open the governor switch. Additional actuator force can be obtained only from greater armature speed which, of course, results in greater beater speed. This type of "infinite" speed control, employing a governor, is also frequently used in stand-type food mixers.

Stand-type food mixers. The stand-type mixer has a comparatively heavy frame with the mixer head at the top and a wide base big enough for large bowls or a bowl turntable. To ensure complete mixing, the rotating beaters either turn the bowl and turntable or circle inside a stationary bowl.

Stand mixers have larger motors than the portable models, and most have variable-speed controls with a regulator to maintain the same speed whether mixing light liquids or heavy batter. The speed controls may have as many as 10 to 14 positions, ranging from very slow to full speed. The methods of speed control most common in these mixers are the tapped-field windings, the governor-controlled, and the solid-state speed controls. Stand model motors are rated about 150 to 400 W.

Servicing Mixers

Common troubles with food mixers include openings in the cordset, problems in the switch or speed control circuits, and lack of lubrication in the bearings. Methods of finding and repairing these problem areas have already been covered in previous chapters of this book. Troubles typical of universal motors and described in Chap. 2 may be encountered, of course.

It is seldom necessary to *completely* disassemble and reassemble a food mixer, since most repair work can be confined to the area immediately affected. But if it is necessary to disassemble a mixer, always follow the advice of the manufacturer as it appears in the service manual. For example, the disassembly and reassembly information for the typical stand mixer shown in Fig. 3-13 is taken from the manufacturer's service manual.

Disassembly

1. Remove the ejector knob assembly by turning it counterclockwise. The ejector spring will be released upon removal of the knob.
2. Withdraw the two rear screws and the two front screws holding the case and trim

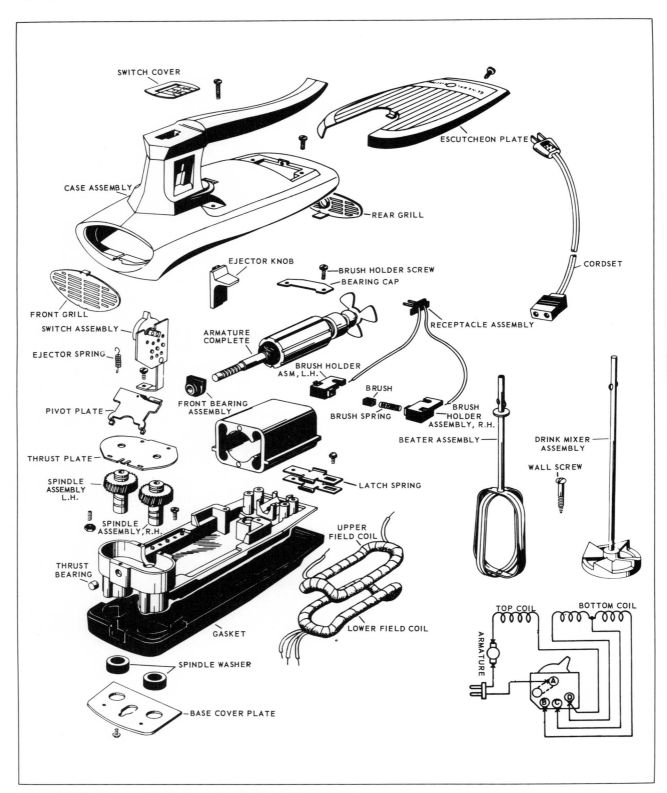

Figure 3-12. Disassembled view of typical portable-type mixer.

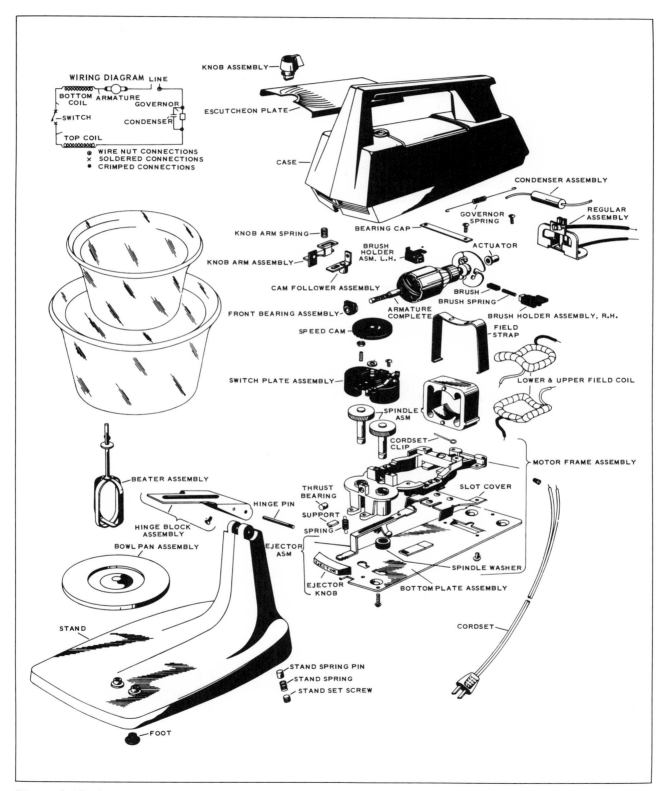

Figure 3-13. Disassembled view of typical stand-type mixer.

plate assembly to the motor base. Lift off the case and trim plate.

3. Withdraw the screw and remove the handle cover by lifting the rear end and sliding it back.

4. The control knob assembly may be removed or replaced by removing the retaining ring from the knob.

5. The grill is held to the case by the grill clip which is pushed onto two pins on the grill. Do not attempt to remove the clip from the grill, as to do so will ruin the clip and/or the grill. To remove the grill and clip assembly, simply lift the clip ends over the stops in the case with a small screwdriver.

6. Note the lead dress.

7. Unhook the governor spring first from the bracket assembly and then from the governor leaf assembly.

8. Remove the two screws securing brush holders. Lift out the brush holders, brush springs, and carbon brushes.

9. Remove the two screws holding the rear bearing strap and thrust plate retainer. Once loosened, lift off these parts.

10. Remove the two screws holding the front bearing strap, camshaft assembly, and bracket assembly. These two assemblies are held together with a small retaining ring. Once loosened, lift off these parts.

11. Removal of the three field screws will allow replacement of the armature or bearings without disconnecting any of the leads. Simply lift the field and armature so that the armature may be withdrawn from the field.

12. Should a field require replacement, use a complete field assembly. Unsoldering and soldering will be necessary, and proper forming and fitting of the leads is important.

13. Spindles may be replaced by removing the governor spring, two screws, front bearing strap, camshaft assembly, and bracket assembly. Remove the retaining ring and washer from the spindle shaft at the bottom of the motor base, and the spindle can be lifted out of the motor base from the top.

Reassembly

Reassembly can be made by reversing the disassembly procedure, but details are also listed below.

1. Lubricate the spindle shafts (with a thin coating of grease) before assembling them into the motor frame. Align the spindle slots with the two dimples on the motor frame. Assemble the washer and retaining ring to each spindle.

2. Place the field with armature in the motor frame, making certain that the spindle slots are correctly indexed to the dimples before and after the armature is seated between the gears.

3. Fasten the three field screws.

4. Place the thrust plate in position at the commutator end of the armature shaft.

5. Place the thrust plate retainer and bearing strap in position at the commutator end of the armature shaft. Secure with two screws.

6. Replace the carbon brushes, brush springs, and brush holders and secure with two screws.

7. Replace the bracket assembly, camshaft assembly, and the bearing strap and secure with two screws.

8. Place the actuator in position in the fan end of the armature shaft.

9. Replace the governor leaf and shoulder screw.

10. Replace the contact bracket assembly and the condenser assembly with a screw.

11. Solder the shunt of the condenser assembly to the governor leaf.

12. Place the switch plate assembly in position and secure with a screw. Be certain the free end of the switch leaf is against the slider.

13. Be certain all solder joints are proper.

14. Connect the governor spring.

15. Be certain all leads are properly dressed according to the wiring layout. The shunt on the condenser must be kept away from the grill and the fan.

16. Connect the motor assembly to the power

source. Be careful of the fan and of possible electric shock.

17. Close the governor and check the high speed. Tap the field iron and front and rear bearing straps gently to remove any binds. Check the wattage; it should be approximately 76 W.

18. Turn the ejector assembly clockwise until the switch just closes (about 36° from the OFF position).

19. Set the minimum speed by bending the contact bracket of the regulator forward or backward. Recheck the high speed by turning the ejector assembly to full ON position. It must be a minimum of 800 r/min at 120 V.

20. Assemble the handle cover, control knob, and retaining ring. With the grill out of the case, assemble the handle cover assembly to the case. To do so with the control knob at position 2, place the handle in the case, down and backward, pressing the control knob against the detent spring until the front tab of the handle cover is all the way down; then push the cover forward and secure with a screw. Check the detent spring operation.

21. Disconnect from the power source; place the receptacle in position in the case and attach the cord. Assemble the case and, with the ejector assembly in OFF position, rotate the control knob to seat into the D of the camshaft assembly.

22. Replace the four screws holding the case to the motor frame. Replace the ejector knob spring, insulation washer, and ejector knob. Tighten the knob securely by hand; it has a right-hand thread.

These instructions from a particular service manual are quite detailed. Unfortunately, not all service manuals are this detailed. For this reason, we include some servicing tips that should be kept in mind when working on food mixers.

For example, when replacing a gear, be sure to clean the gear case thoroughly and discard all the old lubricant, for it is impossible to rid this enclosure of worn or broken gear fragments in any other way. Remember that one stray chip can do serious damage. To clean the gearbox as well as the spindle bearings, use the same solvents as recommended for motor bearings in Chap. 2. Never reuse old gears.

Aligning the spindles is not difficult, because most manufacturers provide some obvious means of positioning the gears on the shafts, such as a countersunk setscrew seat in the spindle shaft and/or mating marks on the gears. It is best to always be sure of exact alignment by trying the beaters in the spindle sockets. After some experience, however, a glance into the sockets as you position the spindles will suffice. Just remember that in nearly all two-beater machines the spindles are timed 45° apart with their indexing pins in this position: /—. In three-beater mixers, the two outer spindles are usually timed parallel and the center beater timed 45° ahead, like this: —/—.

When a spindle shaft seizes in its bearing, remove the setscrew from the gear, improvise a wrench from an old beater for extra leverage, and use this tool to twist out the jammed spindle. Flood the spindle shaft and bearing edge with penetrating oil first, so that as the tight shaft is worked out, oil will follow it into the bearings, thereby reducing the force required as the process is continued. If there is any difficulty in forcing the jammed spindle and its gear part of a revolution against the opposition of the worm in order to gain access to the gear setscrew, check the manufacturer's service manual to see in which direction to apply such force. Once the spindle has been removed, clean it and the bearing, test for free movement, and lubricate as suggested by the manufacturer.

The beaters are locked into the spindles by snap action (usually the ring-and-groove type) in most models; in a few others, by a yoke and setscrew arrangement. Exact radial positioning is ensured by a squared, keyed, or slotted end on the beater shaft which fits into a corresponding socket in the spindle. A beater ejector is a convenient feature found on most models using snap-action beaters. This simple mechanism enables the user, by flipping the handle or by pushing a lever, to partially eject the beaters with little effort.

Before closing the gear case, put the right quantity and type of lubricant into it as recommended by the manufacturer, install a new gasket (if one is used), attach the gear-case cover, and clean the outside of the machine. If the manufacturer recommends the hi-pot test, do it between one of the "hot" prongs on the power plug and a spindle. The speed control should usually be set on LOW.

Analysis of complaints. In order to intelligently analyze a customer's complaint, the service technician should be familiar with the following problems and solutions.

Motor will not run.

1. Check the cordset for continuity using an ohmmeter. Replace if defective.
2. Check the operation of the ON-OFF switch. If defective, replace with a new switch.
3. If the mixer uses a governor switch (usually mounted on or at the end of the armature), see that the points are closed. If necessary, use a point file or crocus cloth and clean the fouled governor contacts. If they are badly pitted, replace them, and test the condenser; replace it if it is shorted.
4. Check for mechanical bind by turning the armature and spindles by hand. If there is binding, try lubrication. If that does not solve the problem, it is often possible to tap lightly on the case with a plastic hammer to line up the armature shaft and the bearings.
5. Check the carbon brushes as well as the springs and leads. If brushes are worn short or hang up, replace the brushes and brush holders.
6. Check the field and armature winding for shorted turns or an open circuit. If a fault is found, replace the motor.
7. Check for a bent fan. Repair or replace as necessary.

Insufficient speed or power. Speed is an indication of power. Check the spindle speed with a tachometer at high and low settings. (A typical reading for a portable mixer is a minimum of 350 r/min on LOW and 800 r/min on HIGH at 115 V or 400 r/min on LOW and 850 r/min on HIGH at 120 V.) If the speed is normal, the customer is expecting too much power for the size of the mixer. If the speed is low, here are some of the possible causes and their remedies.

1. Check for any possible bind. Tap field lightly to left or right to relieve this problem. Be sure the field strap is secure.
2. Check the commutator and carbon brushes for excessive wear or fouled condition. Replace, if necessary.
3. Check for a bind in the bearing or spindles. If necessary, remove the source of the bind.
4. Check the segment-to-segment resistance of the armature and compare it with the proper rating in ohms stated in the service manual (usually 4 to 5 Ω). If the resistance is low, there is a short in the coils; if high, an open exists. In either case, the armature must be replaced.
5. Check the field coil resistance and compare it with the proper rating in ohms stated in the service manual. If any of these resistances is low or high, replace the complete motor.
6. Check the speed control. If faulty, take proper corrective action.

Motor runs hot. Run the motor to confirm an excessive rise in temperature. It may be caused by a bind, in which case the motor will probably also run slow, or by shorted turns in the windings. Check the armature windings as described above under Insufficient Speed or Power and measure the resistance of the field coils separately. Also check for dried-up lubricant; clean and relubricate if necessary. If recommended in the service manual, lubricate the gears and spindles. Use just enough to coat all gear teeth and surfaces, but not so much that the excess will be forced out in the operation.

Motor will not shut off.

1. Check the ON-OFF switch. If faulty, replace.

2. Check for lead wires that may be shorting across the ON-OFF switch.

3. In some governor-type mixers the ON-OFF switch is part of the governor, and in the OFF position the governor points are opened. Check the linkage from the control arm to be sure it is not bent or otherwise not correctly adjusted. Correct as necessary.

4. On any mixer using mechanical linkage between the internal switch and the speed control, check the linkage to be sure that it is not bent or misaligned. Correct as necessary.

Mixer runs only on high speed, will not run on low speed.

1. In governor-controlled mixers check for a shorted capacitor across the governor contacts. Check also to be sure that the governor contacts are not bent, misadjusted, or fused together. Take proper corrective action.

2. In some portable mixers check for a defective switch or an open field winding in the slower-speed positions. If either is faulty, replace.

3. Check control-plate spring. If unhooked, attach to top of control plate and squeeze end.

4. Check operation of control-plate contacts when switch is turned from HIGH to LOW. If contacts do not separate, they are welded together. Replace control-plate assembly.

5. If low speed cannot be correctly adjusted at the control knob, check for excessive end play, actuator tip melted off, or a bent governor body. Take proper corrective action.

6. If the governor flyweights are properly located, be sure the actuator moves freely on the armature shaft. If necessary, correct or replace.

7. This condition will often arise when regulator contacts are closed because the governor leaf is not free or the actuator is out of position from a governor arm. Check the situation and make necessary repairs.

Mixer has no high speed.

1. Check for a bind in the bearings. Remove bind, if one is present.

2. Be sure there is no bind in the gears by turning the spindles. Remove source of bind.

3. Check for shorted or open turns in the armature. If faulty, replace the armature or complete motor.

4. Check for a weak governor spring. Replace if necessary.

Planetary turns, but blades do not revolve. This complaint can usually be traced to a broken pinion gear drive pin. To correct, remove planetary and take off pinion gear. Replace the drive pin. The beater gears may also be stripped; replace beater gears, if necessary.

Operator receives a shock when touching mixer. To check, pull out wall plug, and, with switch on, check for ground with an ohmmeter. Touch one prong of the ohmmeter to prong of plug, and touch other prong to an unpainted metal spot on housing. If the ohmmeter shows little or no resistance, the mixer is grounded. Examine all wiring in order of its accessibility until grounded wire is found, then repair it.

Mixer will not run, although switch clicks and motor hums. Check all bearings in order of their accessibility until frozen bearing is found, then repair or replace it.

Motor surges.

1. Check the actuator for a melted tip or for stickiness on the armature shaft. To correct the latter, polish the end of the shaft with crocus cloth. Never use sandpaper or emery cloth.

2. Check for a loose or a binding bearing. Take necessary corrective action.

Motor runs intermittently.

1. Check cordset for intermittent break. If defective, replace.

2. Check for loose connections. Repair connections as needed.

3. Check the ON-OFF switch. If faulty, replace.

4. Check the motor brushes. Replace if worn.

5. Check the governor points (if used). If defective, replace.
6. Check for binding bearings or bad gears. If defective, replace.

Beaters will not eject or stay in. Check for rusted or encrusted spindles or retaining springs. Clean, lubricate, or replace. Check for loose plate or broken or missing spring. Take necessary corrective action.

Bowl does not rotate (with some stand models). At least one of the beaters should rest on the bottom of the bowl, but you should be able to raise them slightly without lifting the motor. If necessary, correct by raising or lowering the screw as required. The pivot pin on the bowl pan may be lubricated with liquid petrolatum available in any drug store.

Mixer interferes with radio or TV. Check capacitors for opens or shorts. Nearly every governor-controlled mixer is equipped with a radio or TV interference-suppressing condenser which is connected across the line terminals. A third wire leading from this condenser is grounded to the motor body.

Noisy operation.

1. Check for an unbalanced armature by operating the mixer at high speed while holding it by the handle on outstretched fingertips. The vibration from an unbalanced armature will be very noticeable.
2. Check the end play. For most mixers, it should be between 0.002 and 0.010 in.
3. Check the fan to make sure it is tight on the shaft. Inspect it for bent blades which might cause an unbalanced armature. If the fan is faulty, replace it.
4. Check for a loose bearing and replace the bearing cap if necessary.
5. Check for a noise which sounds like the armature striking the field. Such a sound may also indicate shorted turns in the armature and field windings. Turn the armature by hand to feel whether it is striking. If this is the case, loosen the field strap and realign the field.
6. Check for shorted or open turns in the armature by removing the brushes and measuring the segment-to-segment resistance on the commutator. If any one measurement differs markedly from the other, replace the armature.
7. Check for worn bearings, dry gears, or worn or dry spindles. Repair or replace any faulty component.
8. If the motor sounds "bumpy," the actuator may be out of position from a governor arm. Check also for a bent or damaged armature shaft. A "bumpy" sound can be caused by a bad segment on the armature, too. Check segment-to-segment resistance.

As a final test, always check the calibration of the control knob by measuring the spindle speed with a tachometer and compare the readings with those in the manufacturer's service manual. If these figures are not available, the following list can be used as a general guide.

High: 800 r/min minimum,
           1,300 r/min maximum
Medium: 575 r/min minimum,
           975 r/min maximum
Low: 300 r/min minimum,
           700 r/min maximum

There should be a minimum speed separation of about 150 r/min between high and medium and between medium and low. Never send out a repaired mixer with no gear backlash.

Many mixers have available as attachments such items as salad makers, food grinders, citrus juicers, and juice extractors. Some manufacturers offer these devices as separate appliances. Whether they are attachments or appliances, servicing them is the same as servicing the mixer. This also holds true for the multipurpose "kitchen center" which uses a basic motor drive to operate several attachments.

FOOD BLENDERS

The blender is similar in its principles of operation to a food mixer, although its manner of assembly and its end purposes are different. For instance, an electric blender does the same job as the mixer, only in a different way. Food is placed in a glass or plastic container; knives or blades mounted in the bottom, revolving at a high speed, chop it into a fine pulp. In fact, a blender with strong blades and container and a high wattage can crush ice. Power ratings range from 350 to 1,200 W.

A blender runs at speeds that are about three times as fast as those of a food mixer; typical blender speeds are between 3,000 and 14,000 r/min. While both employ universal motors, the blender's motor is vertically mounted, whereas the mixer's is horizontally mounted. The blender's motor is connected to the drive shaft by means of a tooth-reinforced belt or is directly connected to a driver sprocket. In the latter case, the drive member of the cutter assembly fits into the drive sprocket by means of six or eight projecting lugs. The method of speed control, as for the mixer, varies from a simple three-step field winding to a solid-state electronic type.

Many blenders feature both manual and timed blending. In such units, with the timer on MANUAL and the desired speed selected on the control, blending may be obtained by pushing the ON button. When desired blending is obtained, the user pushes the button to OFF. For timed blending, with the timer set at the desired time setting and the proper speed selected on the control, the user pushes the ON-OFF switch button to ON. This slides a cam member over on the timer, mechanically actuating a timer release. When desired blending time is achieved, the timer returns to 0.

In most modern blenders, speeds are controlled by push buttons. In fact, at any selected push-button setting, the speed frequently is obtained through a combination of the diode and/or taps in the field coil windings. To check a multispeed push-button type switch in a diode, let us take a look at a *typical* seven-speed blender circuit schematic (Fig. 3-14). To check the switch continuity, leave all connections intact except for one side of the diode; use an ohmmeter set at R × 1 range. Unsolder one side of the diode—

use care since excess heat will ruin the diode (see Soldering in Chap. 1). Here are the checks that should be made according to a typical schematic.

With No. 1 (Whip) button in, there should be continuity between terminals L1 and 5, 6, 7, 8.

With No. 2 (Puree) button in, there should be continuity between terminals L1 and 6, 7, 8.

With No. 3 (Crumb) button in, there should be continuity between terminals L1 and 7, 8.

With No. 4 (Chop) button in, there should be continuity between terminals L1 and 5, 6, 7, 8, L2, and L3.

With No. 5 (Grate) button in, there should be continuity between terminals L1 and 6, 7, 8, L2, and L3.

With No. 6 (Blend) button in, there should be continuity between terminals L1 and 7, 8, L2, and L3.

With No. 7 (Liquefy) button in, there should be continuity between terminals L1 and 8, L2, and L3.

With the same blender, check the field coils

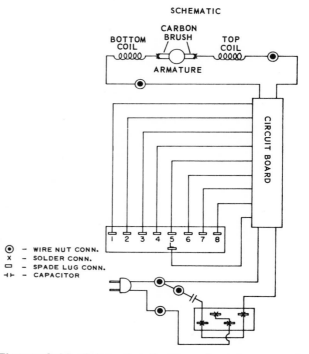

Figure 3-14. Schematic diagram of a seven-speed solid-state blender.

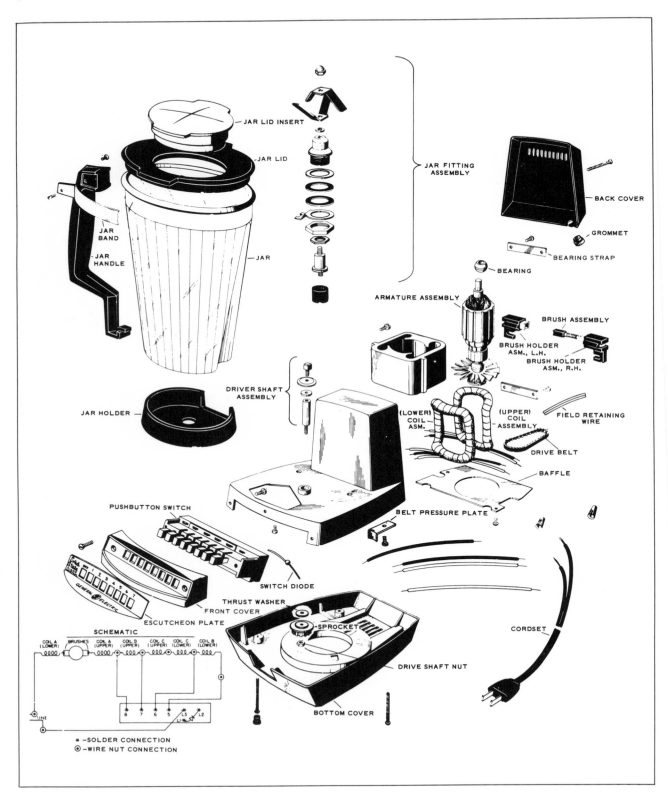

Figure 3-15. Schematic diagram and parts breakdown of a seven-speed field-winding controlled blender.

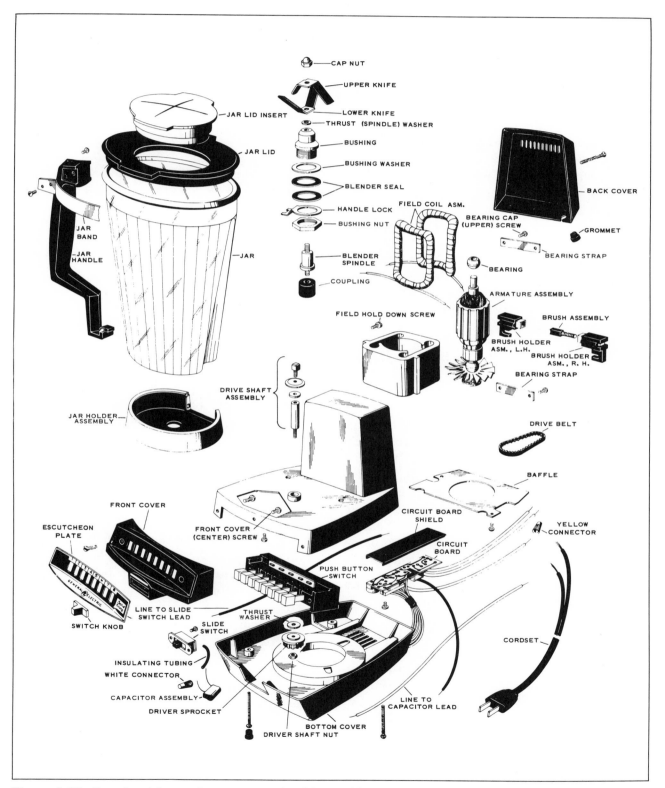

Figure 3-16. Parts breakdown of a seven-speed solid-state blender.

by pushing the OFF button fully in. Test between terminal 5 and the wire nut connection at which the two red field coils are joined. If there is no continuity, replace the lower field coil. Also test between the wire nut connection at which the two red field coils are joined and the brush holder to which the white lead from the upper coil is attached. If there is no continuity, replace the upper coil. Also check from the brush holder with the white field lead from the lower coil to the cordset. If there is no continuity, replace the lower coil. When checking an individual field coil, remember that each coil has three windings.

Upper coil: Resistance in this typical blender should be 1.8 Ω (as stated in the manufacturer's service manual) between one of the double leads in the black tubing and the single white lead. It should be 1.7 Ω between one of the double leads in the black tubing and one of the double leads in the white tubing. It should be 1.7 Ω between one of the double leads in the white tubing and the single red lead.

Lower coil: Resistance in this typical blender should be 1.8 Ω between one of the single white leads and the single black lead. It should be 1.7 Ω between one of the double leads in the black tubing and the other single white lead. It should be 1.7 Ω between one of the double leads in the black tubing and the single red lead. High or low resistance indicates an open or shorted coil, and replacement of the coil is required. If checking an individual switch (no connections), use an ohmmeter set at R × 1 and one probe at terminal L1.

There should be no continuity between L1 and any terminal with red (OFF) button in.

There should be no continuity between L1 and L2, L3, 6, 7, 8 terminals with No. 1 (Whip) button in.

There should be no continuity between L1 and L2, L3, 5, 7, 8 terminals with No. 2 (Puree) button in.

There should be no continuity between L1 and L2, L3, 5, 6, 8 terminals with No. 3 (Crumb) button in.

There should be no continuity between L1 and 6, 7, 8 terminals with No. 4 (Chop) button in.

There should be no continuity between L1 and 5, 7, 8 terminals with No. 5 (Grate) button in.

There should be no continuity between L1 and 5, 6, 8 terminals with No. 6 (Blend) button in.

There should be no continuity between L1 and 5, 6, 7 terminals with No. 7 (Liquefy) button in.

For a final test in this typical blender setup, plug the unit into a 120-V ac outlet through a wattmeter.

With the whip (No. 1) button pushed in, wattage should be 95 W maximum. Speed should be 8,000 ± 1,500 r/min.

With the puree (No. 2) button pushed in, speed should be 8,500 ± 1,500 r/min.

With the crumb (No. 3) button pushed in, speed should be 9,800 ± 1,500 r/min.

With the chop (No. 4) button pushed in, speed should be 10,500 ± 2,000 r/min.

With the grate (No. 5) button pushed in, speed should be 11,300 ± 2,000 r/min.

With the blend (No. 6) button pushed in, speed should be 12,000 ± 3,000 r/min.

With the liquefy (No. 7) button pushed in, wattage should be 175 W maximum. Speed should be 12,500 r/min minimum.

There should be a 500-r/min differential between the speeds of buttons 1, 2, 3, 4, and 5.

There should be a 1,000-r/min differential between speeds of buttons 5, 6, and 7.

Troubles in the blender are similar to those of a mixer, and, in general, their solutions are handled in exactly the same manner as with the food mixer. Such troubles as damaged knife blades are exclusive to blenders, but these blades are easily replaced. Other typical troubles involve defective cords, switches, and speed. Motors can cause some problems, particularly when they have seen more than their share of "heavy" duty. (See Chap. 2.) Since blenders are run only for short periods, they can get by with very little lubrication or none at all.

Analysis of Complaints. Here is an analysis of the complaints usually encountered with food blenders.

Motor will not run.

1. Check cordset for continuity. If faulty, replace.

2. Check the switch for continuity or damage. If defective, replace.

3. Check the control circuit for continuity. If the circuit has a diode in it, check as follows. Unsolder one side of the diode from the switch; use care since excess heat will ruin the diode. With an ohmmeter set at R × 1 range, read the resistance, which should be approximately 4 to 10 Ω in the forward direction and infinity with the probes reversed. If the diode is found defective, replace it.

4. Check carbon brushes to be sure brush is not "hanging up" in brush holder. If brushes are worn, replace.

5. Check the armature for open or shorted windings. Remove armature and check the resistance from segment to segment of the commutator with an ohmmeter and compare with the service manual. If resistance in any coil is low, it indicates a shorted coil. If resistance is high, it indicates an open coil. In either case, the armature must be replaced.

6. Check the field coils for open or shorted turns. Remove field and check the upper coil with an ohmmeter and compare with the service manual. Check the lower coil and compare with the service manual. High or low resistance indicates an open or shorted coil, and replacement of the coil or entire motor is required.

7. On some models, check the timer by using an ohmmeter set at R × 1 range. Remove spade connectors from the timer; touch probes to timer contacts. With knob pointing to MANUAL and switch pushed to full ON, there should be continuity. With knob pointing to any portion of the timed area and the switch pushed to full ON, the timer should unwind and there should be continuity until knob points to 0. If the timer is faulty, replace it.

Blender abnormally noisy.

1. Check the fan blades for clearance. Make certain they are not hitting the bearing strap, brush boxes, bearing boss, or belt. If necessary, take proper corrective action.

2. Check for excessive clearance between the driver shaft and the driver bearing. If there is, the base assembly must be replaced.

3. Check the thrust washer under the driver sprocket to be sure that it is properly positioned. Reposition or replace it, if necessary.

4. Check the switch arm. It may be bent or distorted enough to touch the driver sprocket. If so, repair or replace the switch.

5. Check the baffle for distortion. It may be touching the belt, and it should be readjusted.

6. Check for loose tooth-reinforced belt. Adjust pressure plate to just touch the belt when the belt is in tension. Too tight a belt will result in high wattage and low speeds; too loose a belt will result in a noisy blender.

7. Be certain the hex nut is securely tightened against the blades and the cap nut is securely tightened against the hex nut. Minimum torque on each nut should be 45 in-lb. Do not overtighten.

Jar noisy. The jar hardware may not align with the driver, causing an objectionable noise when the blender is on with the jar in place and empty. This can generally be corrected by unscrewing the large nut on the bottom of the jar and repositioning the jar bushing. If turning the bushing to two or three new positions fails to correct the problem, replace the universal coupling. If noise still persists, observe whether the hole in the jar is off-center. If so, replace the jar.

Motor runs but blades will not turn.

1. Check for a broken tooth-reinforced belt. If defective, replace.

2. Check for excessive wear of the universal coupling or driver shaft. If faulty, replace either or both components.

3. Check driver for damage. If faulty, replace.

4. Check whether the cutter shaft is frozen or broken. Clean or replace as needed.

Blender jar leaks.

1. Check for poor seal between jar and bushing assembly. Replace the neoprene rubber seals or add an extra seal if necessary. Tighten large nut at the bottom of the jar.
2. Check for loose fit between the spindle and bearing. Replace both parts, if necessary.

Loose cap nut. Make certain the cap nut is securely tightened.

Blender runs at only one speed. Visually check to see that all connections are proper and secure and that the switch has continuity. Check armature and field. If faulty, replace the motor.

Blender will not run at one or more speed settings. Visually check to see that all connections are proper and secure. If all connections are proper, check continuity of push-button switch between single spade at rear of switch and other spades—one at a time. If all check all right, replace printed circuit board (with solid-state speed control models only).

Blade edges chipped or rolled over. Replace the blades. Since this may be a result of cracking or crushing large, solid ice cubes, caution the customer about putting such big ice cubes into the blender.

CAN OPENERS

The electrical can opener is mechanically similar to the hand-operated model. In the simplest of models, when the operating lever is depressed with a can in position, the cutting wheel pierces the can and a projecting spring holds the folded lip firmly against the toothed drive wheel. A slight further pressure actuates the switch. The drive wheel turns the can under the cutter. The magnet contacting the can top holds it after the can is cut through. Raising the lever stops the operation and releases the can.

There are several variations of this operation. Some are just as simple, others are more complex. For instance, in one design, when the clamping lever is depressed with a can in position, the can guide, can guide spring, and cutter all move down to hold and puncture the can. In this movement the clamping lever strikes a step on the interlock slide causing that also to move down from its normal position which locks the switch bar open. A light touch on the switch bar then closes the circuit through the motor which activates the driver and causes the can to turn under the cutter. With the magnet resting on the top of the can, this part is held after it is cut through; raising the clamping lever releases the can.

A great many of today's can openers are sold as combination appliances: juicer/can opener, salad maker/can opener, ice crusher/can opener, and knife sharpener/can opener. The latter two are by far the most popular, and the servicing of them is discussed later in this chapter. However, all these combinations are simple devices that require little or no servicing once the operation of a can opener is understood.

While shaded-pole motors were once widely used in can openers, most manufactured in recent years have used the universal type. Many motors used in conjunction with can openers employ a governor-type control to keep the gear speed constant regardless of cutting pressure. This gear arrangement or train, which usually consists of a worm gear on the motor shaft that drives a spindle gear connected to the cutting wheel, reduces the motor speed from about 3,500 r/min to an operating speed of approximately 250 r/min. In addition to reducing the speed, the gear train increases the usable torque proportionally.

Besides the motor, the only other electric components include a switch and cordset. The switches in can openers are all the momentary-contact type; that is, they close to ON only as long as the clamp handle is kept down to hold the rim of a can. Remember this when checking a switch with an ohmmeter for its ability to close and open a circuit.

Since can openers are primarily mechanical devices, most troubles are due to dulled cutting edges, worn gears, etc., which may need replacement from time to time. But because this appli-

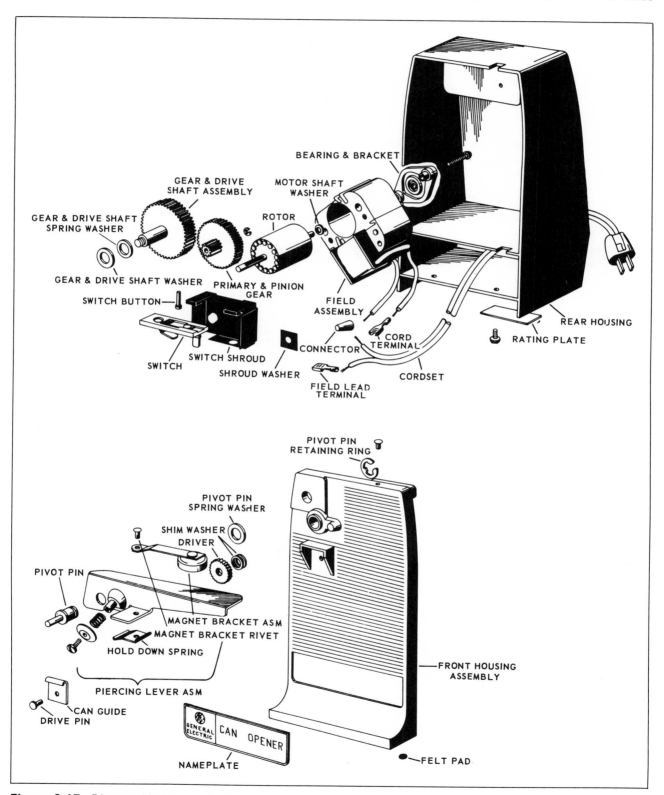

BEARING & BRACKET

GEAR & DRIVE
SHAFT ASSEMBLY

MOTOR SHAFT
WASHER

GEAR & DRIVE SHAFT
SPRING WASHER

ROTOR

GEAR & DRIVE SHAFT WASHER

PRIMARY & PINION
GEAR

SWITCH BUTTON

FIELD
ASSEMBLY

CONNECTOR

CORD
TERMINAL

REAR HOUSING

RATING PLATE

SWITCH

SWITCH SHROUD

SHROUD WASHER

CORDSET

FIELD LEAD
TERMINAL

PIVOT PIN
RETAINING RING

PIVOT PIN
SPRING WASHER

SHIM WASHER
DRIVER

PIVOT PIN

MAGNET BRACKET ASM

MAGNET BRACKET RIVET

HOLD DOWN SPRING

FRONT HOUSING
ASSEMBLY

PIERCING LEVER ASM

CAN GUIDE

DRIVE PIN

GENERAL ELECTRIC CAN OPENER

NAMEPLATE

FELT PAD

Figure 3-17. Disassembled view of a typical can opener.

ance is run for only seconds at a time, it can be expected to give long service. Most models are "lifetime" lubricated; some might need a dab or two of light grease on the reduction gear every few years. If motor troubles are suspected, refer to Chap. 2.

Here are the most common complaints about can openers and some of the solutions to them.

Motor does not run.

1. Check the power at the outlet. If there is none, take proper corrective action.
2. Check the circuit completely including the armature, field windings, cordset, and internal wiring for continuity. Correct or replace as required. Bear in mind that some can openers contain, as an integral part of the field assembly, a thermal protective cut-out which opens the circuit if excessive heat is developed for any reason. If the circuit is open, allow it to cool for 10 min and test again. The cut-out will reestablish the circuit when temperature of the field winding drops within safe limits. Under normal operation, the thermal cut-out should not open in less than 5 min. If cut-out occurs repeatedly, check for mechanical bind or improper lubrication; if no fault is found, replace the field assembly.
3. Check the switch contacts for contamination or wear. Clean or replace the switch as needed.
4. Check the brushes for wear or binds; replace if necessary.
5. Check the governor contacts for continuity, contamination, or wear. Take any corrective action necessary.
6. Check to be sure that the gear train is not bound. Clear the jam-up as necessary.

Motor runs continuously.

1. Check the switch return spring. It must break contact when the button is released. Adjust the spring or replace it as necessary.
2. Be sure the switch is opening. Adjust contacts or replace the switch.

3. Check for a short circuit across the switch. If shorted, replace the switch.

Motor runs but lacks power.

1. Check the drive gear for binding. Clean the shaft and oil the bearings to correct bind if necessary.
2. Check for dull or damaged cutter wheel or blade. If faulty, replace.
3. Check for poor brush contact. Clean the commutator and replace brushes if necessary.
4. Check for defective (shorted) armature or field winding. If defective, replace.

Motor noisy.

1. Check the field and armature alignment by turning armature by hand. If necessary, realign or replace bearings. If situation cannot be corrected, replace armature.
2. Check lubrication of thrust bearing. Lubricate if necessary.

Motor runs too fast or too slow (some models).

1. Check the governor actuator for wear. Readjust or replace as necessary.
2. Check the governor centrifugal arms for proper placement. Readjust arms as necessary.
3. Check for brush wear. If worn, replace brushes.

Cutting edge will not pierce can.

1. Check for a binding cutter assembly. Remove bind or replace assembly.
2. Check for the proper location of the cutter spring. Correct or replace as required.
3. Check for the proper clearance between the driver and cutter. The proper measure is given in the service manual. Adjust as necessary.
4. Check for bent, worn, or dull cutter point on cutter assembly. If faulty, replace cutter.

Can stalls or does not turn.

1. Check the cutter wheel, screw, and spring

for excessive contamination with food, etc. Clean or replace as required.

2. If the drive wheel does not turn, check the gears for broken or missing teeth. Replace spindle gear, if stripped.
3. If the drive wheel turns but the can slips, check for worn or chipped drive wheel and cutter. Replace drive wheel and/or cutter as necessary.
4. Check for cutter gap which should be between 0.002 and 0.010 in for most can openers. (Check the service manual for the exact dimension.) Correct the gap by adding or removing spacing washers on the drive wheel shaft.
5. Make sure that the drive gear is not binding. Remove bind, clean and lubricate shaft.
6. Check for loose cutter assembly. Tighten or replace as required.
7. The large seam on the can may be binding.

Cutter wheel wanders away from the rim. Make sure the cutter turns freely and the spring drives it firmly against the head of the retaining screw. Check for bent or misaligned operating lever and pivot. Take proper corrective action.

Unit will not clamp some cans. Cans are customarily made by setting the top and bottom inside the cylinder and folding the edge of the top and bottom over and down on the outside of the cylinder. It is this edge on the outside that enables the driver of crank-type openers, whether manual or electric, to support the can for clamping the opening. A few products, notably some brands of condensed milk and sardines, are packed in special cans without this folded edge and, therefore, cannot be opened by any crank-type opener.

Clamping lever will not latch down. If the clamping lever will not go to the bottom of the slot, it is probably striking the boss for the adjacent backplate spacer screw. Reach through the slot and file down the boss as required. If the lever reaches the bottom of the slot but will not enter the detent, file the bottom of the slot as required.

Magnet hangs up or fails to hold lid. Check the magnet bracket for free vertical movement in the slots of the housing. Bind may be caused by bent magnet bracket legs, contamination in slots, or excess paint and/or flash in slots. Replace or correct as required. The magnet will not hold aluminum can lids. Advise customer not to cut all the way around, but to "hinge" can.

Opener drops cans. Make sure the clamping lever locks in the detent, that the driver is not chipped or worn, that the cutter washer is not missing, that the cutter gap is within specification, and that the cutter is mounted with the cone face out. If you find no defect or faulty adjustment, replace the case support assembly. In most cases do not file more than 0.040 in as this creates a tendency for the opener to drop cans.

Ice Crusher

As mentioned earlier, many can openers feature an ice crusher in combination. Each manufacturer has its own method of making the combination; the two most popular are as follows.

1. A slide button is provided to hold the switch ON when the ice crusher feature is to be used. A series of rotating and stationary blades shave or "crush" the ice, which then accumulates in a removable ice drawer. A manually operated knob is usually provided in the event the ice jams and stalls the mechanism. This knob can then be turned counterclockwise to release the blockage.
2. A detent is provided to hold the operating lever in the ON position when the appliance is used for crushing ice. The rotating hub causes the pivoted flails to strike the ice in rapid succession, breaking it into small pieces. The rake holds the ice until it is completely broken and the cracked ice is discharged into the storage container.

Complaints about ice crushers operating by method No. 2 will generally involve the following.

1. See if flails remain in the hub. If faulty, replace hub.

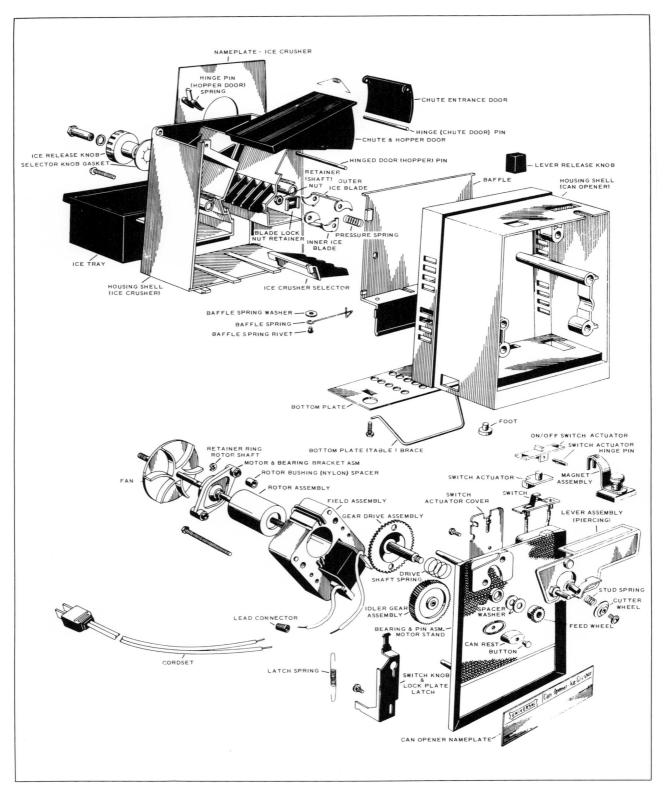

Figure 3-18. Disassembled view of a typical can opener-ice crusher combination.

2. Flails hit rake on side toward can opener end. This also indicates that the hub is defective. Replace as above.

3. Gray residue is noticeable in melted ice. This residue, although harmless, may be cause for some complaint. It is from wear of the slots in an incompletely processed hub in some of the earliest units. Replacement hubs have been specially treated to avoid this effect. Replace hub.

4. Flails hit rake on side toward operating handle. This indicates that the retaining plate is improperly assembled. Check whether the plate is secured on the gasket and the bearing bump on plate is toward hub.

A few manufacturers produce separate ice crushers. The basic servicing of these units is the same as for can opener/ice crusher combinations.

KNIFE SHARPENERS

Most electrical knife sharpeners are made in combination with can openers. However, some manufacturers still produce a knife and scissors sharpener as a separate appliance. These sharpeners usually employ a small shaded-pole induction motor to drive one or two grinding wheels. The other parts of the unit usually include an ON-OFF switch and a fan attached to the motor shaft.

To operate, the knife blade is placed gently on the wheel(s) so that the handle is closest to the user and the knife is drawn toward the user for its complete length. On long knives, light finger pressure is necessary throughout the entire sharpening stroke to prevent "scalloping" or uneven sharpening of the blade. It may be necessary to repeat the sharpening process several times.

Electrical knife sharpeners are extremely easy to service. The major problems are the cutter or grinding wheels needing replacement.

Remember that specific details on parts replacement are usually contained in the manufacturer's service manuals relating to the make and model being serviced. If you suspect motor trouble, refer to Chap. 2 for information on motor troubleshooting and repair.

The servicing problems of the knife-sharpening portion of a sharpener/can opener combination unit and that of a single unit are the same. As a rule, a detent is provided to hold the operating lever in the ON position when the appliance is used for sharpening knives.

Here are the common complaints received about knife sharpeners and what the service technician can do about them.

Motor stalls.

1. Check for grinding wheel interference. If the grinding wheel has worked loose from the shaft, it could bind against one side of the case. To repair, follow service manual instructions.

2. Be sure abrasive disk is not bent and binding against the guide bushing. If the disk has been bent sufficiently to bind against the guide bushing, replace it. If it is slightly warped, straighten with light finger pressure. If necessary, relocate and adjust the abrasive disk.

3. Check for tight gear mesh. If the fit between the gear and worm is tight, carefully bend the bracket away from the worm.

4. Check the gap between abrasive disk and guide bushing to be sure that it is not too narrow. Adjust, if necessary.

5. Check for seized bearings. *Note:* If you find nothing wrong with the knife sharpener, the motor stalling could be caused by the user applying too much pressure while sharpening knives or scissors. This happens as part of the design and is a safety feature.

Sharpener is noisy.

1. Check for loose or binding parts. Repair or replace any faulty components.

2. Check for loose gear mesh. Adjust mesh as needed or replace.

3. Check both ends of driver gear shaft for lubrication. Lubricate if necessary.
4. Check for stripped gear. If defective, replace.
5. Check for driver gear hitting the guide bushing. Repair as necessary.
6. Check for fan hitting the baseplate or motor. Repair as necessary.
7. Check for abrasive disk hitting the guide bushing. Repair as necessary.
8. Check for foreign material in the case. Remove the foreign object.

Sharpener will not start (motor does not run).

1. Check for binding parts. If found, remove bind.
2. Check cordset. If defective, replace.
3. Check all electric connections. Repair as needed.
4. Check power at outlet. If there is none, take proper corrective action.
5. Check the switch. If defective, replace.
6. Check motor windings. If faulty, replace motor.

Sharpening wheel scallops knives. Check for excessive grinding-wheel wobble. Tighten wheel or replace it as necessary. Also check the shaft and bearing and take any corrective action that is necessary.

Motor runs slow.

1. Check for bound gears. Take necessary corrective action.
2. Check for dry bearings. Relubricate as needed.

Motor hums but wheels do not turn.

1. Check to be sure that the armature shaft is not bound. Realign field assembly and lubricate bearings as needed.
2. Check for jammed grinding wheel. Remove source of bind.
3. User may employ excessive pressure when sharpening. Caution user on maintaining a light touch.

Sharpener will not shut off. The switch is usually the problem with this complaint and should be replaced if faulty.

Pencil Sharpeners

Pencil sharpeners are frequently incorporated into knife and scissors sharpeners or may be a separate appliance. In either case, most pencil sharpeners are designed to sharpen pencils by "sanding" them to the desired shape with a tungsten carbide abrasive disk.

In addition to the problems already discussed for knife sharpeners, here are some complaints that may arise with pencil sharpeners.

Pencil fails to rotate or rotates intermittently.

1. Check for warped abrasive disk. If it is warped, straighten with light finger pressure. If it is excessively bent, replace the disk.
2. Check the gap between the abrasive disk and guide bushing to be sure that it is not too wide. Adjust position of the abrasive disk, if necessary.
3. Check for gears broken off the driver. Replace, if necessary.
4. Check for excessive motor end play. If it is excessive, replace the motor assembly.
5. Check the location of the guide bushing. If it is out of adjustment, repair as instructed in the service manual.

Pencil takes too long to sharpen.

1. Check the abrasive disk for accumulation of crayon, etc. Clean the disk as follows:
   a. Set the unit on the bench in an upside-down position.
   b. Remove the scrap box and turn the unit on.
   c. While the sharpener is running, place a small brush, preferably a hard-bristle toothbrush, against the abrasive disk using light pressure.
2. Check guide bushing cap to see if it is in need of adjustment. Adjust according to the instruction manual, if necessary.
3. Check the gap between the abrasive disk and guide to make sure that it is not too

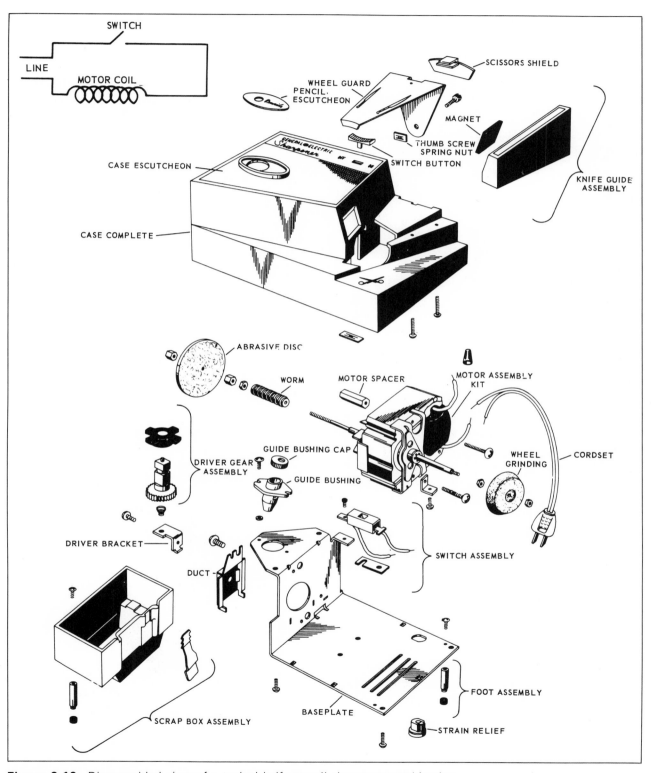

Figure 3-19. Disassembled view of a typical knife-pencil sharpener combination.

wide. Adjust as the instruction manual states, if necessary.

Electric letter openers work on the same basic principle as knife sharpeners and are generally serviced in much the same manner.

SLICERS

Electrically and mechanically a knife sharpener and an electric meat slicer are very similar. The electrical parts of the slicer consist of a universal motor, an ON-OFF switch, and a cordset. The mechanical linkages between the motor and the blade that does the actual cutting is a gear train that retains the motor's rotary motion, but changes the speed and effective torque.

One of the major problems with an electrical meat slicer results from the failure by the user to clean it properly. Contaminants build up and the unit does not cut properly. Scrubbing the assembly with hot detergent water and a tooth-brush will usually solve this problem. Remove more stubborn contaminants by carefully scraping the cutting assembly with a sharp knife.

While the power units of many electric slicers have a lifetime lubrication, from time to time the gear assembly should be given a small dab of gear grease on the moving parts, and the motor should then be run for a few minutes. But never over-lubricate the gear train; excessive lubricant can splatter into the motor and contaminate the brush assembly. When it is necessary to clean a gear train assembly, wash the components with kerosene or isopropyl alcohol. Once the gear assembly has been cleaned, flush it with a lightweight oil to remove any remaining cleaner, and then apply a cream-type gear grease.

Here is a summary of the complaints often received about the electric slicer and ways to correct them.

Slicer will not start.

1. Check the power at the outlet. If there is none, take proper corrective action.

2. Check for binding parts. If found, remove bind.
3. Check the cordset. If defective, replace.
4. Check the ON-OFF switch. Replace if necessary.
5. Check all electric connections. Repair as needed.
6. Check the motor windings. If faulty, replace motor.

Motor runs slow.

1. Check for bound gears. Take necessary corrective action.
2. Check for dry bearings. Relubricate as needed.

Motor hums but cutting blade does not turn.

1. Be sure that the armature shaft is not bound. Realign field assembly and lubricate bearings as needed.

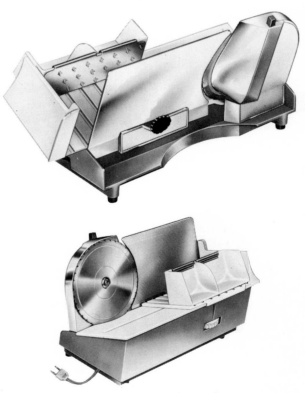

Figure 3-20. A typical meat slicer unit.

2. Check for a jammed grinding wheel. Remove source of bind if necessary.
3. User may employ excessive pressure when cutting. Caution user on maintaining a light light touch.

Motor stalls.

1. Check for cutting-blade interference. If the cutting wheel has worked loose from the shaft, it could bind against one side of the food platform. To correct, follow service manual instructions.
2. Check for tight gear mesh. Lubricate or realign gears as necessary.
3. User may employ excessive pressure when cutting. Caution user on maintaining a light touch.

Slicer is noisy.

1. Check for loose cutting blade. Tighten blade as needed.
2. Check for worn or dry bearings. Relubricate or replace as required.
3. Check for loose or worn gears. Adjust mesh or replace as required.
4. Check for foreign material in the case. Remove foreign object, if found.

SEWING MACHINES

The electrical sewing machine is a fairly simple device, both electrically and mechanically. Most sewing-machine motors are of the universal type, and their speed is controlled by a foot pedal or a knee lever. This control mechanism may be either a step or carbon control.

The step control changes the speed in a series of steps or intervals, usually numbering from five to eight, from slow speed to top speed. With some step controls, the first step does not provide the slow speed the user sometimes wants in sewing. The carbon control, however, adjusts the speed from slow to fast smoothly and uniformly, especially when starting, and at very slow speed.

On either type of control the power is reduced at low speeds. The three ways of controlling speed in use on today's machines are the following.

1. Withholding electric current. This will reduce the power of the machine as well as the speed.
2. Transmission with a gear-reduction system, which provides slower speed with increased power. This is similar to the gear system in an automobile. Full or greater power and slow speed are sometimes needed at the same time.
3. Solid-state electronic control system. This maintains full power at various speeds.

Most of the electrical troubles found in sewing machines will generally be simple ones: broken wires, worn brushes in the motors, and loose connections on the speed control circuit. Always check the condition of the cordset and wires between the switch and the motor. Sewing machine motor problems are the same as those of any other universal type and are fully described in Chap. 2.

All sewing machines have the same basic components to control their ability to stitch properly: the takeup lever, needle, feed dog, and shuttle (hook). In a typical sewing machine, when the takeup lever is in its highest position,

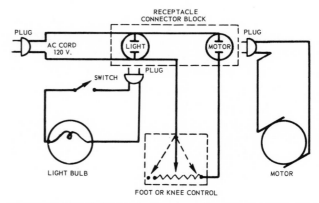

Figure 3-21. Schematic diagrams of a typical resistance (current-withholding) sewing machine control.

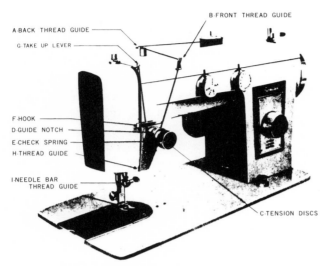

A-BACK THREAD GUIDE
B-FRONT THREAD GUIDE
G-TAKE UP LEVER
F-HOOK
D-GUIDE NOTCH
E-CHECK SPRING
H-THREAD GUIDE
I-NEEDLE BAR THREAD GUIDE
C-TENSION DISCS

Figure 3-22. Thread and adjustment controls of a typical sewing machine.

the needle bar and needle have started downward to pierce the material and carry the thread to the shuttle mechanism. At this point, the feed dogs have dropped below the needle plate, and the material is stationary. As the needle starts to rise, the takeup lever moves downward, releasing the thread so that the shuttle (hook) can take the thread and pass it around the hook to make a lockstitch.

When the needle leaves the material, the feed dogs move upward, contacting and moving the material into position for the next stitch. As the needle reaches its highest point, the takeup lever moves upward to carry the bobbin thread into the material to make the lockstitch and also to pull additional thread through the tension assembly from the spool in preparation for the next stitch. If these components fail to operate in proper relation to each other, the machine will not stitch correctly.

The belt in modern machines is usually made of polyurethane and provides its own operating tension. However, if the belt is too tight, the motor may overheat. Proper belt adjustment allows sufficient slack so that the belt can be pinched together easily just above the motor pulley. It is necessary to remove the handwheel when changing belts. But remember that many

of the newer motors are equipped with oil-impregnated bronze bearings and no lubrication is ever necessary. Sometimes brush noise on new units leads the operator to believe that there is need for oil. This sound is natural and will subside as the machine is used.

As a rule, the belt tension is adjusted by loosening the motor bracket screw and moving motor and bracket assembly downward to increase tension. Tighten only enough to eliminate belt slippage. A belt which is too tight will overload the motor. To align the belt, loosen the motor pulley setscrew and position the pulley on the motor shaft so that the belt runs true.

Thread tension is usually automatic and rarely requires adjustment, even when sewing material of different thicknesses. The needle and bobbin threads should be locked in the center of the material thickness as shown here. The tension of needle thread (top) in most machines is usually regulated by lowering the pressure bar and turning the tension-regulator knob clockwise to increase, or counterclockwise to decrease, tension. In ordinary use the bobbin tension need never be changed from its factory-adjusted setting. If, however, the tension does need to be changed, follow instructions given in the service manual.

Advise your customer, when selecting the correct needle for the type of sewing to be done, to be sure it is straight and sharp and that the eye is large enough for the thread being used. A perfect needle can be determined by placing the flat side of the shank on any flat surface. The point of a perfect needle will be in line with the shank.

The presser foot usually holds the material firmly against the feed dogs; the sides of the foot should be parallel with the slots in the needle plate and resting level on the feed dogs. This alignment can be checked by placing a piece of white paper behind the machine and tilting the head so as to sight between the bottom of the foot and the feed dog while slowly lowering the foot on the feed dog. Bend the foot slightly to the left or right to correct leveling. If the foot is not level, the material will feed crookedly.

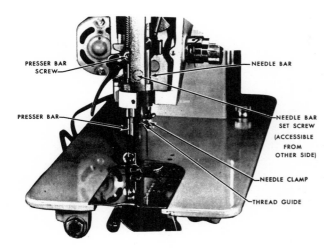

Figure 3-23. Needle guides and adjustments of typical sewing machine.

Figure 3-24. Typical bobbin case assembly.

The takeup lever pulls the thread up after the stitch is formed and locks the knot, much the same as is done in hand sewing when the thread is pulled through the material and brought up tight by hand. It also pulls enough thread off the spool for the next stitch. It is accurately timed through its connection with the takeup cam so that it corresponds with the downward stroke of the needle and the movement of the feed dogs. To test whether it is bent vertically, check to see that the takeup is past its lowest position and starting up just as the needle bar reaches its highest position. If it fails to meet this specification, do not attempt to bend the takeup but install a new one. If bent sideways, it may rub on the face plate. You can straighten it by bending it with caution. A damaged takeup lever will cause the following problems.

1. Skipped stitches
2. Thread breakage
3. Possible needle breakage
4. Poor stitching

The feed dog under the presser foot does the actual moving of the material. These dogs have positions from the full drop to the full rise for proper feeding of all materials. The high position is for heavy materials, and the low is for delicate materials. Position of the feed dog is usually manually selected by the operator with the knob on the bed of the sewing head. The feed dog must move up and down freely in the slots in the needle plate; any friction here would cause noise and hard running.

Practically every complaint about poor sewing can be traced to one or more of the following: wrong needle, improper thread, thread tensions. Always check these three items when servicing any sewing machine. Of course, there may be other problems, and here are some of the most common ones.

Machine operates noisily.

1. Be sure that there is not too much presser foot pressure. Check owner's manual and advise customer.
2. Check for feed dogs binding on the needle

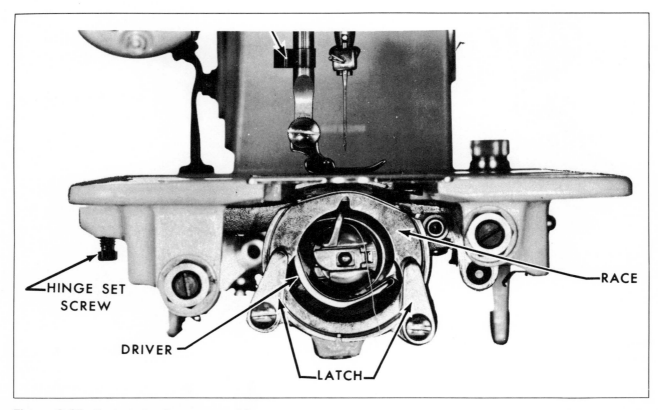

Figure 3-25. Typical shuttle race assembly.

plate. Adjust as directed in service manual, if necessary.

3. Check for lint, dirt, or thread in the hook or race assembly. Remove obstruction and lubricate the race assembly, when needed.

4. Check for looseness of parts such as the rock shaft fork, rocker shaft, zigzag drive assembly, etc. If necessary, tighten, repair, or replace any faulty component.

5. Check for wires hitting the undercarriage causing vibration. Staple the wires to the case or cabinet if necessary.

6. Check for failure by the consumer to oil vital bearing points. Check owner's and/or service manual for this information.

Machine runs hard.

1. See if the motor belt is too tight. Readjust the motor bracket, if necessary.

2. See if the thread is accidentally wound around the main shaft at the hand wheel under the belt guard. Remove the hand-wheel and clean out the cause of the bind.

3. Check for lint, dirt, or thread in the shuttle assembly. Clean, if necessary.

4. See if the machine is lubricated with unsuitable gummy oil. To clean, pour a few drops of kerosene or isopropyl alcohol into each oil hole and run the machine for a few minutes. Then wipe off all assemblies and lubricate with the proper sewing-machine oil.

5. Make sure that the main shaft is not bent. If faulty, send the head to the manufacturer for repair.

6. Check for tight or binding connecting tie rod, rocker shaft, feed fork rod, etc. Adjust as directed in the service manual.

7. Check for damaged or broken cushion spring in the hook rocker. Replace, if necessary.

8. See if the bobbin winder is released and is

running while the machine is sewing. This is a matter of customer education.

9. Check for failure of the owner to oil bearing points. Check owner's manual and/or service manual and advise customer as to what must be done.

Motor runs hot.

1. Check for lint, dirt, or thread in shuttle assembly. Clean assembly, if necessary.

2. Check for overoiling of the motor. Check service manual and/or owner's manual for exact oiling procedure.

3. Be sure that the drive belt is not too tight. Adjust as the service manual instructs.

4. Check the zigzag assembly and the rocker shaft for binding. Adjust, repair, or replace as needed.

5. Make certain that the main shaft is not bent. Send the head to the manufacturer for repair, if necessary.

Foot or knee control runs hot.

1. The machine may have been run for too long at one period. Advise the customer to reduce or shorten the running periods.

2. When the control fails to shut off the machine, check and repair as necessary.

Machine will not operate on zigzag.

1. See whether the stitch-width regulator lever has been set. Instruct the customer on its use.

2. Make sure that the width-regulator locks are loose. Secure after the width lever has been set. Take necessary corrective steps to repair.

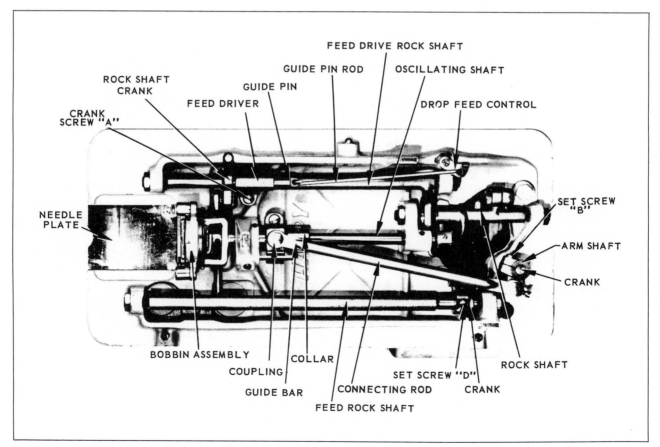

Figure 3-26. Typical method of connecting rock arm to the feed driver.

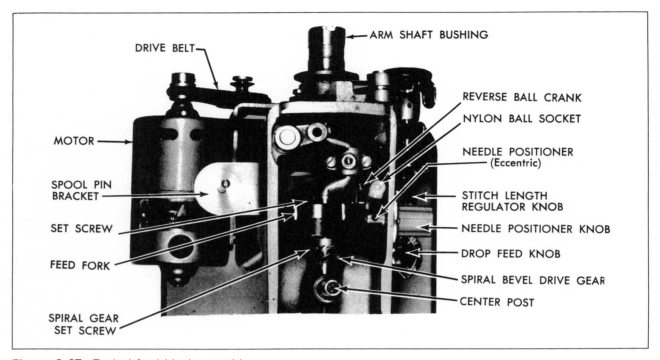

Figure 3-27. Typical feed-block assembly.

3. Check whether the geared cam is worn. Replace, if necessary.
4. Check for a loose or broken needle bar drive rod. Tighten or replace as needed.
5. Check the upper shaft gear for wear. Replace, if necessary.
6. Check the vertical rack shaft to see if it is broken. If faulty, replace.

Machine skips stitches.

1. Customer may use incorrect length of needle, or the needle may be incorrectly inserted or improperly threaded. Also the thread may be too heavy for the needle. All these possible causes can be corrected by customer education.
2. Check for improper adjustment of the upper tension check spring. Readjust or replace as needed.
3. Check for a bent or blunt needle. Replace the needle, if necessary.
4. Check for dirt or lint in the shuttle assembly. Clean thoroughly.
5. Check the point of the shuttle to be sure that it is not blunt or rough. Replace.

6. The pressure of the presser foot may be insufficient. Check service manual and follow manufacturer's instructions.

Machine makes loose stitches.

1. Check for a bent top-thread check spring. Replace the spring, if faulty.
2. Check the upper thread tension to be sure that it is not too tight or too loose. Regulate for the proper thread tension as needed.
3. Check for a damaged needle point. Replace the needle, if necessary.
4. Be sure that the needle thread is not too coarse for the material. Replace the needle if necessary.
5. Be sure that the bobbin thread is not too coarse. Replace if necessary; it should be the same as needle thread.
6. Be sure that the thread on the bobbin is not unevenly wound or that the thread does not pile up on one side of the bobbin. Correct the situation as the service manual instructs.
7. Be sure that the needle plate is not rough.

Remove burrs or replace.

8. Check for accumulation of lint or thread in the shuttle assembly. Clean, if necessary.
9. The setting of the presser foot may be improper or presser foot may not be down completely. Check the lower presser foot lever; also check the snap-lock darner. Repair or replace, as necessary.
10. Check for a broken or damaged feed dog. Replace, if necessary.

Needle becomes unthreaded. This is usually caused by insufficient thread in the needle. Review the threading and operating instructions with the customer.

Machine wrinkles the material.

1. Check for a rough needle plate. Remove the burrs or replace the plate.
2. Make certain that the feed dogs are not set too high. Make proper adjustment, if necessary.
3. Presser foot pressure is too great. Check service manual.
4. Make certain that the tension of the bobbin thread or of the needle thread is not too great. Take necessary corrective action.
5. Check for improper needle for the size of the thread. Replace with proper needle, if necessary.

Cloth will not feed properly.

1. Check for broken or damaged feed dogs. Replace, if necessary.
2. Pressure on presser foot may be insufficient. Increase pressure by depressing the snap-lock darner.
3. Presser foot may not be down on material. Presser foot must be on the cloth during sewing period, except when darning or embroidering. Instruct the customer, if necessary.

Bobbin does not wind properly.

1. Be sure that the machine is threaded correctly. Check owner's manual and advise customer.
2. Check whether the thread is jumping out of

the thread guides. Review the way the bobbin winds in the service manual and make all necessary adjustments.
3. Be sure that there is no misalignment of the bobbin thread guide disks. Make necessary adjustment.

Upper thread breaks.

1. Check for incorrect needle length or a needle too fine for the thread used. Replace, as needed.
2. Check for a needle that is bent, has a broken point, or is inserted incorrectly. Replace or install properly.
3. Be sure that the needle is threaded properly. Check owner's manual and advise customer.
4. Be sure that the tension of needle thread is not too tight. Adjust as instructed in the service manual.

Lower thread breaks.

1. See if the hole in the needle plate is rough or sharp. Replace, if necessary.
2. Be sure that the bobbin is inserted correctly and that the bobbin thread is brought up correctly. The bobbin could be wound too full. Check owner's manual and advise customer.
3. Make sure that the bobbin thread tension is not too tight. Make the proper adjustment.
4. See if the hole in needle plate is rough or sharp because of the needle striking the plate. File or sandpaper the hole to ensure a smooth entrance, or replace the plate and needle.

Needle breaks.

1. Check for a bent needle. If faulty, replace.
2. Check for incorrect length of needle, or wrong size of needle or thread for the material being sewn. Replace the needle or thread.
3. Check for a loose needle clamp. Replace the needle and retighten the thumbscrew.
4. Pulling material from behind the needle while sewing may be the cause. With most modern machines, it is not necessary to

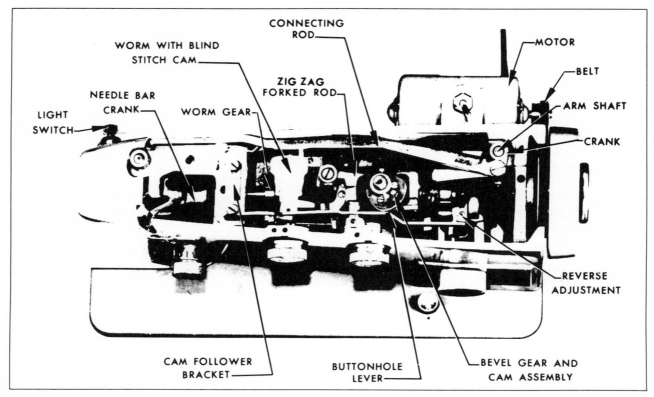

Figure 3-28. Typical underside parts of a sewing machine.

feed the material. Tell the customer this.

5. Be sure that the presser foot or attachments are securely fastened to the presser bar. Be sure that the needle goes through the needle hole plate without touching the sides of the plate or attachment. Tighten the foot attachment if necessary.

6. Check for a rough needle plate hole. Smooth out the rough spots, or replace the plate if needed.

7. See if the needle strikes the needle plate. Adjust according to the service manual.

8. Be sure that the cushion spring in the rocker is not bent or broken. Replace the spring, if necessary.

9. Check the feed rocker shaft to see if it is out of time. Adjust according to the service manual.

10. Be sure that the bobbin case is inserted properly. Replace in proper position.

Machine stops while sewing. With many machines, a loose stop motion knob will cause this problem. To correct in most cases, tighten the stop motion on the inner rim of the hand wheel.

TYPEWRITERS

The servicing of electrical typewriters does not normally fall within the bailiwick of an appliance service technician. In fact, when an electrical typewriter does come into shop, it is best to undertake only a limited service procedure. Leave major problems to typewriter and business machine technicians. By limited servicing, we mean, for example, the replacement of a cordset. As in most other small appliances, it is the line cord that usually gives problems long before the typewriter itself develops any. There is little bottom clearance on most machines, and this frequently results in the cord being crushed by the weight of the machine.

Limited servicing can also include inspection for possible loose screws and cleaning. In fact, cleaning is the keynote to typewriter maintenance, according to most manufacturers. A great many troubles are created by accumulations of bits of paper mixed with eraser rubbings. The latter are, of course, abrasive, and in time can cause problems for the bearings. But overoiling can cause problems, too, because oil holds these abrasive materials in the machine.

The electrical typewriter is a fairly complicated machine. It gets its power from a small, high-speed motor that turns the equivalent of a compact flywheel. When a key is touched, a portion of the kinetic energy of the wheel is employed to release the strike bar, the end of which presses the letter to the ribbon and the paper underneath.

KNIVES

Electrical knives come in two models: standard or 120-V line-operated type, and the battery-operated or cordless type. Basically, either type of this compact appliance consists of a powerful small motor, a gear drive that converts the rotary motion of the motor's shaft to a reciprocating (back and forth) motion, and a knife that is actually a pair of blades. One of the blades is stationary, the other moves back and forth a fraction of an inch. The edges are serrated, or scalloped. The principle of cutting is a shearing action which occurs as the scallops on one blade pass the scallops of the other.

Standard Cord Models

In the standard cord models, the rotary motion of the motor is converted by the gear train, cam or eccentric transmission, and slide box assemblies to a dual reciprocating motion which is applied to the blades. The blades,

being in close contact, produce a shearing effect, cutting neatly through meat or any other appropriate food. A high-torque series-wound universal motor provides ample power for any momentary heavy cutting that may be encountered. Generally, a fan on the rear of the armature is used to create an air flow through the case for cooling while the knife is being used for extended periods. The switch is usually of the momentary contact type which cuts off power as soon as pressure on it is released. Also, a safety position is generally provided to prevent the switch from being actuated in casual handling. In most designs, the blades can be inserted only in the correct way. The blade release button must be pushed in to insert or remove the blades.

The electric circuit of a standard cord model is simple; it consists of the universal motor, a switch, and a cordset. If the motor is laboring or the knife blades running slow, check the gear train for lubrication. Make certain that the moving parts are not tight or binding. These parts should be packed with a cream-type lubricant, never with a thin oil. Oil will run out of the gearbox and create a problem by dripping onto the food; the heavier cream lubricant will stay in the box. Incidentally, the motors used with electrical knives seldom require oiling; most are equipped with oilless bearings, made of sintered (fine-powder) metal. This metal has millions of tiny pores which hold oil, releasing it if the bearing gets hot. If the motor seems to need oil, put a single drop of fine oil on each bearing. If the motor speeds up while doing this, you know that the bearing did need lubrication. Never overoil.

Analysis of complaints. Here are some of the major complaints received on the operation of electrical knives.

Motor will not run.

1. Check the switch; it may be contaminated or damaged by overheating. Clean or replace as necessary.
2. Check the circuit including the armature, field, and cordset for continuity. If any components are found faulty, replace.

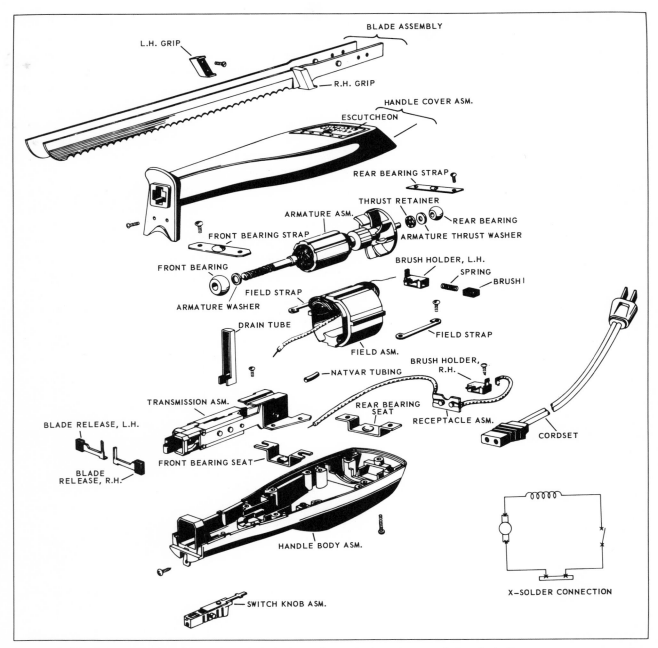

Figure 3-29. Schematic drawing and exploded view of a typical cord type of electric knife.

3. Check as for Insufficient Power below.

Power is insufficient (sluggish).

1. Check the brushes and commutator for excessive wear. Replace any faulty parts.
2. Be sure that the armature is not striking the field. Insulate carbon brushes from the commutator (a $\frac{1}{4}$-in-wide plastic or paper strip); check armature resistance from segment to segment on the commutator. On most units, it ranges from about 8 to 12 Ω. If the resistance in any winding is noticeably

low, there are probably shorted turns in that winding, and the armature should be replaced. Excessive resistance is an indication of open winding in the armature, and the motor should be replaced.

Motor runs hot. Run the motor to confirm an excessive rise in temperature. It may be caused by either a bind or shorted turns in the windings. Check the armature windings as above under Insufficient Power and measure the resistance of the two field coils. Take what corrective action is necessary.

Appliance is noisy.

1. Check for worn or loose bearings, a warped fan, a lead striking the fan, or an armature striking the field. If the motor sounds "bumpy," there may be a defective armature or excessive end play. End play in most units should be anywhere from 0.002 to 0.035 in.
2. Check for bent blades, worn collar, and/or worn drive rivet blade tang. Replace blades if necessary.
3. Check for worn holes in the latch springs which drive the blades in the transmission assembly. Repair or replace as necessary.

Excessive vibration. Run the knife with and without blades to learn whether the vibration is caused by the motor or the blades. Unusual blade problems such as bent blades, tight rivet, and blades "toed in" at the points will cause excessive vibration. Replace the blades if necessary.

Double cutting. When the gap between the blades is too wide, food becomes wedged between the blades and occasionally emerges through the gap as a thin slice. There should be no gap or a very small one between the blades at the scalloped edge. To check the gap, first insert the blades into the handle with the unit positioned on the bench, scalloped edge up. Put a $\frac{1}{4}$-in strip of writing paper (0.004 in thick) between the blades at the handle behind the grips; run the paper along the entire length of the blades. If the paper moves without a

drag, the gap is too wide; replace the blades. If it is very difficult to move the paper because of heavy drag, the gap is within tolerance, and the problem may result from the user exerting too much pressure on the knife while cutting around a curved area. Repair or replace as necessary.

Lack of cutting ability or blades dull. Such a complaint may be caused by any of the following conditions.

1. One blade is not reciprocating because it is not latched. If the condition of the latch spring is doubtful, replace the latch spring.
2. There is poor alignment of the scallops. Check with the blade assembly inserted in the handle. Line up the scalloped edges so the points of one blade are opposite the hollows of the other blade. Simply start and stop the motor until the lineup occurs. The points of one blade should generally be at least one-half the height of the points of the other blade (Fig. 3-30). If they are less, replace the blades.
3. Dull blades indicate that the edges have been in contact with a dish or pan, or that an attempt has been made to resharpen the blades. In such cases, the blades must be replaced. Instruct the customer on proper care of the blades.

Blades will not latch.

1. Be certain that the shoulder rivets on the blades are properly seated and not worn. Correct or replace as necessary.
2. Check the latch springs for worn latching

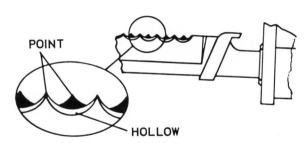

Figure 3-30. Proper heights of typical blades.

hole and other possible damage. Replace any faulty components.

Cordless Models

The principal differences between cord and cordless electrical knives are in the motor and source of power. The cordless knife utilizes a dc motor with a permanent magnet field. The motor obtains its power supply from nickel-cadmium or silver-cadmium batteries (usually five) connected in series with a total output of usually about 6 V. The actual operations of the two models are basically the same: the motor drives two counterreciprocating scalloped-edge blades through a cam or an eccentric transmission. The principle of cutting is a shearing action which occurs as the scallops on one blade pass the scallops on the other. Cordless knives are normally supplied as two units: a handle assembly and a battery recharge or base assembly. The handle assembly usually contains the transmission, motor, switch, and batteries. When the knife's power handle is not being used, it should be placed in the battery recharge unit which is,

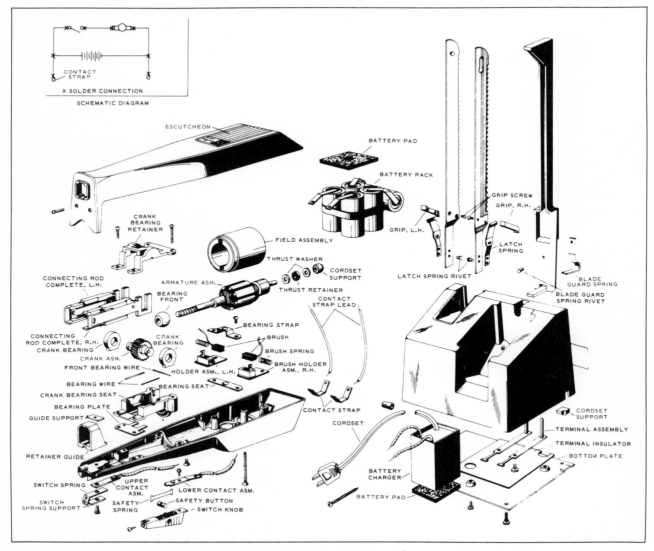

Figure 3-31. Schematic drawing and exploded view of a typical cordless type of electric knife.

in turn, connected to a 120-V ac outlet.

As a rule, the batteries in the handle are connected with welded straps. This eliminates the possibility that the intercell connections will corrode or get dirty. Remember that with a fully charged power handle, the open-circuit voltage of the five-cell unit will measure approximately 7 V on a dc voltmeter. It can be checked without taking the handle apart by measuring across the charging terminals at the back of the handle; the batteries are connected directly across these points. To get an accurate reading, connect the voltmeter to the terminals and place the ON-OFF switch in the ON position. This will give a voltage reading under full load, with the motor running. The voltage should be above 5.7 V. If it drops to below 4.0 V, the battery needs recharging. More on a typical battery test procedure is given under Battery Test below.

While the basic principle of operation of most electrical knife power handles is about the same, the design and operation of the battery recharger can vary from one make and model to another. Here are the two most common recharger arrangements.

1. As shown in Fig. 3-32 (top), the charger unit consists of a transformer which reduces the 120-V ac line voltage to about 4 V alternating current; two silicon rectifiers which change the alternating to pulsating direct current; an electrolytic condenser to smooth out the dc pulses to produce a steady dc voltage of about 6 V to charge the batteries; and a bleeder resistor to discharge any electric energy that may be held in the capacitor should the charger be unplugged. In other words, in a charger of this design, the charging current from the charger base unit flows to the batteries through the electric contacts that are built into both the charger and power handle assemblies.

2. Another very popular type of base is the so-called "induction" recharger (Fig. 3-32 bottom) which incorporates a sealed induction coil connected to the 120-V ac power

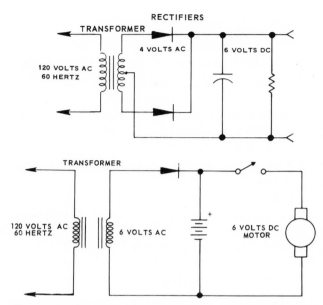

Figure 3-32. Standard transformer recharger schematic (top) and induction recharger schematic (bottom).

outlet. This coil induces a small current in a mating coil contained in the power handle. This alternating current is rectified and smoothed out by a silicon rectifier and filter circuit; thus, in this process the low ac voltage is changed into a smooth direct current source of approximately 6 V which serves to continuously recharge the battery, also in the handle. The handle must be stored in the well of the recharger, and the recharger must be connected to a continuously energized outlet to maintain a full battery charge. Thus, in an induction recharger there is no direct electric connection between the recharger and the power handle. If no electric contacts are noted, it is safe to assume that the recharger is of the induction type.

More on a typical test procedure is given later in this chapter.

Analysis of complaints. While complaints about the knife's blades will be the same as those discussed above, here is an analysis of power problems that may arise with the cordless type of knife.

Motor will not run.

1. Check the switch (there may be a short between the positive battery terminal and the brush wire directly below it). If the unit runs, the switch is defective.
2. Rotate the armature by hand and check for binding armature. If you find a bind, correct the problem.
3. Check the battery (refer to Battery Test below).
4. Check all electric connections and battery welds. Do not repair broken welds; replace the battery.
5. Check with the customer to confirm that a continuous energized outlet is being used. Instruct the customer on the correct procedure.

Motor runs slow.

1. Check the brushes and armature for excessive wear. Repair if either is found faulty.
2. Check the field-mounting screw for tightness. Repair, if necessary.
3. Check battery. (Refer to Battery Test below.)
4. Check charger. (Refer to Charger Test below.)
5. Check motor speed with one blade. (Refer to Motor Test below.)
6. Check with customer to confirm that a continuous energized outlet is being used. Instruct the customer on the correct procedure.

Noisy or excessive vibration.

1. Run the knife with and without blades to determine whether the noise is caused by the knife or the blades.
2. Check for worn or loose bearings.
3. Check the cam or eccentric drive mechanism for worn or loose parts. Repair or replace as necessary.
4. Check the blades for warpage and/or worn collar and rivet. Replace blades if necessary.

The following are *typical* procedures for conducting charger, battery, and dc motor tests. We

state *typical* because the information given here is to illustrate how such tests are to be conducted. Exact information on how to conduct charger, battery, and motor tests is usually given in the service manual.

Charger Test

Input Test

Conditions of test: Plug the charger to be tested into the outlet in the test panel. (Most small appliance manufacturers supply a test panel as an accessory or give instruction which permits easy construction of one. It is of value for servicing most cordless appliances.) Plug the test panel into a 120-V ac 60-Hz outlet. This places a 100-mA ac ammeter in series with the charger and power supply.

Specification

1. With the knife body placed in the charger rack, the reading must be 50 mA ± 20 percent alternating current.
2. With the knife body out of the charger rack, the reading must be less than the reading in specification No. 1 above.

Interpretation

1. If no reading appears on the meter, the charger circuit is open in either the cordset wiring or the charger transformer.
2. If the reading is out of specification, it could mean a shorted transformer or poor contact from knife to pins.
3. Finding no substantial drop in the meter reading with the knife in the charger or out indicates poor contact between the knife and charger pins.

Output Test

Conditions of test: Place the test handle (part of the test panel) in the charger rack to be tested, making sure that the toggle switch is in the TEST position and the charger and test panel are plugged into a 120-V ac outlet. (The test handle must be fully charged.)

Specification: The charger must supply 100 mA ± 10 percent of direct current 30 min after the test handle is inserted in the charger.

Interpretation

1. Observing no reading indicates an open transformer, open line cord, open connecting wires, the absence of contact with the test handle, or the test switch in the wrong position.
2. A reading out of specification indicates a defective charger unit, or a test handle not fully charged.

Battery Test

Conditions of test

1. Charge the battery to a full charge (overnight: 16 h).
2. Place the knife with the fully charged battery pack in the test panel rack.
3. Turn the discharge switch to the ON position. This places a 30-Ω resistance load across the battery. Record the time.
4. Leave on discharge $3\frac{1}{2}$ h. Immediately at the end of this time (while still on discharge) check the voltage at the measuring access pins on top of the rack. (This measurement must be taken exactly $3\frac{1}{2}$ h after start of discharge to obtain an accurate reading.)

Specification: The voltage must be a minimum of 5 V direct current.

Interpretation: If the battery does not meet the above specifications, it is defective and must be replaced.

Motor Test

Conditions of test: Place a dc ammeter (10 or 12 Ω full scale) between the positive terminals on the battery power pack and the brush contact immediately below the positive terminal. This will short out the switch and place the meter in series with the battery pack and motor.

Specification: Without the blades, the battery drain should not be more than 3 Ω direct current or less than 2.2 Ω.

Interpretation

1. If the battery drain is less than 2.2 Ω, check for loose solder joints or high-resistance connections.
2. If the battery drain is more than 3 Ω, connect the meter as indicated in the Conditions of Test and tap the crank-bearing retainer and seat for better alignment and lower battery drain.

Conditions of test: The knife must be completely assembled with a fully charged battery pack and the blades in place. Hold a vibrating-reed tachometer, or similar equipment, in the left hand and the knife in the right. Place the upper edge of the left blade against the side of the tachometer with the vertical plane of the blade at approximately a 30° angle from the vertical plane of the tachometer. Press the switch and record the speed.

Specification: The speed of the knife must be a minimum of 1,350 r/min with the blades in place.

Interpretation

1. If the minimum specification above is not obtained, repeat the test with one blade in place. If one blade passes the test, substitute a new set of blades and test again for speed.
2. Tap the crank bearing retainer and seat for better alignment to improve speed.

TOOTHBRUSHES

Electrical toothbrushes operate in a manner very similar to electrical knives. Both appliances employ a small motor that converts electric energy into mechanical motion. Both use an eccentric transmission or a cam arrangement to change this mechanical rotary motion into the desired kind of oscillations. Most modern electrical toothbrushes permit the user to select either back-and-forth or up-and-down brush motion.

While a few models have a cord arrangement, the vast majority are of the cordless type. The assembly of a cordless toothbrush, like that of a cordless knife, consists of two parts, the power handle and the battery recharger base. The handle must be stored in the well of the recharger, and the recharger must be connected to a continuously energized outlet to maintain a full battery charge. *Note:* Some bathroom fixtures are so wired that the outlet is controlled by the light switch which results in an inadequate battery charge and either poor or no operation.

The power handle should be thoroughly rinsed under running water to remove toothpaste accumulation. The recharger base should be disconnected from the power outlet and wiped clean with a damp cloth. Heavy accumulation of toothpaste either on the handle or in the recharger well will result in poor seating of the handle and prevent proper charging.

Unfortunately most power handles and rechargers used with toothbrushes are hermetically sealed. This is probably for watertightness, but it makes them nonrepairable. The cases are plastic and are cemented together so that they can not be disassembled. The only recourse for the service technician if the handle or recharger goes bad is to send it back to the manufacturer for exchange. Thus, the troubleshooting procedure for most electrical cordless toothbrushes is limited to just determining whether the power handle or the recharger base is bad. Here is how to test these two units of a typical electrical toothbrush.

Recharger. Plug the recharger to be tested into an energized 120-V outlet. Insert a steel screwdriver into the charger well and up against the center metal post. A magnetic vibration should occur. If there is no vibration, replace the recharger. Check for excessive heat after the recharger has been plugged in for at least $\frac{1}{2}$ h.

Power handle. Because of the induction system used in most electrical toothbrushes, the battery condition cannot be measured directly. However, the following procedures will enable you to determine whether the handle is defective and

should be replaced or the battery merely needs to be charged.

1. If the handle does not run, operate the switch several times to be certain it is indexing properly.
2. Turn the switch to OFF (in models with an ON-OFF switch).
3. Place the handle in a recharger unit known to be good.
4. After 1 min, turn the switch ON, or press the push-push type once.
5. If you see no motion of the shaft of a model with an ON-OFF switch, the handle is bad and should be replaced.
6. The push-push type handle, however, should be tested further. Again press the switch firmly—once. Leave the handle in the recharger for another minute.
7. If you see no motion of the shaft, the handle is bad and should be replaced.
8. If motion detected in steps 4 or 7 is sufficient to operate the shaft at least once, turn the switch off and leave the handle in the recharger for an extended period before deciding on its condition. An overnight charge should result in full capacity, power, and speed.

Analysis of complaints. Here are the major complaints you will encounter for electrical toothbrushes.

Handle will not run, runs slowly, or produces inadequate power.

1. Test the recharger and handle as detailed above.
2. Ascertain whether the energized power outlet and proper cleaning instructions are being followed.

Handle is noisy. With the switch OFF, check the main shaft for looseness. Slight side-to-side movement is normal. Excessive movement indicates a defective assembly. Replace the handle.

Toothbrush runs with normal speed but stalls in use. Check the gear mesh by grasping brush and stalling it. If the motor also stalls, check for a poor battery recharger as detailed above. However, if the motor continues to run, the gear mesh is bad, and the handle should be replaced.

Bellows or case seam of handle is open. Replace the handle.

Recharger overheats. The center post in the well should be warm to the touch but not hot. If heat is excessive, replace the recharger since this indicates a defective coil.

Recharger hums. This indicates a poor mechanical/magnetic circuit. Replace the recharger.

WATER PULSER OR PICK

A water pulser or water pick is a dental hygiene aid which consists of a power unit containing a motor and adjustable pulsating pump and a handle, which is attached by a coil tube, containing a water shutoff valve. When not in use, the reservoir receptacle containing a check valve also doubles as a cover for the handle and tips.

With the reservoir receptacle in place and containing the proper amount of water (usually about $1\frac{1}{2}$ pt) and the pressure-selector switch and water shutoff valve in the handle (both in the OFF position), the water pulser should be connected only to a 120-V 60-Hz ac outlet. A water tip should be properly seated in the handle and aimed into a suitable container. When the pressure selector switch is turned on, the pulsating pump forces the water through the tube to the handle. Opening the valve in the handle allows the water to pass through the tip at various pressures (dependent upon the setting of pressure-selecting switch) at a maximum rate of about 2,000 pulses a minute.

Like the electrical toothbrush, if any problems are encountered with the sealed power unit assembly, the complete appliance, as a rule, should be exchanged through the manufacturer. For this reason, no attempt should be made to service the power unit assembly. To replace a

handle assembly, however, the following steps are usually followed.

1. Slide the hose fitting back from the handle onto the hose.
2. Pull the hose off the handle fitting.
3. If necessary, replace the hose fitting onto the hose.
4. Cut $\frac{1}{4}$ in from the end of the hose.
5. Slide the hose onto the handle fitting full depth.
6. Push the hose fitting over the hose onto the handle fitting full depth.
7. Run the unit, using a small amount of water in the reservoir to test the connection at the handle for leaks.

SCISSORS

Electric scissors operate in much the same manner as an electric knife. The electrical circuit for scissors is simple and can be serviced in the same manner as a knife. Parts for a typical two-speed scissors are shown in Fig. 3-33.

The blades of an electric scissors should be oiled occasionally with light machine oil. Make certain all screws are in position at all times; if they are loose, excessive vibration will cause uneven cutting. Scissor blades can be sharpened by any reliable scissor sharpener.

SHOE POLISHERS

Shoe polishers are available in hand-held ac-powered types and cordless battery models, as well as small ac floor models. The floor model is intended to be fastened to a solid surface and has a dual-shaft motor with a brush or buffer on each end. The battery-operated type uses five nickel-cadmium cells in series and a dc permanent magnet motor. The motor drives a small gearbox, with reduction-gearing to slow

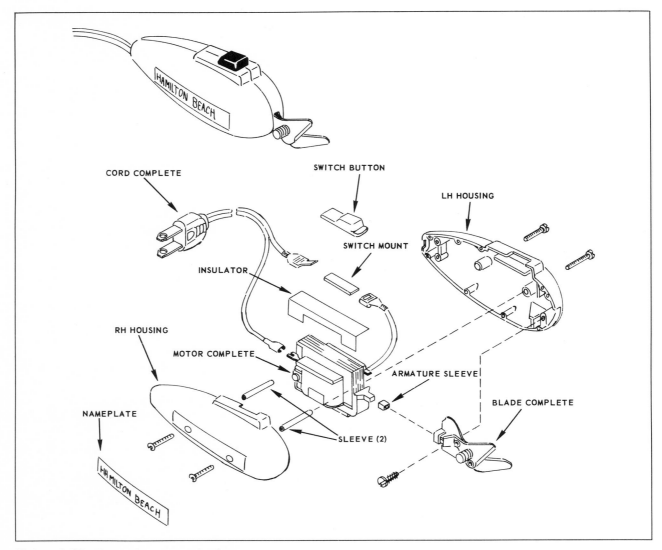

Figure 3-33. Parts of two-speed scissors.

the speed and give more torque. Motor units of this type are usually completely sealed and cannot be taken apart for repair. The gearbox is underneath the motor and can be taken apart for repair. The battery recharger units, which are similar to those used with cordless knives, are also hermetically sealed, and thus cannot be repaired.

The ac-powered hand-held units work in the same way as the cordless models, exept that they are usually more powerful. The power handle, which is hand-held, is driven by a series-wound universal motor which operates a recessed shaft into which the applicator/polishing brush is threaded. The direction of rotation of the shaft is controlled by a three-position switch. The ON or forward position is used to connect the attachments automatically. The RELEASE or back position is used for disconnecting the attachments automatically. The center position is OFF. If the motor refuses to run in either direction, check the switch first.

Analysis of complaints. Here are the complaints generally received about an ac-powered shoe polisher, including the floor type.

Motor will not run.

1. Check the cordset for continuity and replace entire set if defective.
2. Check the switch. Replace if defective.
3. Check all soldered connections. Take necessary corrective action.
4. Check for broken leads. Repair as necessary.

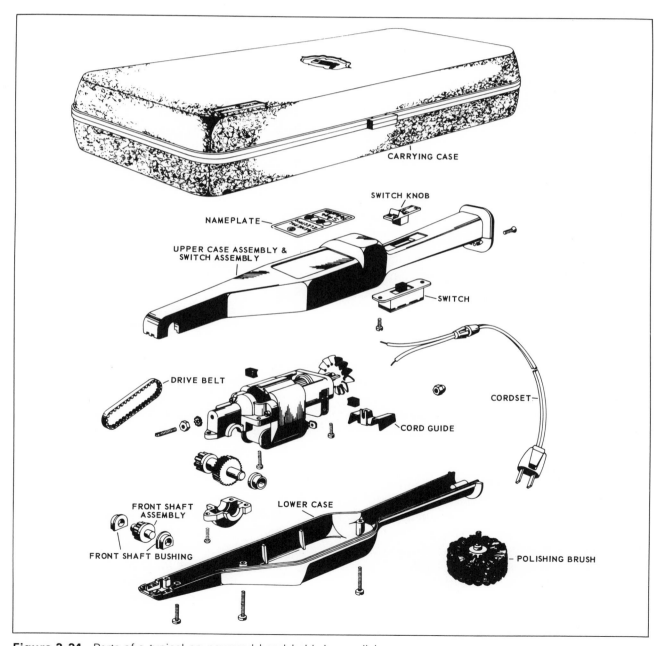

Figure 3-34. Parts of a typical ac-powered hand-held shoe polisher.

5. If above checks do not disclose any defect, you can assume that the motor itself is defective. Replacing the entire power handle is usually required.

Motor runs but shaft will not turn.

1. Check for broken or worn drive belt. Replace belt, if defective.
2. Check for stripped or broken gears. Replace any defective parts.

Handle is noisy.

1. Check the front-shaft bushings for wear. Replace, if necessary.
2. Check front shaft for damage or wear. Replace, if necessary.
3. Check whether the fan is striking the case. Rotate by hand; realign to correct.
4. Check drive belt for "flat." Replace, if faulty.

Appliance runs slow or stalls easily. Check for misalignment of parts, worn bushings, or defective motor. Free-running front-shaft speed should usually be not less than 750 r/min, measured with a tachometer.

CLOTHES BRUSHES

Most electrical clothes brushes are of the cordless type and are designed to brush lint, dirt, and foreign material from clothing. This battery-operated clothes brush consists of a recharger rack which holds the power handle, makes contact to it, and maintains the battery supply at full charge. In use, the dc motor, which operates on two nickel-cadmium batteries, and gear train assembly drive the rotating brush at high speed, creating a slight vacuum which draws lint and dust into the dirt storage cavity. The storage rack, designed as a rule for either counter top or wall mount, has a battery charger built into it so that the battery automatically makes contact with the charger when the charger is plugged into a 120-V ac outlet.

Analysis of complaints. Here are two complaints that are generally made about cordless clothes brushes and ways to service them.

Motor will not run or runs slow.

1. Check with customer to learn whether a continuously energized outlet is being used.
2. Check condition of charger and handle contacts. Clean with a soft eraser.
3. Check switch for positive contact. If the switch is faulty, replace.
4. Perform battery, motor, and charger tests outlined below.

Brush deposits line of dust on clothes. This indicates a loaded dirt storage cavity. Clean out the cavity, and instruct the customer about proper maintenance. A rotating brush in poor condition could also cause this effect; replace the brush.

The following battery and motor tests are *typical* of those made on these components of clothes brushes and similar cordless appliances. For exact testing procedures, check the service manual for the specific appliance.

Battery Test

1. Inspect the battery assembly for a white deposit, indicating electrolyte leakage. If found, replace the battery power pack. Remove the white deposit by scraping the inside of the handle. Then wash it with a boric acid solution (20 to 100 percent). Rinse with clear water, or wipe with a wet cloth, to remove the boric acid. If deposit cannot be removed by this method, replace handle.
2. A quick indication of the battery condition (charge capability) may be obtained by following these steps:
   a. With the switch off, insert the handle in the recharger for 5 min or longer.
   b. With a volt-ohm-milliammeter (VOM), read the open-circuit battery voltage (side contact to end contact). A reading of less than 2.0 V indicates that one or both cells are bad. If so, discontinue the test. A reading of 2.0 to 2.6 V indicates

that the cells are probably good and the test should be continued as in step 3.

3. Charge the battery to full charge (overnight: 16 h).
4. Place the brush handle with the fully charged battery in a battery discharge rack such as shown here. This places a 1.6-Ω resistance load across the battery.
5. Leave on discharge for 36 min and check the voltage.

Specification: The 36-min discharge voltage must be 2 V or more.

Interpretation: Voltage lower than 2 V indicates low-capacity cells; replacement of the handle is necessary.

Motor Test

Conditions of test: Place a dc ammeter between the switch contact (with the switch in OFF position) and the lower brush contact. This will place the meter in series with the battery and motor.

Specification: For most motors of this type, the motor current should not be more than 2 A or less than 1 A.

Interpretation

1. If the motor current is less than 1 A, check for a loose or high-resistance connection.
2. If the motor current is more than 2 A, check for a motor out of position.
3. Check the brush and brush retainer for proper placement. Check the motor for proper alignment and for binding. Take whatever corrective action is necessary.
4. If the unit draws more than 2 A with the brush in position, repeat the test with the brush removed. If it then passes the test, replace the brush.

Input Test

Conditions of test: Plug the recharger to be tested into the outlet of the test panel (see Charger Test above). Plug the test panel into a 120-V 60-Hz ac outlet. This places a 100-mA

ac ammeter in series with the recharger primary circuit.

Specification

1. With the clothes brush handle in the recharger stand, the meter reading must be 28 mA $\pm$ 20 percent.
2. With the brush handle removed from the rack, the meter reading should not change more than 5 mA.

Interpretation

1. If no reading appears on the meter, the recharger circuit is open.
2. If reading is excessive, the transformer could be shorted. Check for excessive heat.
3. No change in the meter reading with the clothes brush in or out of recharger would indicate poor contact or no contact between the clothes brush and recharger.

Output Test

Conditions of test: Place the test brush (fully charged) in the recharger. Use a standard volt-ohm-milliammeter set on the milliampere 120-V full-scale setting or higher. Place the probes into the test brush; maintain proper polarity and place the test brush switch on TEST.

Specification: The recharger should generally supply 110 mA direct current $\pm$ 10 percent 5 min after the test brush is inserted into the recharger.

Interpretation: A reading out of the specifications would indicate a defective recharger unit, the test handle not fully charged, or the test brush switch in the wrong position.

MANICURING SETS

While there are many different electrical manicuring sets on the market, all work on the same basic principle. Most have the same basic components: a case or base which contains a transformer, full-wave rectifier, and fusible element assembly which supplies power at a low voltage to the power handle that drives the attachments at the correct speed for the application. Developed power is deliberately limited to a safe value. With most models, the attachments may be removed and installed while the motor is running without danger or harm.

To better understand the operation of a manicuring set, let us take a look at a typical unit and trace a series of the most common complaints. (Remember that the voltage and resistance readings given below are only *typical*; check the service manual for exact readings.)

Unit will not run. Make sure the customer is using the appliance in an energized outlet; some outlets in the home are controlled by either a wall or light switch. Check to see that the switch is placed in the ON position. Check plug on retractable cord to confirm that it is fully seated on the connector pins, and that "key" on the plug mates with the slot in the case. If these items are all right, localize the trouble as follows.

1. Connect the line cord to a 120-V ac outlet and pull out the retractable cord plug.
2. With a VOM on the 50-V dc range, read the voltage at the connector pins. If the switch, cord, and power supply are all right, the voltage should read approximately 15 V.
3. If the voltage reading is very low or zero, disassemble the case and make the following tests to localize the defect.
   a. Disconnect the line cord from the outlet and pull out retractable cord plug from the deck connector.
   b. With the blade of a screwdriver, pry off the shield and insulator fastened on the switch. This will expose the switch terminals for testing.
   c. With the VOM (ohmmeter portion) set on the R × 10-Ω range, trace the continuity of the primary circuit from one male prong to cordset through the switch, through the transformer (about 95 Ω), wire nut connector, and out through the other male prong of the cordset. Any open-circuit reading will show a defective part.
   d. Snap the metal shield and fishpaper

shroud back onto the switch.
   e. If *c* above is all right, set the VOM on the R × 10-Ω range and read the resistance across the deck connector pins. Reverse the ohmmeter leads and read again. One direction should read continuity; reverse direction should read open. If not, the transformer pack is defective and must be replaced. *Caution:* Continuity and resistance checks above must be made with a VOM. Do not use a 120-V series light continuity tester because damage to the low-voltage components will occur.
4. If the proper voltage is read, the trouble is in the power handle which should be tested as follows.

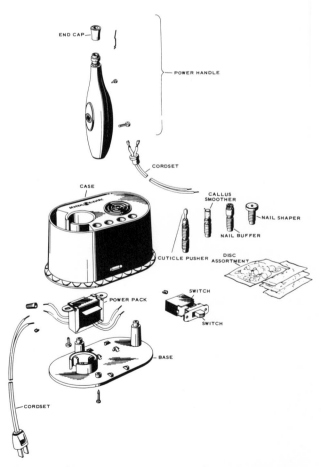

Figure 3-35. Disassembled view of a typical manicure set.

a. Pull the retractable cord plug out of the socket in the case.
b. With the VOM on R × 1 range, insert the probes into plug openings.
c. Install the callus smoother attachment onto the nosepiece.
d. Slowly turn the drum by hand while watching the ohmmeter. Do not spin it fast or the countervoltage developed from the motor will give erratic readings.
e. The reading on the ohmmeter should be approximately 12.5 Ω; at no point in a revolution should it be less than 10 Ω nor more than 20 Ω.

Interpretation

1. If the reading is uniformly about 12.5 Ω, the motor is all right.
2. If the reading dips at times to less than 10 Ω, the motor is shorted; replace.
3. If the reading is over 20 Ω, the motor segment is open. Replace.
4. If an open circuit is indicated, the cord is defective. Replace.

Attachment will not hold on. Check for flattened, bent, or missing bow spring in the slot of the nosepiece.

Power handle has full speed but attachments run slow. Clean the attachments in isopropyl alcohol and relubricate them with petroleum jelly applied with a toothpick for application. Use a minimum amount of jelly.

Power handle tests defective or is noisy. Replace the entire power handle assembly.

As a final operation test, connect the manicurer to a 120-V ac supply. Insert the cuticle pusher and test for a handle speed of 2,000 to 3,000 r/min with a vibrating-reed tachometer held lightly against the tip of the cuticle pusher.

SHAVERS

While electrical shavers—for both men and women—vary in shape, size, and color, their basic operational principles are the same. Most of the shavers on the market today have three general forms of motive power: vibrator, universal motor, and battery.

Types of Shavers

Vibrator-driven shavers. Most of the early models of electrical shavers were of the vibrating-motor type which operates from the 60-Hz electromagnetic field of a coil. In this type an alternating current from the house line passes through an electromagnet, near the ends of which is suspended an iron bar called a *vibrating armature.* As the alternating current varies in direction and strength, it attracts the vibrating armature and repels it in rhythm with these variations. (Some vibrating shavers move the cutters at about 7,200 strokes per minute.) The vibrating armature is connected to a set of small cutting blades, generally shaped like a comb, which is enmeshed with a fixed set. Thus, as the head of the shaver is pressed against the hair, it snips off the hair when it gets trapped between the fixed and movable cutting blades. The actual movement of the movable blades is very small.

Motor-driven shavers. In this type of shaver, a small motor agitates or drives the cutting blades. An eccentric transmission or cutting blades convert the motor's rotary motion into high-speed oscillations in the same manner as for electric knives and toothbrushes.

While for many years shavers used shaded-pole motors, today most use small, high-speed universal motors.

Cordless shavers. The cordless electric shavers usually employ a low-voltage (usually about 2.5 V) universal motor which operates from a set of rechargeable batteries built into the shaver unit. There is also a recharger which is used to keep the shaver's batteries charged. The method of servicing the recharger unit is the same as for the clothes brush's recharger described earlier in this chapter.

Some shavers have dual-voltage motors so that the unit may be operated on either the house power line or from an automobile power supply.

Many shavers have so-called "lifetime" lubrication; with others, motor lubrication is important. But remember that excess oil can "gum up the works" by picking up fine bits of hair, dust, etc. When oiling is a must, be sure to use a thin oil, and put only one drop on each end of the motor shaft. To lubricate the cutting heads, rub one drop of oil between your thumb and forefinger, then rub the parts to be lubricated with your fingertip.

A great many makes of shavers cannot be serviced by local technicians because many manufacturers will not supply the necessary replacement parts, preferring to have the shaver returned to their authorized service centers for repair.

Analysis of complaints. It is rather difficult to establish solutions to various complaints received from owners because of differences in designs of shavers. Here, however, is a general listing of complaints and possible causes and solutions.

Vibrator-type Shavers

Shaver does not operate.

1. Check for defective cordset. Almost all cords are of the detachable type, which makes them easy to roll up and pack with the shavers in their containers. If the plug or wire becomes damaged, replace the cordset.
2. Check for loose wires. If any are found, resolder.
3. Check the motor winding. Replace motor, if defective.
4. Check the switch. Replace switch or motor as required.
5. Check for correct adjustment of the motor. Refer to the service manual; if this is not available, remember that most shavers of this type will operate when the rotors overlap the stators by between 0.045 and 0.055 in. The air gap between the rotors and stator should be approximately 0.003 in (Fig. 3-36). Use a feeder gauge to check these adjustments.

6. Check for a faulty head or inner cutters. Replace if found defective.

Motor hums but blades do not move.

1. Be sure that the vibrating armature assembly is not broken or jammed. Correct, or if the armature is defective, replace it.
2. Check the shaver head to see if the comb assembly is dented or broken. Replace, if defective.
3. Check the blade assembly to be sure that its shaft is not bent, broken, or badly worn. Replace, if defective.
4. Check the transmission system for mechanical problems such as a broken or bent connecting rod or a cutting blade shaft jammed by excessive accumulation of cut hairs or dirt. Clean, correct, or replace faulty components as needed.

Shaver is noisy.

1. Check for loose parts. Tighten.
2. Be sure that the shaver head is not bent or dented. Replace, if necessary.
3. Check the vibrator assembly. If faulty, replace.

Shaver gives poor shave

1. Be sure that the motor is properly adjusted. Refer to the service manual for the correct method or adjust as previously mentioned.

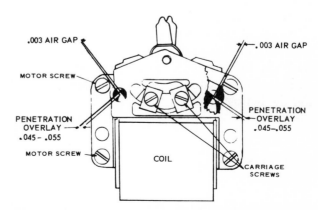

Figure 3-36. Air gap and overlap adjustments on a typical shaver motor.

2. Check for a faulty head or inner cutters. Replace if found defective.
3. Check the throw of the inner cutters. Replace the inner cutter springs, if defective.

Shaver runs erratically.

1. Check for an intermittent cordset. Replace, if faulty.
2. Check for a loose wire. If one is found, repair.
3. Check for a defective motor. If faulty, replace.
4. Check the switch. If faulty, replace the switch or motor.

Shaver draws too high current. (Vibrator-type shavers usually draw less than 100 mA or about 8 to 10 W.)

1. Check for shorted motor winding. If motor is faulty, replace.
2. Be sure that the motor is properly adjusted. Remember that if tension, gap, and balance are not correctly set, the shaver will not run fast enough and may draw excessive current. Refer to the service manual for the proper amount or adjust as previously mentioned.

Motor-driven Shavers

Shaver does not operate.

1. Check continuity of the cordset. If defective, replace.
2. Check for a broken or loose wire. Repair.
3. Check for a shorted capacitor or a defective coil. Replace faulty component or components.
4. Check for broken or worn motor brushes. If defective, replace.
5. Check for open motor windings. If found, replace the motor.
6. Check the ON-OFF switch. If defective, replace.

Motor hums but blades do not move.

1. Be sure that the motor bearings are not too tight. Take proper corrective action.
2. Check the transmission system for mech-

anical problems such as a broken or bent connecting rod, broken or loose eccentric cam on the motor shaft, or cutting-blade shaft or transmission jammed by excessive accumulation of cut hairs or dirt. Clean, correct, or replace faulty components as needed.
3. Check for partially shorted field winding. Correct or replace the motor as needed.
4. Check the shaver head to see if the comb assembly is dented or broken. Replace, if defective.
5. Check the blade assembly to be sure that

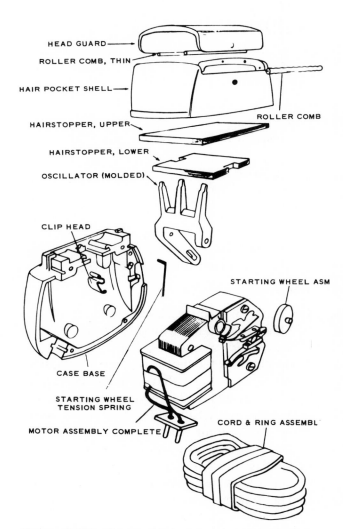

Figure 3-37. Disassembled view of a typical motor-driven shaver.

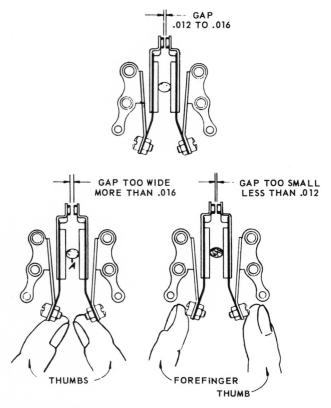

Figure 3-38. Method of adjusting the point gaps on a typical shaver motor.

its shaft is not bent, broken, or badly worn. Replace, if defective.

Shaver operates slowly.

1. Check for a dirty shaver head or switch. Clean and oil according to instructions in the owner's manual.
2. Be sure that the motor is properly adjusted. When adjusting a shaver motor, be sure to set the points as the service manual recommends. As a rule, these points are never set closer than 0.012 in (Fig. 3-38). The shaver motor may run faster with a closer spacing but will stop running within a few weeks because the cam will have worn out. The points should not be spaced wider than 0.016 in, however.
3. Check for burnt or worn contacts. If defective, replace contacts or motor.

4. Be sure that the inner cutters are not too tight. Replace, if faulty.
5. Check for motor binding. Clear the cause of binding.
6. Check for worn brushes. Replace the brushes.
7. Check for defective motor bearing. If found, replace.
8. Determine whether the motor field is partially shorted. Correct or replace the motor as needed.

Shaver gives poor shave.

1. Determine whether the motor is running too slow. See above.
2. Check for defective heads or inner cutters. If found to be defective, replace them.
3. Check for weak cutter springs. Replace, if faulty.
4. See whether the oscillator is low or worn. Replace or shim into proper position.

Shaver runs erratically.

1. Check for an intermittent line cord. Replace, if faulty.
2. Check for dirty, burnt, or worn contact points. Clean or replace contact points.
3. Check for sticking brushes. Form brush leads or replace.
4. Check for loose wires. Tighten and resolder.
5. Check for poor adjustment. Refer to the service manual for the correct method or adjust as previously described.

Shaver draws too-high current. (Most motor-driven shavers draw about 135 mA or about 15 W.)

1. Check for poor adjustment. Refer to the service manual for correct method or adjust as previously described.
2. Be sure that the heads are not too tight. Replace the heads, if necessary.
3. Be sure that the heads are not dirty. Clean and oil the heads.
4. Check the bearing to be sure that it is seated properly or there is no bad bearing. Repair or replace.

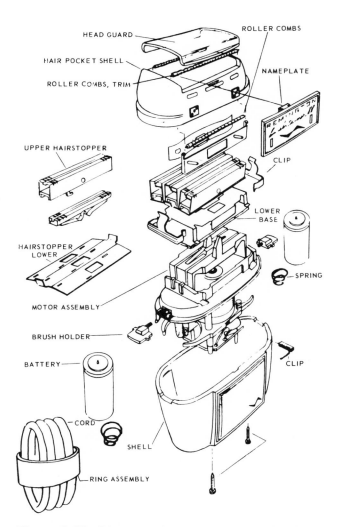

HEAD GUARD
ROLLER COMBS
HAIR POCKET SHELL
ROLLER COMBS, TRIM
NAMEPLATE
CLIP
UPPER HAIRSTOPPER
LOWER BASE
HAIRSTOPPER LOWER
SPRING
MOTOR ASSEMBLY
BRUSH HOLDER
BATTERY
CLIP
CORD
SHELL
RING ASSEMBLY

Figure 3-39. Disassembled view of a typical cordless shaver.

5. Check the oscillator or motor armature for defectives. Replace any defective component.

Shaver is noisy.

1. Check for loose motor screws or mechanical parts. Tighten.
2. Check for a bad bearing or bent armature shaft. Replace motor, if faulty.
3. Check for a weak cutter spring. Replace springs, if necessary.
4. Be sure that shaver head is not bent or dented. Replace, if necessary.

5. Check for worn oscillator. Replace, if faulty.

Cordless Shavers

In addition to the complaints about motor-driven shavers, here are some that are frequently made for the cordless, battery-operated ones.

Motor does not run. Check the batteries (see Battery Test above) and replace, if necessary.

Shaver does not run on accessory recharger.

1. Check for a defective cordset. Replace if faulty.
2. Check for an open transformer winding in recharger stand. Replace the recharging stand.
3. Check for a defective rectifier. Replace, if necessary.
4. Check for a defective switch. Replace, if necessary.
5. In some models, there is a fuse in the transformer network. See if this is blown. Replace it if it is.

Shaver runs slow.

1. Check for one cell that may be defective. Replace cell.
2. Check for defective rectifier. Replace, if found to be the case.

CLOCKS

All electrical clocks, whether they be of wall, digital, alarm, or timer types, cord or cordless, operate on the same basic principles and consist essentially of a shaded-pole motor which runs through a gear train. This gear train, in turn, transfers the rotating motion of the clock hands. Unlike mechanically wound clocks and timers, there are no hair springs and escapements in electrical clocks—just the step-down gear trains.

The shaded-pole induction motor used in all clocks, except the quartz type, operates in synchronism with the generator located at the local public utilities power plant. Because of the great multiplicity of present-day electrical and electronic equipment in the home and in industry, generating plants produce power carefully speed-controlled to exactly 120 alternations, or 60 Hz, per second.

The quartz-type clocks use a precisely cut quartz crystal which, when activated by electricity, vibrates at a specific rate (generally about 262,150 vibrations/s). These vibrations, converted into electric impulses, are brought into manageable proportions by binary dividers (the final step is 64 vibrations/s). The end stage activates a motor, which, in turn, drives the hands. Although the vibrations of a quartz crystal are incredibly constant, they are subject to variations under different temperatures and barometric pressures. A hermetically sealed and evacuated container protects the quartz. Most quartz clocks use a synchronous inductor-type motor that operates on approximately 300 μW (3/10,000 of a watt). Under normal and consistent barometric pressure and temperatures, an accuracy of ± 1 min per year can be obtained.

Cordless clocks generally use a single non-rechargeable type C cell flashlight battery to power their shaded-pole motors. A cell will last for about a year. Therefore, the first check to give a cordless clock is of the battery.

There is little in the way of servicing that

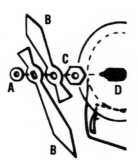

Figure 3-41. Method of disengaging a clock movement from a frame.

can be done to a clock motor when it goes bad. Since many electrical clocks are relatively inexpensive, it is usually not advisable to spend a great deal of money on repairs. However, for more expensive clocks, a replacement motor unit is available from appliance supply dealers or directly from the manufacturer. Be sure to copy the make and model number from the clock unit. This information is usually stamped on the clock case or on a small metal plate fastened to it somewhere. When factory service is needed, return the complete clock mechanism, fixation nut, hands, and pendulum (if it is a pendulum clock) unless otherwise directed in the service manual. Be sure everything is packed securely.

A typical clock movement can be disengaged from a frame (Fig. 3-41) as follows. Cover the jaws of a pliers with adhesive tape to protect the finish of the clock. Remove the knurled nut (A) and slide the clock hands (B) up and off the mechanism stem (D). With the pliers, carefully remove the flat hexagonal nuts (C) located directly under the clock hands. Doing this will usually release the entire mechanism from the clockface. Always check the service manual for exact details on disengaging the clock mechanism from the frame or case.

The motor's condition is easy to test. Remove the movement from the frame. Plug it into an ac line or use new batteries and check the back of the motor unit. There is usually a small hole through which an aluminum disk can be seen. Since this is perforated, determining whether it is turning is easy. If the disk is turning, the

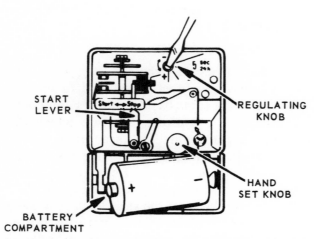

Figure 3-40. Controls and adjustments of a typical cordless clock.

motor is operating; if it is not, the motor coil may be open. Check with a VOM or test lamp to make certain that there is power at the clock wire terminals. If there is power and the disk is not turning, the diagnosis is complete: the motor is defective.

When replacing the clock mechanism, just fit the new unit into the place where the old one was. Make certain that the pinion meshes with the teeth of the driven gear, but never force teeth into place. When installed properly, it will slip into position very easily. Be sure that the flanges of the case are down flat on the frame and that screw holes are lined up. With the screw-holding screwdriver, put the screws back and tighten them.

The wires will be hooked up to a "terminal board" on the clock frame. It is a good idea to make a rough sketch of the wires as they were originally connected, before disconnecting any of them. Note wire size, color, or any other distinguishing feature. If they should be all of the identical color, take pieces of tape and secure one to the end of each wire. Write a number, or some other identification, on each. When replacing the wires, be sure that each is on the original terminal and that they are secure. Also, be certain that the insulation on each wire is tight up against the terminal lugs. Do not leave any bare places on the insulation which might permit two wires to touch each other, or to contact the frame of the clock. If necessary, tape these places or use wire nuts so that there

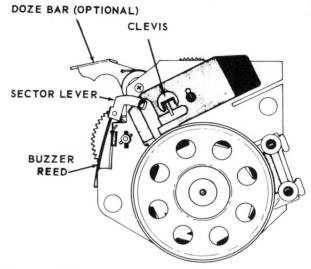

Figure 3-42. The internal mechanism of a typical electric alarm clock.

will be no chance of their making electric contact. The wires which are fastened with wire nuts require no soldering. Make certain that the insulation of the wire leads is well up inside the insulating cover of the wire nut so that there will be no problem of a short circuit.

Because of their sealed construction, electrical clocks seldom if ever really require lubrication. Once in a great while, should a clock buzz or rattle as it runs, it means that there is a worn or dry motor-shaft bearing. Take the clock out of the frame and put one or two drops of a light oil on each end of the motor shaft. Do not use a heavy oil; it could cause the clock to be slow because of the added drag on the motor.

Servicing portable power tools

Portable power tools are now considered small appliances. They are just as important in doing construction and repairs around the house as the electric can opener, mixer, or blender is in preparing meals. With the increase in number of power tools in our homes, there has been, of course, a steady increase in the number of service calls for the appliance technician.

All portable power tool problems, like those of all other small appliances, can be divided into two categories, electrical and mechanical.

Electrical Problems

The electrical system of all portable power tools is simple. It usually consists of only the motor, a switch, a cordset, and the necessary internal wiring. In other words, a portable power tool can best be described as a universal motor connected in series with an ON-OFF switch.

Series-wound universal motors are used for most portable power tools because of high torque ratings. Complete information including servicing these motors is given in Chap. 2. In recent years a few tools such as electrical drills, screwdrivers, grass shears, and hedge trimmers have been made in cordless models.

Like most other cordless small appliances, these power tools use nickel-cadmium batteries to drive a permanent-magnet dc motor. Keep in mind that nickel-cadmium or silver-cadmium batteries must be cycled approximately five times times to develop full capacity. This cycling is known as *conditioning*. Thus for the first five times the cordless power tool is charged, discharged, and charged, it will not provide maximum running time. If a power tool refuses to take a full charge or runs down rapidly, particularly after long periods of storage, the batteries may need conditioning. To do this, completely charge the batteries, then run the portable power tool until the batteries are exhausted. Repeat this cycle about three times. The power tool should run longer each time until full power is restored. Remember that both nickel-cadmium or silver-cadmium batteries

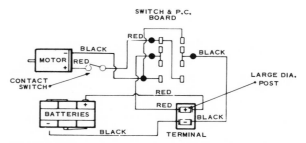

Figure 4-2. Typical schematic diagram for a cordless electric screwdriver.

work best under a full charge–full discharge type of use. Incidentally, the normal charging time for portable tool power packs is about 10 to 16 h. Silver-cadmium batteries can usually be quick-charged in approximately $3\frac{1}{2}$ to 4 h. But always follow the charging rates recommended by the manufacturer since it has some effect on the life of the battery.

The silver-cadmium cells provide more power for a given size than nickel-cadmium cells. However, the silver-cadmium cells are more expensive. The output of either of these cells is dependent on temperature. The output increases as temperature increases and decreases as temperature decreases. Over the range of normal operating temperatures, you will notice very little difference in operation.

Here is a typical charging procedure for a cordless power tool.

1. Place the power tool switch at OFF.
2. Plug the charger into a 120-V 60-Hz ac outlet that is continuously supplied with electricity.
3. Plug the connector end of the charger into the receptable on the top of the power tool handle. There is a ridge on the connector end of the charger and a corresponding groove in the power tool receptacle. You must align this ridge with the groove in the receptacle when connecting the portable power tool to the charger.
4. Leave the power tool on charge for approximately 16 h to ensure a full charge.

Separate power-pack chargers are used with

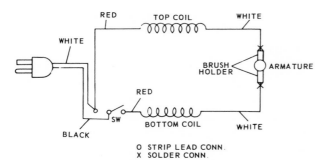

Figure 4-1. Typical schematic diagram found in most portable power tools.

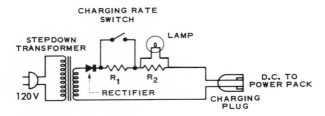

Figure 4-3. Simplified schematic of power-charger circuit.

some portable cordless tools. To prevent charging a power pack at the wrong rate, these chargers are designed so they will accept only one type of power pack. That is, the 6-V power battery pack will not physically fit into the 9-V charger.

Figure 4-3 is a schematic diagram showing the principle of operation of power pack chargers. A transformer steps down the ac line voltage. Then a silicon diode rectifier converts the low voltage ac to half-wave direct current. Series resistors and a lamp limit the current to the power pack. The charging-rate switch shorts out one of the resistors for the fast-charging rate.

The charger requires little or no maintenance. If the lamp burns out, be sure to replace it with a lamp of exactly the same type. Otherwise, the charging rate will be affected. If the charger will not charge a discharged power pack, make sure the power pack is not defective before attempting to repair the charger. A defective (internally open) power pack will not draw any current from the charger so the charging lamp will not light. Check to be sure that there is line voltage available at the plug where the charger is plugged in.

When you determine that a charger is definitely defective, unplug it from the line and from the power pack. Check the line cord for breaks and use your ohmmeter to check for continuity across the line cord plug. Remove the charger base plate to expose the parts inside. Check the electric connections. Check for continuity of the secondary winding of the transformer. If the charger has a line switch, check it for continuity with your ohmmeter. Measure the

resistor values. Check the rectifier with an ohmmeter. The rectifier should measure a very high resistance in one direction. Reverse the ohmmeter leads and take another reading with the leads reversed. The reading should be a low ·value—only a few ohms. Very little can go wrong with a charger unit unless it has been physically damaged. Replace any defective parts and reassemble the unit.

Some power tools such as electrical drills and screwdrivers have a two-speed feature. As a rule, tools employing this feature use either a tapped field-winding motor control or a rectifier-type control. The selection of speed is usually made by a trigger ON-OFF switch.

A few power tool manufacturers recommend a hi-pot test for their equipment. It is usually accomplished by applying 1,100 V for about 1 min with the leads connected between one of the "hot" prongs on the power cord and the tool's metal case. The power switch should be ON and the tool run sufficiently to warm up the motor. Always check the service manual before conducting a hi-pot test for any specific instructions. Incidentally, the hi-pot test should not be given to cordless power tools.

It is good servicing policy to always check the tool's power consumption and current drain before returing it to the customer. This can be done, of course, with a wattmeter and an ammeter, and all readings should be within 10 percent of nameplate rating when the tool is working under a moderate load.

Mechanical Problems

The purpose of the mechanical portion of a portable power tool is to convert the electric energy into the mechanical work of drilling, cutting, or sanding. There are two basic linkages between the motor and the work end of the tool. In one, the linkage retains the motor's rotary motion but reduces the motor's revolutions per minute and increases the effective torque. In the other, the linkage changes the motor's rotary motion into an oscillating motion. Portable power tools that employ rotary motion include drills, circular saws, belt sanders, screwdrivers, routers, edgers, and lawn mowers, while saber saws, hedge trimmers, grass clippers, and orbital-finishing sanders use oscillating motion.

Now let us take a look at some of the mechanical problems that the service technician will find with specific tools.

DRILLS

The electrical drill is the most common of all portable power tools and the one that will come into the service shop the most.

Common Problems

Stripped gears. The symptoms of this problem are that the motor turns but the chuck does not. In addition, a grinding sound is usually heard. In most instances, both gears will be stripped. To check, remove the screws that hold the gear case to the drill housing. On most portable drills, there is a thin casting between the gear case and the drill housing. Do not pull the gear case off yet. (Should the intermediate casting be pulled off with it, the armature will probably be pulled out also. If this occurs, the spring-loaded carbon brushes will pop into the center of the tool and the armature will not go back until they have been removed.) Hold the casting against the drill housing with one hand and with the other gently pry off the gear case until it is almost free. Turn the drill so the chuck points up. In this position, complete the removal of the gear case.

If the drill has a single train gear (most $\frac{1}{4}$- and $\frac{3}{8}$-in drills have), a small gear will be found that is either cut into or screwed onto the end of the armature shaft. This is the pinion gear that protrudes through the intermediate casting. The gear case holds the second or jack-shaft gear that is connected to the chuck. Examine both gears. If teeth are missing or are worn, the reason for the slipping of the chuck or the grinding sound is quite obvious. If the pinion gear can be removed from the end of the armature, it should be replaced along with a new jack-shaft gear.

However, if the pinion gear is permanently connected to the armature, it will probably not pay to repair it. (The cost of a new armature with a cut-on gear and jack-shaft gear is about the same as a new drill.)

If the drill has double reduction gears (as better slow-speed $\frac{1}{4}$-in and most $\frac{3}{8}$- and $\frac{1}{2}$-in drills have), there is a double, intermediate gear as well as the two gears mentioned above. This intermediate gear consists of one small and one large gear on one shaft. The intermediate gear reduces the speed and increases the power (torque) of the chuck. Usually the smaller gear of the two is the one that goes bad. If this occurs, the jack-shaft gear that it drives also requires replacement. As a rule, this is not too difficult a task. Generally, the intermediate gear can easily be removed by lifting it out of its position. In most cases, there is a spacer or thrust washer in one or both ends of its gear shaft. Most often the washer does not stay on the shaft when the gear is removed but clings to the casting because of the grease on the casting. Incidentally, this is a good time to clean out the old grease and put in fresh gear grease. Wash out the old grease with any good solvent such as kerosene or isopropyl alcohol. The old grease has small metal particles in it and if not removed will soon grind up the new gears as well. When repacking the gear case, pack it only halfway, since too much grease is as damaging as too little. Grease expands when it gets hot. Excess grease resulting from overpacking will overflow onto the armature and field coil, damaging the windings and clogging vent holes, and thus causing the tool to overheat.

Bad bearings. Another very common mechanical fault found in electrical drills is bad bearings. The symptoms of bad bearings are various. The moving parts may be frozen, very stiff, or hard to rotate. There may be a screeching or grinding sound. A bad bearing at the commutator end of the armature can cause excessive sparking.

Three types of bearing are used with electrical drills (as well as most other portable power tools): oil-impregnated or sleeve (used on the most reasonably priced drills), ball bearings (used

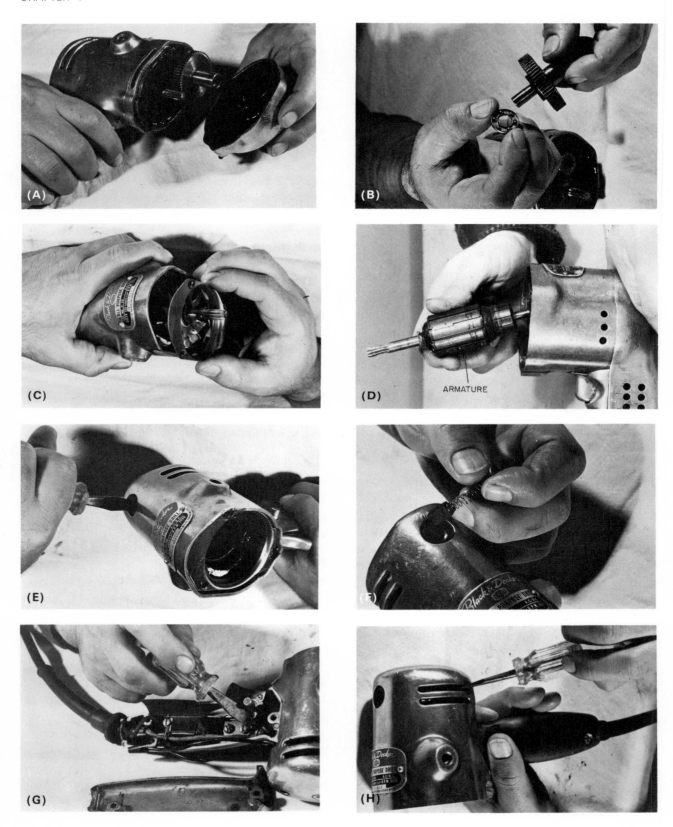

(A)

(B)

(C)

(D)

ARMATURE

(E)

(F)

(G)

(H)

in conjunction with the sleeve in the better drills), and needle bearings (used in conjunction with ball bearings in the better drills). We would not suggest the changing of any bearings unless the proper equipment is available, such as an arbor press (a drill press works well on small tools) and a small gear or bearing puller. It is not difficult, for example, to press out a bad sleeve bearing, but when you press in the new one, very often the bearing constricts, changing its inside diameter. This must be reamed out to a precision fit, or the tool will bind or overheat at the new bearings. We therefore suggest you leave the changes of sleeve bearings to those with proper equipment (see Motor Bearings in Chap. 2).

Some drills have clamshell construction in which the tool casing is made in two halves like a clamshell (see Brush Troubles in Chap. 2). Changing bearings on this type tool does not require special equipment, but the problem is in opening up the tool without the parts falling out and keeping the parts in place when closing the two halves of the casing.

If the motor has been disassembled (or even sometimes if dropped), it may run slow because of misalignment of the bearings. Usually, misalignment can be cured by simply turning the motor on and tapping near the bearings with a small hammer. Keep tapping around the bearing

Figure 4-4. Steps in checking an electric drill. (A) The cover of the front case (right) holds the grease that lubricates the drive and worm gears. It should be filled only halfway. (B) This is the sealless type of bearing contained in many power tools. Note that the ball bearings are exposed. It fits on the backside of the drive gear. (C) This metal-type gasket separates the front case and the case holding the armature assembly. It keeps grease from getting on the armature assembly. (D) Remove the armature assembly from the case and inspect for wear or damage. If in very poor condition, replace with a new one. (E) Most power tools have brushes that can be removed externally by unscrewing the brush holder caps. (F) The brush is easily removed from the holder after the cap is removed. (G) In this electric drill, the wiring setup, consisting of the line cord and switch, is contained in the handle. To get at the wiring, simply unscrew the handle cover. Most power tools have similar setups. (H) Vent holes must be kept clear of dirt so air can flow unobstructed through the tool. Clean with a flat stick or tool.

location until the motor is turning normally, and if using an ammeter or wattmeter in the line, until the least amount of current is drawn.

Anytime a portable power tool such as an electrical drill is opened up, inspect the bearing for lubrication. If both sides of the bearing are sealed with plate or felt seals, hiding the ball bearings, do not clean in solvent. This type of bearing is permanently lubricated at the factory and has sufficient grease packed into it to last for its lifetime. Do not break the seal. However, carefully wipe dirt from the seals with a clean cloth. The other type of bearing is exposed. This type should be cleaned in solvent and relubricated with a bearing grease.

Faulty chuck. To remove a faulty chuck, determine whether the chuck is threaded to the drill or on a tapered shaft. On most reversing drills, the chuck is on a tapered shaft, and it will not come off when turning counterclockwise. But as a rule, most drill chucks are threaded to the drill with a conventional right-hand thread. To remove this type of chuck, place the chuck key in one of the chuck holes with the jaws tightly closed. Then put the chuck on the edge of a solid workbench with the drill in a horizontal position extending off the workbench. Hold the drill with the left hand. With a wooden or plastic mallet or a sturdy stick, strike the chuck key to drive the chuck in the same direction it turns. A few such sudden strikes should crack the chuck loose.

If the chuck is on a tapered shank, you must remove it with a pair of wedges. If the proper steel wedges are not available, removing this type of chuck will be almost impossible. A few reversing drills have a conventional right-hand-threaded chuck with a retaining screw in the center of the chuck, usually left-hand threaded. This screw must be removed before the chuck can be unscrewed.

Before reassembling any power tool, be sure that all the ventilation openings are clear. Most motors have holes in each end bell for ventilation, and some have channels down each side of the frame behind the field coils. Blow the dirt out of these channels, if necessary, but do not push a wire or rod through to see if they are clear—if

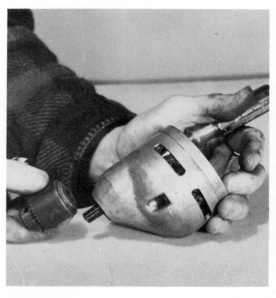  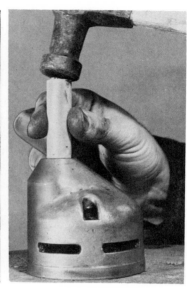

Figure 4-5. To remove a bad bearing: (left) Remove the gear housing, jam the gear with a screwdriver and remove the chuck by turning it to the left. The gear may now be slipped from the housing. (Center) "Pressed fit" bearing is removed by filling a dowel to size and tapping bearing out. Replacement parts are available from factory. (Right) Tap the new bearing into place with help of wood scrap and the large gear, reversed, after first ascertaining that new bearing is smooth and free of burrs.

there are wires in there, they can easily be broken. Also clean any dirt from the case and field coil by immersing it in a solvent, such as kerosene. Let it air-dry before reassembling it.

PORTABLE CIRCULAR SAWS

One of the most common mechanical problems of portable saws is an inoperative saw guard. The guard will retract but then will not spring back to cover the blade after the cut is completed. Most often, the return spring is broken or missing or there is an accumulation of dirt on the guard pivot. If it is an internal spring, the guard must be removed. Most guards are held in place by a spring-type retaining ring. This ring can be removed with a small screwdriver or fine pair of needle-nose pliers. With the spring (and any other parts that interfere) removed, the guard should slip off its pivot, revealing the return spring. Most often these return springs are intact, but an accumulation of dirt, gums, resins, or just sawdust gets into the pivot casting and binds the guard. This dirt can sometimes be removed, without taking the guard off, by washing it out with a solvent such as kerosene, perchloro-ethylene, trichloroethane, or trichloroethylene. An air hose, if it is available, can be used to remove a great deal of sawdust without taking the tool apart at all.

Another common problem with portable circular saws is worn or improper arbor washers. This problem, if not corrected, can cause more extensive and expensive problems. Improper or loose-fitting washers do not grab the blade firmly. This not only cuts down on the ability of the blade to cut properly but also wears a groove in the jack shaft. The result is a poor fit between the blade arbor hole and the shaft, and any blade mounted will be off-center. Running the saw under these conditions will eventually destroy the bearings. Therefore, the arbor washers should be changed at any signs of wear. Remember that with most circular saws, the blade mechanism is driven directly from the motor shaft.

If the jack shaft is worn or if it has any side play, it can easily be changed on most circular saws. To remove the jack shaft in most models, remove the screws that hold the jack shaft bearing casting. Put a scratch line on the top side of the jack casting before removing it, so that when reassembling, you can put it back in exactly the same position. Pry up this casting with two thin screwdrivers. Generally, the casting, bearing, jack shaft, and jack-shaft gear should come out in one piece. Once the casting is out, remove the jack shaft in order to free the jack-shaft gear. This gear is usually retained with either a spring washer or a nut. Remove the washer or nut. Remove the gear by using a puller, wedges, or an arbor press. The gear is generally keyed to the shaft with a Woodruff key. Do not lose the key. Press the jack shaft out of the bearing by pressing from the gear end. Before installing the new jack shaft, inspect the bearing carefully and change it if it feels rough when turned or if it appears worn.

If the gear shows signs of wear, also check the pinion gear into which it engages. If the pinion gear is worn, too, and is not removable from the armature, a complete new armature is required as a replacement. If the bearing and gears are in good condition, repack the gear case chamber with new grease in the same manner as described for electric drills, and reassemble all the parts in the reverse order as they were removed.

If the saw is of clamshell construction, the jack-shaft bearing and gears will all be exposed when the tool is open. Changing the parts requires dexterity; in most cases, the cost of replacement of the armature, gear, bearing, and shaft is more than the tool is worth. Most clamshell saws are designed for light duty only.

Electrical lawn edgers work in much the same manner as circular saws, and many of their basic problems are the same.

SABER SAW

As was stated earlier in the chapter, in the saber saw, the motor shaft drives an eccentric gear that converts the rotary motion into back-and-forth

motion. The same principle is used in many electrical hedge trimmers and grass shears.

Most saber saws in the handyman class are of clamshell construction. This makes them difficult to service, and because of the reciprocal action, they are somewhat self-destroying. If the motor goes but the blade does not reciprocate, or the user cannot guide the blade on a line, the reciprocating mechanism is either badly worn or broken. While a new reciprocating mechanism can easily replace the defective one, its cost usually prohibits such a move with a saw in the handyman class. With the so-called professional or craftsman-type saber saw, it is a different story. In such a tool, the reciprocating mechanism should be replaced as directed in the manufacturer's instruction manual.

FINISHING SANDERS

While the transmission of the saber saw allows for only back-and-forth movement, that in the orbital sander permits the full offset oscillation motion of the eccentric gear which drives the sanding pad. To prevent most of the vibrations from being transmitted to the main assembly, a counterweight is fastened on the inside surface of the sanding pad. This principle of full offset oscillation is also used in some personal-care small appliances such as massagers and vibrators.

The most common service with an orbital sander is the replacement of the felt or rubber pad. On most of these tools, four screws hold the lower pad to four rubber mounts at the base of the tool. Remove the four screws and the pad will come off. (Some tools have a fifth screw that holds the pad to an eccentric weighted flywheel.) The new pad can be purchased from the manufacturer or supply center.

Older and slow-speed sanders of this type are usually belt driven, and frequently the belts slip or break. To replace the belt, remove the lower pad as described previously. (If there is a fifth screw and it cannot be removed because it turns with the eccentric gear, it can be stopped from turning by inserting a nail or allen wrench in a predrilled hole in the base of the pad.) Once the pad has been removed, the belt will be exposed and can easily be replaced.

If the sander binds, there is a bad bearing on either the eccentric or the armature. With the drive mechanism removed, the armature should turn freely. Plug the tool in and turn it on. The motor should turn at a high speed, quietly, with no vibration. If the motor runs smoothly, the binding is caused by a bad bearing(s) on the eccentric gear. The eccentric is usually bolted to the tool casting with one bolt. With the eccentric out, the bearings can be removed by the use of wedges or a small puller. When replacing the bearings, be certain to use exactly the same type as those removed, and be sure they are of the double-sealed design. Open bearings on a sander will be destroyed quickly by dust and dirt.

ROUTERS

The router is a simple tool with only two bearings and one major moving part. The most common problem is bad bearings. Remember that the lower bearing cannot be changed until the armature has been removed from the motor. In order to do this, you need an impact wrench for removing the collet chuck and special equipment to hold the armature while you remove the collet chuck. Check the service manual for complete instructions before attempting to remove any bearing from a router.

POWER LAWN MOWERS

While electrically powered rotary lawn mowers are not as common as the gasoline-driven ones, nevertheless the service technician should know how to service them. With most electrical lawn mowers, a series universal-type motor is vertically mounted on rubber bushings to reduce vibration,

and it is energized when the ON button of the switch is depressed. The armature shaft usually has a helix gear on the end which meshes with the main drive gear. The main drive gear generally has a multiple disk-type clutch built into the hub to protect the motor in the event of blade stoppage when the motor is running. The blade on the shaft driven by the main gear and clutch assembly is usually attached by means of an adapter key and a mounting bolt.

Here is an analysis of common electrical rotary lawn mower complaints.

Motor will not run.

1. Remove the cover and with the extension cord plugged into a 120-V power source (or 110-V source with test cord), check for voltage at the terminal block at the top of the motor. Depress the ON-OFF switch to check for incorrect button installation. (With most models, the mower with the blade installed must be tested at a voltage not to exceed 110 V when using a short test cord. If the blade is removed, the voltage should not exceed 70 V. Check the service manual for complete information.)
2. Check the brushes for excessive wear. Replace if necessary.
3. If voltage is obtained at the terminal block and the brushes check all right, remove the complete motor assembly and replace it.
4. If no voltage is measured at the terminal block, disconnect the unit from the power source and check the following for continuity:
    a. Cordset from the switch to the motor
    b. Switch in the ON position
    c. Cordset from the switch to plug
    d. Extension cord
5. Check the proper source for a control switch in the OFF position or blown fuses.

Mower vibrates.

1. Check the motor-mounting bolts for tightness.
2. Check for a bent or damaged blade.
3. Check for loose parts such as the hood, handle, wheels.

Mower will not cut properly.

1. Check for proper blade installation (refer to the service manual).
2. Check the blade sharpness. Sharpen if necessary.
3. Check for a slipping clutch. Replace the motor and gear assembly if necessary.

Motor is noisy. All complaints of a noisy motor should be checked. If abnormal, replace the complete motor and gear assembly.

Handle grip is loose or switch binds. Check for handle grip slippage or loose tubing on the upper handle. Be sure to inform your customer as to a proper lubrication schedule. With most models, use SAE-20 or SAE-30 motor oil to lubricate the following.

1. Upper motor bearings: fill hole twice a year under normal usage and during reassembly or repair of mower.
2. Handle pivot pins: oil frequently.
3. Wheel bearings: oil frequently.

Be sure to warn your customer not to use excessive oil on the upper motor bearing because the lubricant may run into the motor and cause premature deterioration of electrical parts. The lower motor bearing and gearbox are lubricated at the factory with a special gear lubricant; therefore they should not require any further attention.

Sharpening of rotary blades. When a service technician is called upon to work on an electric rotary lawn mower, it may be wise or necessary to sharpen the blades before returning the machine to the customer. The blades can be sharpened by using a honing stone mounted on an electrical grinding wheel. Hold the blade at a 20 to 30° angle to the stone and take one long, steady stroke across. Then turn the blade around and do the same for the opposite side.

Failure to sharpen both sides can unbalance the blade, which may lead to damage from the violent vibration the imbalance can cause.

Before leaving the subject of sharpening blades, it may be wise to mention it for other power lawn equipment, too—specifically edgers, hedge trimmers, and grass cutters or shears. Not all need sharpening in the sense we have been using it here. Most grass shears, for example, are equipped with self-sharpening and self-cleaning blades. If they seem to lose their ability to cut, all that has to be done is to tighten down on the tension screws. The rubbing action of the blades keeps them sharp and clean. In time though, blades will have to be replaced when badly worn. Some hedge trimmers are double-bladed but may not be self-sharpening. Single-sided hedge-trimmer blades are not self-sharpening, whereas two-sided blades are.

To sharpen a blade of the non-self-sharpening type, unbolt the movable blade. Sharpen only the movable (top) blade by putting it in a vise and holding a file at a 20 to 30° angle to each cutting surface. Give it one long, steady stroke, moving the file toward the pointed end of the cutting edge. Taking more than one stroke raises the danger of rounding off the blade.

The only attention required by self-sharpening hedge-trimmer blades is that they be kept clean. Place the blade in a pan of water and scrub with a brush. Dry well and oil lightly to prevent rust. If the hedge trimmer is chain-driven, as some are, the chain has to be oiled through the lubrication hole which is provided. Use SAE-30 oil.

Edgers usually have rotary blades. More often than not, they are damaged before they require sharpening. If the blade is badly gouged or nicked, replace it. If the blade seems all right and needs sharpening, clean it off and hone with a motor-driven stone.

Resistance-heating small appliances

CHAPTER

5

As was discussed in Chap. 7 of *Basics of Electric Appliance Servicing*, the heating effect of electricity makes possible many modern small appliances, such as toasters, coffee makers, irons, electrical blankets, hair dryers, and space heaters. The heat developed in these appliances is obtained by passing current through a special type of wire known as *resistance* wire. It has a higher resistance to the passage of electricity than the ordinary wire used in cordsets. Overcoming this resistance causes heat—and the process is known as *resistance heating*.

HEATING ELEMENTS

As was mentioned in *Basics of Electric Appliance Servicing*, the resistance of wire depends on its material (copper, aluminum, etc.), cross section (diameter), length, and temperature. The typical cordset used with most small appliances has a resistance of approximately 4 Ω per 1,000 ft. Thus, a 6-ft cord would have approximately 0.024 Ω of resistance, which is too little to measure with ordinary shop test equipment. In other words, for all practical purposes, this wire at this length would read on an ohmmeter as a short circuit.

On the other hand, a typical resistance wire similar in size to that of the cordset just mentioned, but in a different material, would have a resistance of about 0.404 Ω per inch of coil. This is 4,848 Ω per 1,000 ft, or about 1,200 times as great as the resistance of the first cordset mentioned. Another major advantage of the resistance wire described in this example is that the heat is confined to a relatively small area.

Most resistance wire used today in heating elements is made of an alloy of nickel and chromium called *Nichrome*. Unlike many other wires of different metals, Nichrome can be operated at near white-hot temperatures without weakening or collecting an oxide coating. This wire is available in both coiled and flat (ribbon) forms. As a rule, automatic grills, waffle irons, space heaters, rotisseries, and similar appliances employ coiled wires, while toasters and irons usually use ribbon-type resistance wire. A few

Table 5-1. Approximate length of resistance (heating) wire for given wattage.

Watts	36 Gauge $\frac{1}{8}$ in OD (in)	30 Gauge $\frac{3}{16}$ in OD (in)	25 Gauge $\frac{3}{16}$ in OD (in)	20 Gauge $\frac{1}{4}$ in OD (in)	22 Gauge $\frac{3}{16}$ in OD (in)	22 Gauge $\frac{7}{32}$ in OD (in)	22 Gauge $\frac{5}{32}$ in OD (in)	23 Gauge $\frac{1}{16}$ in OD (in)	18 Gauge $\frac{1}{4}$ in OD (in)
10	6.3								
15	4.2								
20	3.2								
25	2.5	13.6							
30	2.1	11.2							
35	1.8	9.6							
40	1.6	8.6							
45	1.3	7.5							
50	1.2	6.8							
60		5.6							
70		4.8							
80		4.3							
100		3.4							
125		3.0							
200			14.3						
250			10.7						
300			8.5						
350			7.1					13.6	
400					14.5	15.0	20.5	11.5	
500					12.1	11.5	16.5	9.5	
600					10.1	9.0	14.0	8.0	
660				13.9	9.2	8.5	12.5	6.5	
700				13.1	8.6	7.5	10.5		
800				11.5		6.5	9.5		
900				10.3					
1000				9.2					19.1
1200				7.6					15.9
1300									14.7
1500									12.7

appliances employ coiled wire wrapped in an insulating sheath or, in some cases, a coiled Nichrome wire located inside a metal sheath (generally copper). The latter, sold under the trade name Calrod, is sometimes round and sometimes flat. It has the advantage of being rugged physically, easy to keep clean, and shock-free to the touch. In addition to being employed in small appliances such as ovens, rotisseries, irons, and kettles, it is used as the heating element in major appliances such as electric ranges and clothes driers.

Table 5-2. Standard resistance of coiled wire.

Gauge No.[a] and size (in)	Coil OD (in)	Resistance (Ω/in)
28	3/32	7.2
26	3/16	7.5
25 (0.0175)	7/32	6.2
24 (0.020)	9/64	2.45
23 (0.0226)	9/64	1.87
22 (0.0253)	3/16	1.8
22 (0.0253)	7/32	2.1
22 (0.0253)	1/4	2.45
20 (0.032)	15/64	1.1
20 (0.032)	1/4	1.2
19 (0.036)	13/64	0.631
18 (0.040)	3/16	0.404
18 (0.040)	9/32	0.652

[a]Brown and Sharpe (equivalent to American Wire Gauge).

While coiled resistance wire used in heating elements is frequently assumed to have a higher resistance than the ribbon type, it just seems that way because there is more wire in an inch of coil than in an inch of ribbon. Actually, the total length of a resistance wire is its stretched-out length. Manufacturers frequently form resistance wire into a coil or spiral shape to get more wire into a smaller area. While this coiled length might be only a foot or so, the total length could be several times greater.

The current drawn and the watts consumed by a heating element can be determined from the ohm value and the length of the wire. For instance, here are some examples, based on the wire previously mentioned that has a resistance of 0.404 Ω per inch of $\frac{1}{4}$-in coil.

Length (in)	Total ohms	Amperes	Watts (at 120 V)
78.2	31.2	4.2	500
39.1	15.6	8.3	1,000
23.8	9.6	12.5	1,500

As can be noted, there is an inverse relationship between resistance and current: the lower the resistance, the greater the current drawn and, of course, the wattage consumed.

It is also important to remember that the watts of heat energy developed depend on the voltage. Here are some typical examples.

Resistance of element (Ω)	Line voltage	Amperes	Watts
15.6	105	6.67	700
15.6	110	6.98	768
15.6	115	7.35	846
15.4	120	8.33	1,000
15.6	125	8.69	1,090

The above data illustrate how low line voltage can cause improper operation of resistance-heating appliances. For instance, it would take about 30 percent more time to toast a slice of bread at 105 V than it would at 120 V. Slow heating is a frequent complaint of resistance-heating small appliances; more often than not, it is caused by low line voltage rather than a faulty unit.

THERMOSTAT HEAT CONTROLS

Nearly all small electrical appliances which employ heating elements also require a thermostat of one style or another. These thermostats may be either fixed or adjustable. Fixed thermostats either hold the temperature at one particular value or function as protective devices should the appliance get too hot for some reason. Adjustable thermostats provide user control of the amount of heat.

As was stated in Chap. 6 of *Basics of Electric Appliance Servicing*, a thermostat is simply a switch that is controlled by heat. All thermostats used in small appliances are of the bimetal type. The heart of such a thermostat is, of course, the bimetal blade, a strip of metal made from two different metals, one with a high rate of expansion when heated, the other with a low rate. As heat is applied, the two strips expand or grow at different rates, causing the bimetal blade to warp or bend toward the side that has the smaller expansion or growth rate. When the heat is removed, the blade returns to its normal position. By varying the metals or alloys of the two strips, thermostats can be designed to give almost any desired temperature control. Most small-appliance thermostats have an upper limit (the temperature at which the contacts open) and a lower limit (the temperature at which they close).

In fixed-heat thermostats the bimetal blade generally has one of the switch contact points fastened directly to it. In adjustable thermostats the bimetal blade is usually just an actuating arm. This makes the bimetal blade separate from the electric circuit and eliminates heating caused by the passage of current through it. However, in

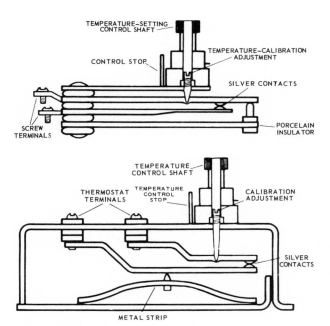

Figure 5-2. Two types of adjustable thermostats: (top) Open frame bimetal thermostat. (Bottom) Typical electric iron thermostat.

the design of some thermostats this is a safety factor since excessive current could cause the thermostat to cut off.

The operating temperature of a thermostatically controlled small heating appliance is varied by altering the distance the bimetal blade must travel to close or open a set of contacts. Many adjustable-type thermostats have two adjustments: one that permits the user to set the appliance's actual operating temperature, and one —usually a setscrew arrangement—that permits the service technician to bring the operating temperature into line with the scale on the user's control knob. The latter adjustment is usually calibrated by the manufacturer at the factory and should not be changed unless you are absolutely sure that it is out of adjustment. The technician's adjustment is nearly always hidden from view so that the appliance will have to be partially disassembled to locate it. For example, on electrical coffee pots, the bottom normally has to be removed, but sometimes it is possible to reach the "second" adjustment

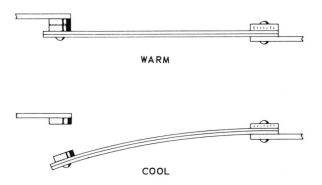

Figure 5-1. Thermostatic action of a fixed bimetal blade.

through a hole provided or through the center of the control shaft after removing the control knob. As a rule, if a higher maximum temperature is desired, the adjusting setscrew is turned so that the thermostat contacts are pushed closer together (or farther away from the bimetal actuating arm). If the minimum temperature needs to be lower, the contacts are adjusted so there is less tension on the contact points or so they are moved closer to the bimetal actuating arm.

By sliding or turning a control knob, the user adjusts the heat of the appliance only within specified limits; that is, through a linkage from the user's control knob the distance between the thermostat's bimetal blade and its associated electric contacts are varied. The user's adjustment does not determine the amount of current that flows into the appliance but usually controls the ON-OFF cycles of the unit. For instance, an electric iron may have a 1,200-W heating element, but on low-heat settings that element may be turned on by the thermostat for as little as 15 percent of the time after the initial warm-up period. On the other hand, at high-heat settings the element may remain on for as much as 80 percent of the time.

HEATING CIRCUITS

Some appliances such as simple space heaters, various cooking devices, hair setters, and dry irons have a so-called "single-element" circuit. In it the complete circuit consists of a cordset, heating element, connecting wire, and simple ON-OFF switch or thermostat—all connected in series. Many other small appliances, however, have two or more heating elements connected in various circuit configurations: in series, in parallel, or in a combination of series and parallel. The schematic diagram in the service manual will illustrate the circuit configuration used for any given appliance.

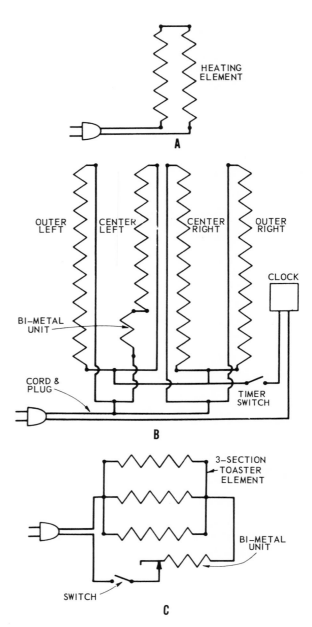

Figure 5-3. (A) Simple single element; (B) series-connected elements; (C) series parallel-connected elements.

Single-Element Circuit

Troubles in single-element appliances are easy to diagnose. In most resistance-heating small appliances the vast majority of troubles can be traced to one or more of the following.

1. Open circuit in the cordset or loose connections at terminals
2. Defective heating element or elements
3. Faulty thermostat or switch

Here is the basic troubleshooting procedure for a single-element appliance. Begin by making a visual check of the cordset, plug, and terminals connecting the cord to the appliance, for any sign of defects. Replace the cordset if necessary. If you note no visual defects, make a continuity check of the cordset with an ohmmeter. If you find an open or short, replace the cordset.

If you locate no defect in the cordset or plug, check the element for continuity with an ohmmeter. If the heating element is open or shorted, repair or replace as described on p. 137. If the element has continuity, check the contact points on the thermostat; they may be dirty or pitted and not making a good connection. In some cases the contacts can be cleaned, while in others the thermostat should be replaced (see p. 138).

The ohmmeter can tell you more about the appliance than just whether there is continuity. For example, if the resistance of a 800-W iron is about 19 Ω, we can use the formula for Ohm's law, $R = E^2/W$, and find that the resistance should be

$$R = \frac{120 \times 120}{800} = \frac{14,400}{800} = 18 \, \Omega$$

Because our reading of 19 Ω is well within the 10 percent accuracy we can expect from an ohmmeter, it is safe to assume that the element is all right.

Another test that should be made is a reading between one post of the heater element and the frame of the appliance. There should be no connection between these two—in other words, a reading of infinity or open should be obtained. If in our circuit we get a reading of 20 Ω, this indicates that the element is grounded. Of course, an internal short or ground of the element may cause the fuse to blow, may create shock hazard, or may cause the iron to get too hot and/or draw too much current.

Speaking of current, another good test is a current reading. Our circuit reading was 6 A. Using the formula for Ohm's law, $I = E/R$, we find that

$$I = \frac{120}{18} = 6.6$$

This again, is within the 10 percent accuracy of the instrument, so everything on this score can be considered good. Besides, since our resistance read a little high (19 instead of 18), we would expect the current to read a little low. Incidentally, if there is a high current drain but low heat, chances are very good that there is a partly shorted heater element.

Series-Connected Elements

In this type of circuit, two or more elements are connected in series, but since there is only one complete path for the current, the amount flowing through each element is equal to the total current of the circuit. The actual power consumption of each heater element is equal to the voltage dropped across it, multiplied by the total circuit current. Keep in mind that the voltage across each element in a series circuit depends upon the relative power ratings of the elements; the element with the smaller power rating takes on a proportionally larger portion of the source voltage. While the total power consumption of the circuit is equal to the sum total of each element, the total power consumption is less than the total amount of power each element would consume if each had 120 V applied across it.

The thermostat may be connected in series with all the elements or connected in parallel with one or more of them. When connected in series with the elements, the thermostat shuts off all the heating elements when it is open. But when the thermostat or switch is placed in parallel with one of the elements, it completely shorts out that element when its contacts are closed. Under these conditions, the circuit is the same as a single-element one. If element No. 1 is, for example, a 500-W unit, the appliance will consume about 500 W when the contacts are closed and will draw about 4.2 A from the lines. But when the contacts of the thermostat or switch are open and element No. 2 is effectively in the circuit, we now have two series-connected heaters. If both have the same power rating (500 W apiece, in this example), the line voltage is divided between them equally. Therefore, if the nameplate wattage rating is 800 W, each element would consume half the power (400 W) and the appliance would draw approximately 6.7 A. Of course, some appliances have series-connected elements of different power ratings. In such cases, as previously stated, the element with the smaller power rating takes on a proportionally larger portion of the supply voltage. That is, because of the larger voltage drop across the smaller-rated element, this unit consumes more power and generates more heat than the element with the higher power rating.

It is very important to remember that the total power rating as appears on the resistance-heating appliance's nameplate is not equal to the sum of the individual power ratings. Also keep in mind that the power ratings for small appliances employing series heater elements are generally rated according to the power they consume when 120 V is applied to them.

Troubleshooting series-connected elements is very much the same as troubleshooting single-element circuits. Of course, when checking continuity, each element should be checked if tests show no continuity between the terminals. If there is uneven heating, the cause is generally a short in one of the elements. To check this, take a voltage reading across the elements with full line voltage applied. If one of the heater elements is shorted, no voltage will be read on the voltmeter.

Parallel-Connected Elements

Several small appliances have circuits in which the heater elements (two or more) are connected in parallel. In such an arrangement, the current flows in separate paths through each of the elements, and its total drain is equal to the appliance's power rating divided by the line voltage. The current through each element should be equal to its power rating divided by the line voltage.

The voltage across all the elements of a parallel-connected circuit is equal to its 120-V rating, which also means that the voltage across all the elements connected in the circuit is equal to the line voltage. But regardless of the power ratings of individual parallel-connected elements, the total power consumption of the appliance is equal to the sum of the power ratings of each heating element connected into the circuit. Keep in mind that the appliance's nameplate rating generally indicates only the maximum power consumption. For instance, a broiler/skillet rated at 1,500 W might consume 1,000 W at its LOW setting, 1,350 W at MEDIUM, and 1,500 W at HIGH. But in this case the nameplate would indicate only the maximum power rating of 1,500 W.

The control switch and/or thermostat is usually connected in series with one of the elements. In some instances, such as in the broiler/skillet described above, this switch may control two or more elements. When the switch or thermostat is open, no current flows through the element, and it is an open circuit. When checking parallel-connected heaters for continuity, be sure to disconnect each element from the circuit and check each separately. Of course, if one element does not heat up and others are hot, this is a good indication that the cold element is open. You can double-check by disconnecting one end of the suspected faulty element and running a continuity test. If the ohmmeter shows infinity or a tester indicates lack of continuity, the element is bad.

If one of the elements seems to be glowing hot in some portions of the heater and cold in others, it is almost safe to assume that some part of the element contains a partial short circuit. The shorted point usually occurs where the element appears to have changed from hot to cold.

Series-parallel circuits

A few appliances—mainly some electrical blankets, heating pads, and a few toaster models that have special thermostat control units built into them—contain elements connected in both series and parallel. The actual circuit arrangement will determine the basic Ohm's law that will apply in a series-parallel element circuit. In general, the sum of the voltage in series arrangements is always equal to the voltage across the parallel elements. Of course, the actual voltage across each of the series-connected elements depends upon the relative power ratings; the element with the smaller power rating uses a proportionally larger portion of the supply voltage.

Repairing and Replacing Heater Elements

If a heater element is faulty, it should be replaced. But before attempting to replace one, consult the manufacturer's service manual to determine what else, if anything, will be needed to do the job. With some makes it may be necessary to order related parts along with the element, such as gaskets or spacers. Although in many makes the installation of a new heater element is relatively simple and can be done with ordinary tools, for a few models special tools are needed to do this work satisfactorily. For instance, percolator element burnouts are so rare that in a small shop it would hardly pay to buy this special equipment if it is costly unless you expect to service a large volume of one make. Thus if you repair various brands and your volume of business amounts, to say, thirty to forty service transactions weekly, you may not encounter more than two or three percolator-element failures a year. It would seem advisable, therefore, to farm out to your authorized service station the element replacement jobs for which costly special equipment is needed—at least until experience dictates otherwise.

As a temporary expedient, mending sleeves—small hollow metal tubes of Nichrome or a similar alloy—may be used to connect broken resistance wires. (Soldering such a connection, even with silver solder, is generally a poor service procedure because frequently the resistance wire runs hotter than the melting point of solder.) To make a good splice, clean the resistance

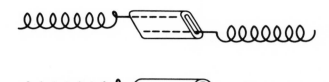

Figure 5-4. Broken heating element wire repaired with a mending sleeve.

wires thoroughly with fine sandpaper before inserting the ends into the mending sleeve. Once the sleeve is in place, crimp it tightly and wrap it with at least one layer of asbestos or fiberglass tape which is made just for this kind of work.

Occasionally, the connecting wires between pairs of heating elements break. This is common in appliances like automatic grills and waffle irons which have two heating elements in different locations. Do not bother trying to splice them; get new pieces of wire and replace them. Some appliances use heavy-stranded wire called *copper rope*; others employ stranded wire covered with a very tightly wrapped, or even braided, asbestos. If the connecting wire is inside a heater-type appliance, the connecting wires must be very well insulated to keep them from shorting to the case.

Speaking of shorts, they can usually be spotted visually. One of the most frequent problems is that of two different sections of heater wire touching each other. Visual inspection may also lead to finding a portion of an element which touches another electric component or the case of the appliance. The solution here is to remove the grounding condition.

Thermostat Problems

The bimetal blade of the thermostat seldom is a source of problems. Most often trouble occurs because the contact points become pitted and oxidized through repeated use. We must remember that thermostats, as employed in small appliances, are switches. They turn the appliance on when it becomes too cool and off when it gets hot enough. After any switch has been turned on and off a thousand or so times breaking a heavy current, it becomes a little dirty; so does a thermostat. Each time the contacts open, they cause a small arc which leaves a deposit of oxide on the contact surface. After several hundred hours of service the points can develop a great deal of resistance at the contacts, creating additional heat which, in turn, creates additional oxidation and an even greater increase in resistance. Eventually, this oxide builds up to a point where the contacts will not complete the circuit. This can occur even though they may seem to be touching each other.

Dirt and grime can be removed from the electric contacts by spraying them with a little aerosol contact cleaner or by wiping them with a cotton swab soaked with isopropyl alcohol or similar solvent. If the contacts are covered with oxidation or are pitted, they can sometimes be cleaned by slipping a piece of fine sandpaper between them, holding the contacts together with the fingers, and pulling out the sandpaper. This process should be repeated until the contact surfaces are clean and shiny. At this point, complete the task by pulling a piece of cardboard (of postcard thickness) between the contacts points several times. Cardboard is just abrasive enough to give the metal surface the desired polish.

In most cases, it does not pay to attempt to clean badly pitted and burned contacts on a thermostat. It is usually best to replace the faulty thermostat with an exact replacement part. In fact, many of today's small-appliance thermostats are sealed in plastic housings and cannot be repaired.

Because of the various designs of thermostats, when it is necessary to recalibrate one, be sure to follow the manufacturer's instructions exactly. But before concluding that an adjustment or a replacement of the control is required, be sure that every other part of the appliance is in good working order and the operating instructions are being followed. When you are assured that the thermostat itself is at fault, study the manufacturer's service manual for the make in hand and follow these directions carefully.

Checking temperatures. To service resistance-heating small appliances, special testing equipment is needed to check temperatures to at least a 5 to 10 percent tolerance. To check the temperatures of irons and some cookware, for example, a thermocoupled-type tester is usually employed. A thermocouple (described in *Basics of Electric Appliance Servicing*) is a device made of two dissimilar metals. By connecting the thermocouple to a meter, the amount of heat can be determined quite accurately.

To check appliances such as coffee pots and sauce pans that contain a liquid when in use, an immersion-type mercury thermometer, especially designed for use in hot liquid, is generally used. Other temperature checking equipment is discussed in the next chapter. But in all temperature tests be certain to allow any appliance three to five cycles (one cycle is each time the thermostat turns off and on) before making a temperature reading or adjustment to ensure that the appliance has heated evenly. For instance, checking the temperature near the element may indicate a high temperature, but the sensing element (bimetal strip) of the thermostat may be located away from the element and may depend on the conduction of heat through metal and air to function correctly. Permitting the appliance to cycle a few times will make certain that the thermostat is working under normal conditions and the temperature test device measures the actual ambient heat in the appliance.

Servicing resistance-heating small appliances

Resistance-heating appliances convert electric energy into heat which may be used to press clothes, toast bread, grill hamburgers, cook stew, brew coffee, and warm a room. This heat is also used in such personal care products as hair curlers, hair dryers, heated combs, lather dispensers, hair stylers, and massagers. As you can see, many important household functions are performed by resistance-heating small appliances. Let us take a look at how they perform and how to service them.

IRONS

Modern electrical irons have lightened ironing and pressing chores in millions of homes and have made possible a variety of operations on fabrics, including raising the nap on velvets and corduroys and blocking knits.

Electric irons can be divided into four general classes on the basis of their construction: dry irons, steam irons, steam/spray irons, and travel irons.

Dry Irons

The automatic dry iron is simple, inexpensive, and dependable. It is used chiefly on heavy, pre-moistened fabrics. Its electric components include a heating element, a thermostat (temperature control), the terminals, the terminal insulator, a cordset, and, in some makes, a pilot lamp and its resistor. The low-voltage pilot (shunt) lamp, which indicates when the iron is heating, is usually connected in parallel with a resistor, while the resistor is connected in series with the heating element.

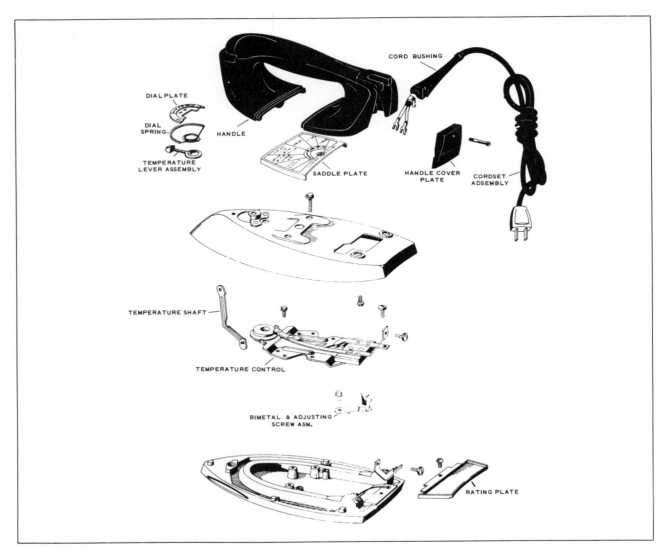

Figure 6-1. Disassembled view of a typical automatic dry iron.

Mechanical parts of a dry iron include the soleplate, the pressure plate, the cover, the heel rest or other type of stand, the handle, and the strain relief for securing the cordset in the terminal enclosure. The first part—and in some makes the first two parts—may be combined with the heating element to form a unit. In still other makes the element may be cast into the soleplate, in which case the pressure plate is not used.

The element heats the soleplate, which may be made of either aluminum or stainless steel, and the amount of heat is controlled by a thermostat. While thermostats used on irons are of many types, the general principle of operation is this: A bimetal blade (described in Chap. 5) fastened at one end bends proportionately as the temperature of the soleplate rises, and this motion opens the contacts of the thermostat. As the temperature falls, the bimetal blade begins its return toward its original position and the switch closes again. When the temperature-control knob, which is conveniently mounted on the iron, is turned toward a higher temperature, the distance which the bimetal blade must travel to open the switch is increased; when turned toward a lower temperature, the distance is decreased. Of course, the way this is accomplished may vary with different manufacturers, but the principle is essentially the same.

Servicing dry irons. The principal causes of trouble in an iron are in the (1) cordset; (2) connections to the element and thermostat; (3) thermostat; and (4) heating element. As pointed out in Chap. 1, except for replacing the attachment plug, more labor is expended in repairing a cordset than in replacing it. Furthermore, even more labor is saved if genuine cordsets are used in preference to either the fit-all variety or those assembled from bulk supplies in the shop. Those supplied by each manufacturer for its own make and model are fitted with the proper eyelets or other terminals, together with the strain relief and/or cord guards if required. And although these assemblies may cost more than any make-do cordset, the labor saved offsets the difference, and the finished job presents a more professional appearance. But if a fit-all cordset or one assembled in the shop is used, it should equal in current-carrying capacity and in every other quality the cordset originally supplied with the iron.

In most irons, some type of insulating bushing is employed where the terminals pass through the cover. Not only must the terminals be bright and clean to ensure a good electric connection, but be sure also that the insulator is intact and that the terminals are precisely in the center of the cover opening so that the insulator will not be cracked when the cover is tightened.

If the thermostat or heating element is faulty, good servicing practice is to replace it. Never try to splice an iron's heating element, if open. As for the thermostat, the troubles are usually mechanical. It may get bent or may get old and lose its resilience; it may be stuck shut or stuck open. While sometimes cleaning the contacts will solve the problem, generally the thermostat should be replaced if it is faulty.

When replacing internal parts, do not fail to provide safe clearance between the current-carrying parts and between these and other metal parts to ensure against short circuits and grounds. Test for short circuits and grounds before and after installing the cover. Some manufacturers recommend a hi-pot test for this. Follow the service manual recommendations when conducting this test, but most suggest 1,100 V for 1 min at maximum operating temperature. The probes of the tester should be connected between one prong of the cordset and the soleplate assembly. Heat the iron fully when conducting the hi-pot test. Remember that grounds

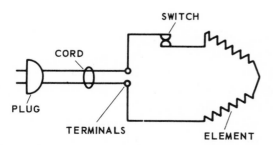

Figure 6-2. Schematic diagram of a typical dry iron.

can occur as the temperature rises which would not show if the iron were tested cold, for the thermostat parts move as the iron heats and any misalignment or distortion of these parts could result in a ground at a higher temperature. Bear in mind that a grounded appliance subjects the user to a serious accident hazard; for this reason spare no effort to make certain that no grounds exist at any temperature.

Thermostats may be adjusted up or down the entire scale if they are provided with a regulating device. For example, if a thermostat is good in every other respect and the heat test reveals that the temperature is, say, 120° too high in every stage tested, it follows that an adjustment is advisable. To test the calibration of an iron and to adjust the thermostat, a test stand such as the one shown here is necessary to measure the temperature. The iron must be placed on the stand so that its toe rests flat on the button; otherwise a true temperature reading cannot be obtained. The readings obtained should be compared with the manufacturer's soleplate specifications given in the service manual. Failure of the thermostat to perform within the range specified will require its adjustment. Thermostat adjustments should not be attempted prior to a careful reading of the manufacturer's service instructions, since they determine the exact adjustment procedure. Here are adjustment instructions for the iron illustrated in Fig. 6-1 as taken from its service manual.

1. Place the iron on the iron test stand and connect it to a 120-V ac source. Adjust the temperature-control knob to the highest setting. Wait until the temperature-recording meter needle stops climbing before taking a reading. Always disregard the first reading because of the tendency of the thermostat to overshoot. The second and following readings will be correct.

2. By removing the saddle plate, the adjustment screw can be reached through a hole in the top. Turning it clockwise decreases temperature, and turning it counterclock-

Figure 6-3. An iron test stand is the best way to measure the soleplate temperature of an automatic iron.

wise increases temperature. By means of this setscrew, adjust the temperature so that all the following conditions are fulfilled.

   a. The iron should be off when the control lever is in the OFF position.

   b. With the control lever at the middle of WOOL, the ON point should be 340°F ±35°F.

   c. With the control lever at maximum heat, the ON point should be 465°F ±50°F.

   d. The amplitude, difference between ON and OFF, at any setting should be at least 20°F but not more than 50°F.

   e. The amplitude plus overshoot should not exceed 80°F.

While the exact adjustment procedure and temperature ranges may vary with different manufacturers, the basic temperature tests are fairly typical for most dry irons. Always follow the service manual, but the chart below will serve for general use whenever this information for a specific make is not available.

Average temperatures, °F

Synthetic 220
Rayon 275
Steam 300
Wool. 340
Cotton 400
Linen 455
Maximum .500

Note: All temperatures ±50°F; control set in the middle of each range.

Of all the mechanical parts, the soleplate is most often the source of trouble. It should be cleaned as directed in the user's care and maintenance booklet or the manufacturer's service bulletin. Slight scratches in the soleplate will usually cause no problems, but burrs or any other projections above the soleplate surface could damage clothes. To remove such burrs and projections, rub the surface with fine emery. But before returning the iron to the customer, carefully clean and wipe off the soleplate.

When servicing an iron, it is a good practice to protect the soleplate by working on a clean pad. In fact, careful observation and good judgment on the part of the service technician will result in neat and workmanlike mechanical repairs. Handles, control knobs, terminal box covers, and all other miscellaneous parts should be replaced in the same manner in which they originally were installed by the manufacturer. Every part, however insignificant it may seem, has its place and purpose. Neither add nor take away—and never improvise. And although a slightly chipped control lever or a cracked terminal box cover may appear to have no effect on satisfactory operation, renew any such damaged parts so that the finished job will be completely whole.

Analysis of complaints. Here is an analysis of the typical complaints faced when servicing an automatic dry iron.

Iron will not heat.

1. Check for absence of power at outlet. Take necessary corrective measures.

2. Check the cordset and plug for continuity. If faulty, replace the plug or cordset.
3. Check the connecting terminals. Be sure that all terminals are clean and make good connections. If defective, be sure to replace.
4. Check the operation of the thermostat. Clean the contact points, if dirty or pitted. If the thermostat is faulty, replace it.
5. Check the heating element for an open circuit. If you find one, replace the element or soleplate depending on model involved.

Iron gives insufficient or excessive heat.

1. Using an iron temperature test stand (see Servicing Dry Irons above), check the iron for operation within the temperature limits specified in the service manual. Adjust the temperature, or if proper control cannot be obtained, replace the thermostat or the temperature control as required.
2. Check the prongs of the cordset and terminal pins on the iron for burned or corroded condition. If found, caution the customer to turn the knob OFF before connecting or disconnecting the iron. This will minimize the arcing at the terminals and consequent deterioration. Replace any faulty parts.
3. Be sure that the house voltage is not low. Remember that resistance-heating appliances, such as electric irons, are manufactured to operate properly on a voltage variation of approximately plus or minus 10 percent. The standard 60-Hz ac voltage in most sections of the country is usually 110 to 120 V, and most irons are rated at 105 to 125 V of alternating current.

Iron shocks user.

1. Check the cordset and internal wiring for bare spot. If found, replace or repair as needed.
2. Check the thermostat for insulation breakdown. Replace the thermostat if found faulty.
3. Check for grounded heater element. Take proper corrective action.

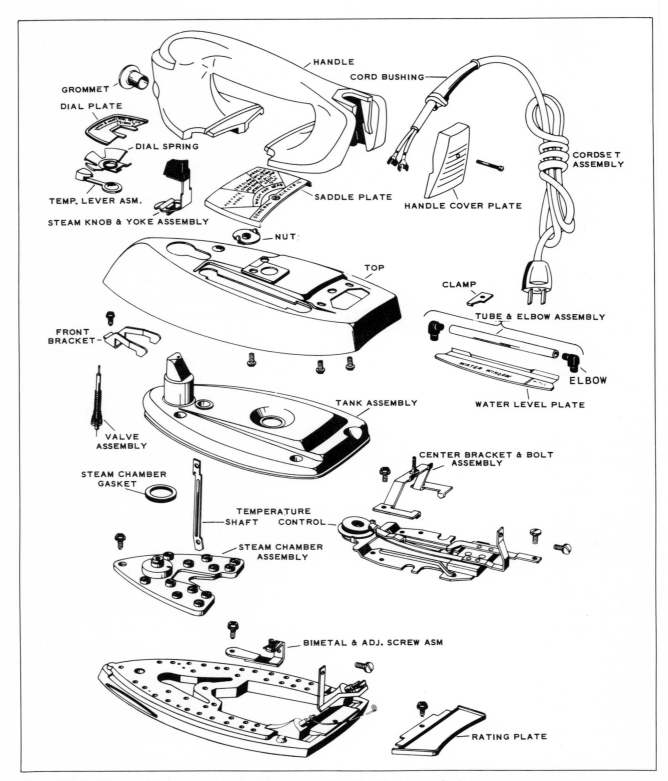

Figure 6-4. Disassembled view of a typical steam/spray iron that features a flash-type steam generator.

Cordset sparks.

1. Check for loose connections. If found, clean and tighten.
2. Check for broken wire. If found, repair or replace.

Blisters on the soleplate. This condition is usually caused by excessive heat. After taking the proper corrective measures to remove the cause of the high heat, replace or repair the soleplate, depending on its condition.

Iron sticks to clothes.

1. Check for dirty soleplate. Clean as directed in user's booklet.
2. Question customer about possible use of excessive starch in clothes. Instruct customer on use of correct amount.
3. Be sure the iron is not too hot for the fabric being ironed. Instruct customer on correct temperature.

No OFF *position.* See if the contacts in the temperature control or thermostat are open when the control lever is in the OFF position. If the contacts are closed, inspect them to see if they are fused together and check the position and operation of the thermostat. Take proper corrective measures.

Steam and Steam/Spray Irons

Steam and steam/spray irons are adaptable to a variety of tasks. With steam on, they can touch up permanent-press clothing, raise the nap of velvets and corduroys, block knits, or press wools. With steam off, they function as dry irons. The steam/spray iron has a nozzle that sprays water to help relax deep wrinkles.

There are two basic types of steam or steam/spray irons, differing primarily in the way the steam is generated. The boiler type has a water tank that is heated by the soleplate heating element to generate steam within the tank. This steam is then directed, by a valve, through holes in the soleplate to the fabric being ironed. When an iron of this type is filled to the proper level, plugged into a 120-V ac power source, and switched to the STEAM range, it operates as

follows. When the steam button is down, the steam button connector depresses the valve stem so as to close the orifice preventing the passage of water from the tank to the boiler chamber. When the steam button is released, the spring on the valve stem lifts the valve above the orifice, allowing water to drip into the boiler forming steam. The pressure created causes steam to fill the dome where it follows two paths. One path is through a metering opening in the steam chamber cover and thence through the soleplate distribution channel and out the soleplate vents. The other path is up the balance tube into the tank. There the slight pressure aids the flow of water through the valve and orifice maintaining a constant full flow of steam. When the button is in the STEAM position, a washer and gasket on the valve stem seal the fill path to the tank, preventing the loss of tank pressure. The spray mechanism frequently is of the standard bellows pump design and is independent of steam pressure. Each stroke of the spray plunger draws water from the tank forcing it through the check valve, spreader, and spray nozzle and producing a fine spray directed in front of the iron.

The flash-type steam generator uses a water tank so located that the tank does not receive much heat directly from the heating elements but, instead, serves as a reservoir. The water from this supply is metered, by means of an adjustable valve, into a steam chamber, which is part of the soleplate. As the water strikes this plate, it is immediately vaporized and is forced out under pressure through holes in the soleplate. In a typical flush-type unit, as the steam button is released to its extreme UP position, a gasket seals the tank at the top and the valve is pulled away from the orifice allowing the water to flow

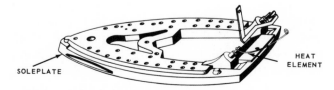

Figure 6-5. A typical soleplate and heating element. The heating element in modern irons is cast into the soleplate and cannot be repaired.

from the tank to the steam chamber. As the water hits the hot soleplate, steam is generated; the pressure control valve opens, and steam flows out the steam holes. Steam also flows back into the tank through the balance tube which creates pressure. A pressure valve between the boiler and the steam passages maintains the boiler and tank pressure required for spray. A small water tube connects the spray nozzle with water at the bottom of the tank, and a shorter surrounding steam tube connects it with steam pressure at the top of the tank. When the spray button is depressed to open the nozzle, a mixture of water and steam is forced through, very much in the manner of a paint spray gun or a household insecticide sprayer. In many models of this type, with the spray knob set in the PERM-PRESS position, a rod remains in the water tube mixer restricting the flow of water which results in a fine mist spray. When set on regular spray, the rod is out of the mixer, allowing full wet spray.

Some steam and steam/spray irons have a so-called "self-clean" feature. With most irons of this type when the self-clean feature is utilized, the iron must first be at maximum temperature to ensure full steam production during the cleaning cycle. Unlatching, depressing, and holding down the self-clean button operates another valve in the steam dome. This valve usually has a large opening and passes a large volume of water from the tank directly into the steam chamber where steam is created, forcing water and steam rapidly through the soleplate distribution channels and out the vents. This action generally loosens and expels mineral and lint contaminations that have accumulated. Releasing the self-clean button seals the "quick-dump" valve restoring action to normal.

Many irons have a safety feature that incorporates an over-temperature fuse link that will completely stop the flow of current to the element should the soleplate temperature rise beyond safe limits through failure of or damage to the temperature control. This fuse link consists of a bus wire soldered to a strip spring with a eutectic. The spring mount is such as to put a bias stress on the connection. If the critical temperature is reached, the eutectic will melt

and the spring will pull the connection open, preventing further heating.

Servicing steam and steam/spray irons. Problems with steam and steam/spray irons are similar to those of dry irons except that a few special problems may be due to the water, the water reservoir, and the water seals.

Distilled or demineralized water is recommended in all steam and steam/spray irons because it has none of the impurities which can cause deposits and corrosion in the iron. When ordinary tap water is used—especially in hard-water areas—the inside of the reservoir or boiler and the steam chamber will develop scale and deposits, and the valve and the escape holes may become clogged. Chlorine, used in many areas as a purifier of water, reacts with aluminum to form aluminum chloride. Thus in irons with aluminum internal parts and soleplates, if a flaky gray substance is present it is usually aluminum chloride rather than hard-water scale. Aluminum chloride deposits do not stick to the steam passages and are frequently carried along by the steam onto the cloth; when ironed into cloth, they leave a black smear that is most difficult to remove.

Sometimes it is possible to scrape or scratch-brush the boiler to clean out the mineral deposits. But a better cleaning method is to boil a tankful of saturated solution of agricultural gypsum through the reassembled iron before considering the job complete. Shake enough gypsum into a container of water so that after stirring or shaking some of it settles to the bottom undissolved. Be sure to run the iron at full temperature and with steam until the tankful is used up. The operation can be speeded up by placing the iron—with its soleplate down—on a block of asbestos-covered wood.

Another good cleaner is white vinegar. Fill the tank with vinegar, set the temperature for maximum heat, and plug in the iron. Put the steam button on and let the vinegar pass through the valve and nozzle. One tankful generally removes the scale. If the reservoir and other parts are badly contaminated with scale and deposits, they should be replaced.

If the soleplate has been removed for any

reason, the contacting surfaces of the bottom of the steam chamber must be painted with silicone adhesive to seal against steam leaks. Coat a small quantity of silicone adhesive around the collar of the filler elbow to provide a seal when it enters the fill hole of the tank.

Operational test. After servicing any iron, give it a complete operational test before returning it to the customer. Here is a typical test recommended for a steam/spray iron.

1. Fill the iron with 4 to 8 oz distilled water with the steam button down.
2. Set the iron on the test stand. The watt-meter should indicate 0 at OFF and nameplate wattage ±5 percent at 120 V when turned to the operating area of the dial.
3. The temperature at the center of the sole-plate at the third or fourth cut-in of the thermostat should be between 290 and 320°F when set at the center of the STEAM area.
4. Pressing and releasing the steam button should result in a good flow of steam from the ports.
5. Pressing the spray button should result in a cone of fine spray from the nozzle.
6. The temperature at the third or fourth cut-in of the thermostat should be between 445 and 500°F when set at the maximum or LINEN position.

Note: This is a typical operational test. Since the technique and measurements will vary with different makes and models, check the service manual for specific temperatures and methods of conducting the test for a particular iron.

Analysis of complaints. In addition to the problems given for automatic dry irons, these are some of the common complaints that service technicians receive concerning steam and steam/spray irons.

Iron leaks water. This complaint may refer to either water or steam and may or may not be justified.

1. Check the soleplate temperature at the steam setting. Water may run through the iron without vaporizing if the heat is low and may be blown through in drops if the iron

is too hot. If the customer tips the toe of the iron down, it puts additional water pressure over the valve and may cause water to enter the steam chamber faster than the heat of the iron will vaporize it.
2. Occasionally, this complaint may be due to water that drips down the vertical edge of the soleplate. This may be water which the user inadvertently trapped between the tank and the soleplate or between the handle and the top when filling the iron, or it may be condensation which drips off the tank or the inside of the top. The latter condition is more in evidence when the iron is first put to use and the tank and top are cold.
3. An apparent water leak may be due to care-less filling or overfilling, trapping water which will be expelled as the iron heats. Similarly, a defective or partially open steam valve may release water to the steam chamber, and if the temperature setting is below the steam range, it will be expelled through the soleplate holes as water drop-lets. To check, fill the tank carefully and stand the iron on a piece of paper for 15 min with the steam button closed (down). Note location of water spots, if any.
4. As a rule, actual water leaks can occur only at the gasket which seals the tank to the steam chamber and at the joint of the steam chamber cover; spray leaks at the spray

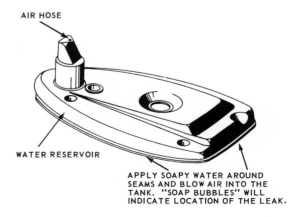

AIR HOSE

WATER RESERVOIR

APPLY SOAPY WATER AROUND SEAMS AND BLOW AIR INTO THE TANK. "SOAP BUBBLES" WILL INDICATE LOCATION OF THE LEAK.

Figure 6-6. One method of checking for a leak is to apply soapy water around the seams of the water reservoir and then blow air into the reservoir while watching for telltale soap bubbles, indicating a leak.

body gasket and at the seal between filler elbow and tank fill hole or the tank itself. Small leaks can be repaired with a silicone adhesive applied as described above. With some irons, check for a bad gasket. When a gasket is worn, cracked, or not sealed correctly, the iron can leak water and replacement is necessary. When you encounter a bad leak in the steam chamber, service is not possible for many makes because the soleplate and chamber assembly are not available for replacement.

5. With some models, water which enters the spray nozzle opening instead of the fill tube when the tank is being filled may cause this complaint. Continuous dripping at the spray nozzle would indicate a tube sleeve assembly which did not close firmly against the nozzle gasket. Correct this condition by lubricating, with a minimum of lubricant, the slots in the temperature bracket assembly where they engage the tabs on the tube sleeve assembly. Do not put any lubricant on the tube or cap.

6. With some models, check for the short yoke which would leave the valve slightly open. With the steam knob locked down for dry ironing, press on the knob while observing the steam control shaft through the fill part. If it moves in response to the pressure, the yoke is short and should be lengthened slightly by bending.

Iron will not steam or gives insufficient steam.

1. Be sure there is water in the tank. If not, fill the tank.

2. Check whether the water tank and the steam passages are completely or partially clogged with hard-water deposits. Clean the parts (see above) or replace, if necessary. Advise the customer to use distilled water to prevent deposit buildup.

3. Be sure that the thermostat is not set too low. Set the control higher or adjust the thermostat as needed.

4. Make certain that the steam valve stem rises and falls when the knob is actuated.

(This can usually be accomplished by looking into the filler hole.) If the stem is not operating properly, disassembly is required to determine whether the steam valve is clogged with hard-water deposit or if the valve stem is defective. Clean or replace faulty part.

5. This complaint may be due to a dry filming effect caused by oil or grease on the valve. Frequently this can be corrected by operating the iron for at least an hour at maximum heat to bake out the oil film. If this does not correct the situation, it may be necessary to heat the valve with a small gas flame to drive off the oil and unclog the valve. Be sure to keep the flame about an inch away from the valve. Do not overheat and melt the steam chamber. Cool naturally since the use of compressed air is apt to deposit a new film of oil.

6. Steaming complaints with some models of irons may be the result of the yoke being too long. Correct by pressing down firmly on the steam knob. The yoke is usually made of soft brass and will bend to the proper dimension.

Iron steams when control is on OFF *or* DRY *position.* This indicates a defective steam valve system. It can be caused by a bent or underformed steam knob connector wire, a weak valve spring, or a corroded steam valve stem. Also, the clean-out wire or pin at the bottom of the valve stem or body may be bent or broken, causing the valve to close improperly. Replace faulty parts as needed.

Iron gives poor or no spray.

1. Check the orifice of spray cap to be sure that it is not clogged. Remove cap, inspect, and clean with a fine needle or wire.

2. Check the gasket to be sure that it has not deteriorated or is misseated. Replace, if required.

3. Check the pump assembly to see if it is defective. Disassemble the handle and replace the pump. *Note*: In most cases, the pump must be actuated several times before

it will start to spray when the iron is first filled. This is normal.

4. With some models remove the spray mechanism and check the bellows for leaks. Replace, if faulty.

Steams but will not spray.

1. Press the spray knob to check for steam pressure in the boiler and tank. If there is pressure, the water tube is probably clogged. Remove the nozzle and connector assembly and clean the water tube and the nozzle opening.

2. If there is no pressure, disassemble the iron and look for a defective pressure valve between the boiler and steam passages.

Steam comes from fill port. This would indicate that the fill tube valve was not closed when the knob was in the STEAM position. If the steam knob is pushed tight against the handle, the steam knob connection is probably underformed and holds the fill valve open. If the knob is loose, look for a defect either in the valve or in its seat at the top of the fill tube.

Iron sputters or spits water from steam ports. Occasional water droplets emerging from the steam portion can be considered normal, but when there is an excessive amount, here is an analysis of things that can cause the problem.

1. Iron is not hot enough. Adjust the temperature control as specified by the manufacturer's service manual.

2. Iron is too hot. Adjust the temperature control as specified by the manufacturer's service manual.

3. If a check of the temperature at the steam setting shows the iron to be correctly calibrated, try the iron for steaming. Treat with one tank of gypsum solution (as described above) if it sputters. Recheck and if it still sputters, scratch-brush the steam chamber for possible contamination with oil or grease and repeat the gypsum treatment. If it does not sputter, the customer possibly is steam-ironing at the fabric setting instead of at the steam setting.

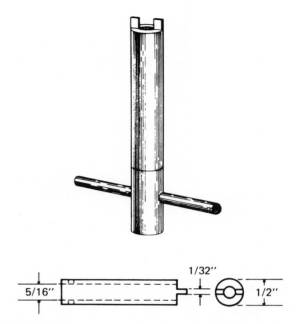

Figure 6-7. A wrench similar to the one shown here is used to remove the steam outlet nut. The manufacturer's service manual usually describes how the tool may be made or tells where you can buy one.

4. This can be caused by a corroded steam valve or, in an extreme case, by a corroded or deposit-filled stem. Disassemble the iron and clean or replace as required.

5. In some models, the inside of the steam chamber has a special coating which, when damaged, usually requires replacement of the soleplate.

Water sloshes from fill port. This malfunction probably means that the iron has been used at a high temperature for a protracted period with water in the tank and that the toe has been tipped down. Caution the customer against both of these practices; suggest an empty tank for dry ironing.

Iron spots material. Complaints of brown spots usually imply charred organic material in the steam chamber and passages. The cause may be lint which works up into the iron or filling with water from contaminated vessels (pop bottles etc.). Clean by filling and flooding soleplate while iron is cold, then turning to steam range with valve still open. The boiling water will flush brown material from the iron. Repeat if necessary. Other contaminants introduced into the steam chamber in the water used in the iron may also cause spots.

Iron burns holes in clothes. Complaints that iron burns holes in clothes usually can be traced to customer buying so-called distilled water from automobile service stations where the water is apt to be contaminated with sulfuric acid. Zippers are sometimes made of two different metals which react electrolytically in the presence of moisture. This may cause a disintegration of material which should not be blamed on the iron or the water used.

Iron drags (*non-stick soleplates*).

1. Advise the customer to do the following.
   a. Clean the soleplate with a mild liquid detergent.
   b. Raise or lower the temperature slightly for those fabrics which create the worst problem.
   c. Avoid using a slick ironing-board cover.
   d. Allow several weeks of use, after which the drag problem tends to disappear.
2. Lightly buffing a nonstick coated soleplate with a buffing wheel that has the excess compound removed from it will tend to make it glide easier. Buff lightly. *Caution:* Do not bear down hard or hold the iron in one place as you may damage the finish. If it is necessary to remove the Teflon coating from the soleplate, use a special 120-grit paper. Buffing and polishing are necessary after the Teflon coating has been removed.

Starch buildup, dirt, or foreign material on soleplate.

1. Clean the soleplate with an antistatic polish.
2. Use a mild metal abrasive for aluminum and stainless steel, baking soda or similar cleaner for other materials.
3. Heat the iron to remove plastic coating and use terry cloth or burlap material to rub it off the soleplate.

4. Starch buildup on the soleplate can be removed by rubbing lightly with a soap pad.

Iron gives no steam or poor steam. This probably indicates a dry-film condition of the steam valve. Fill the tank with 1 oz anti-dry-film solution mixed with 6 oz distilled water. This solution will break down the surface tension created by the dry film, and steaming should start. Lightly bumping the iron on the ironing surface may aid in getting the flow started. Steam out the entire tankful of this dilute solution to ensure complete removal of the dry-film condition. If this procedure fails to correct the problem, disassembly of the iron will be required.

Travel Irons

The travel iron is small and light and has a handle that folds, collapses, or detaches to save suitcase space. Some models have steam or steam/spray capabilities accomplished by including a boiler, a steam chamber, and steam passages and ports in the soleplate assembly. A removable bulb usually serves as a water tank for steam ironing.

In addition to the problems already discussed for dry and steam irons, here are some that occur with the water bulb.

Bulb leaks. Check to confirm and locate the leak. If the check ball is missing or water leaks past it or if there is a leak between the bulb itself and the fitting, replace the bulb assembly. If the leak is through the vent hole when the iron is set back on its heel stand, the bulb assembly is improperly oriented as the vent should be up. Replace the bulb assembly; if this does correct the condition, replace the steam chamber assembly.

Condensation forms on bulb fitting. This is a natural result of exposing cold metal to warm damp air. The condition will be minimized if warm or hot water is used for filling the bulb.

Iron leaks between bulb fitting and steam chamber orifice. Inspect the threads on the orifice and rechase them if they are damaged. If this does not cure the trouble, try a replacement bulb

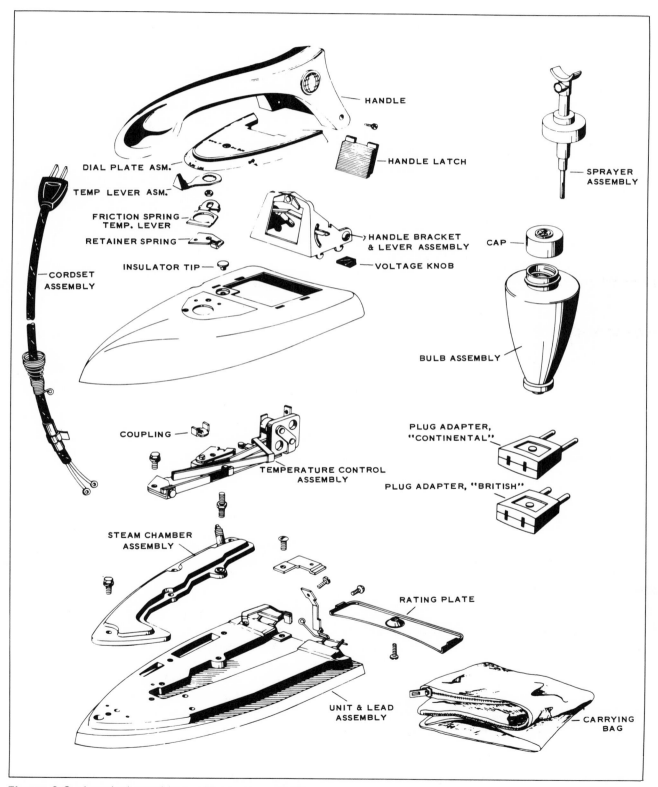

HANDLE

HANDLE LATCH

SPRAYER ASSEMBLY

DIAL PLATE ASM.

TEMP LEVER ASM.

FRICTION SPRING TEMP. LEVER

RETAINER SPRING

HANDLE BRACKET & LEVER ASSEMBLY

CAP

INSULATOR TIP

VOLTAGE KNOB

CORDSET ASSEMBLY

BULB ASSEMBLY

COUPLING

TEMPERATURE CONTROL ASSEMBLY

PLUG ADAPTER, "CONTINENTAL"

PLUG ADAPTER, "BRITISH"

STEAM CHAMBER ASSEMBLY

RATING PLATE

UNIT & LEAD ASSEMBLY

CARRYING BAG

Figure 6-8. A typical travel iron with steam capability.

assembly. If the connection still leaks, the orifice is probably defective or damaged, and the steam chamber assembly should be replaced.

Iron will not steam. Be sure the vent hole in the bulb is open. Pierce it with a hot wire if necessary. Then check the orifice and the steam ports to be sure these are all open. If the trouble has not been located, remove the steam chamber assembly to see that the internal passages are clear. If no stoppage is found, operate the iron for at least an hour at maximum heat to correct a possible dry-film condition in the orifice.

Iron spits or floods. First check the temperature at the steam setting and adjust if necessary. If the temperature is within limits, remove the steam chamber assembly and inspect the boiler area for surface treatment. Replace the screen mesh if required, and boil a bulb full of saturate solution of gypsum, if further boiler treatment is needed.

Some travel irons will operate on 120 V alternating or direct current, and 240 V alternating current. As shown in Fig. 6-9 a typical circuit for such an iron consists of two heating elements in series with a shorting switch across one of them. Thus, when the 240-V setup is in use, the following occurs.

1. The shorting switch is open.
2. The current flows through both elements A and B.

When the 120-V setup is in use, the following occurs.

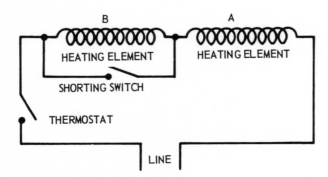

Figure 6-9. A 120/240-V arrangement.

1. The shorting switch is closed.
2. The current flows through element A, then through the shorting switch. Element B is out of the circuit.

AUTOMATIC TOASTERS

A great variety of automatic toasters with somewhat different operational and construction principles are on the market. In all instances, however, the heat from the electric current in the heater element toasts the bread. Actually, automatic toasters may be classed as either upright types that hold slices of bread vertically or horizontal types in which the slices lie flat.

Upright Toasters

Upright models are either conventional toasters that primarily toast or combinations with ovens that can toast, warm, and bake. The conventional upright toasters are built in two- or four-slice sizes. They are often referred to as automatic pop-up toasters. A few manufacturers call them wall toasters. They toast bread, frozen waffles, and thin pastries without toppings or fillings.

Basically, conventional upright toasters are simple appliances with three working parts.

1. A bread carriage that moves up and down inside the toast wells and usually has an external control knob. The carriage operates a switch that turns the toaster on and off.
2. Heating elements, made of resistance wires, positioned on both sides of each toast well. When electricity flows through them, they radiate heat for toasting.
3. A thermostat inside the toaster linked to a toast-color control outside. The control enables the user to adjust toasting time for different types of breads or to suit the user's preference.

Combination uprights have features of both

conventional toasters and small ovens. They have toast wells, bread carriages, and toast-color controls plus oven-type doors and controls for baking and warming. Toasting operations are similar to those in conventional uprights, and some models have the same type of heating elements. In others, toasting heat is provided by wires electrically shielded by metal, glass, or ceramic tubes.

Oven compartments, usually heated by tube heating elements, can be used for warming or for baking biscuits, pastries, small casseroles, and potatoes. Oven temperatures, usually selected on a separate control, range from approximately 200 to 500°F, and are governed by a separate oven thermostat. There may also be a setting for top browning of snacks and muffins.

Horizontal Toasters

Horizontal toasters are either reflector models that toast and warm or toaster ovens that toast, warm, and bake. Both are built in two- and four-slice sizes and have a front door or opening. Both can accommodate most bread sizes and can warm rolls and pastries.

Reflector toasters are named after their toasting method. While one side of the bread is toasted by direct heat from heating wires similar to those in conventional upright toasters, the other side is toasted by heat from the same wires reflected around the bread by a sheet of mirrorlike, polished metal. Because the two sides are toasted in different ways, they may brown to different shades. Reflector models come with or without doors. Some have starting and toast-color controls; with others the user must watch and remove the bread when it is browned.

Horizontal toasters with ovens perform the same functions as combination uprights, but they differ from them both in bread position and variety of controls. To toast, you put the bread on a horizontal rack that slides into the oven. Some models have controls that are fully automatic and stop toasting at a preselected brownness, open the oven door, and slide the toast out. On other models, controls stop toasting but do

not open the door. Some models have no automatic toast controls.

In this portion of the chapter, we shall concern ourselves with only the toaster part of the combination uprights and horizontal toaster/ ovens. The oven operation and servicing is basically the same as for table ovens, and their problem will be discussed later in the chapter.

Servicing Conventional Upright Toasters

Automatic upright toasters are subject to both electrical and mechanical problems. As a rule, the electrical troubles are not very difficult, but sometimes the mechanical ones can be quite trying. Let us first look at electric circuits employed in a typical two-slice automatic upright toaster (Fig. 6-17).

Most ON-OFF main or power switches used today are of two-pole type, and this toaster switch is normally in an OPEN position. The switch is closed when the user puts the bread into the bread well and depresses the handle on the bread carriage. (A few models have an automatic-carriage movement in which the bread, when placed in the bread well, closes the main switch; see the description below.) When the toast is done and pops up, the switch is again in the OPEN or OFF position.

The two-slice uprights have either three or four heating elements: one on the outside of each slice and one double-faced element in the middle, or two single-faced ones. These elements are generally wired in parallel. Consequently, an opening in a heating element would not render the toaster inoperative, as would be the case in a series-wired device. If one element is opened up, the others still work, and it is easy to spot the one that is at fault. This servicing technique would not be possible in the few models that have their outside heaters connected in series.

The heating elements should never be spliced nor any portion of the resistance wire unwound to make slack for reconnecting to a terminal. When ordering new elements, it is important to note whether a rating is stamped on the old element and, if so, to include this information on the purchase order along with the location of the element—that is, center, outside, or any other peculiarities—for in some models the center and outside elements are of unequal resistance; still others may have an opening in one element for the thermostat. Following these precautions will ensure uniform toasting after the installation of new elements.

While all makes and models of toasters differ from each other to a certain extent, most operate on the thermostatic principle of bimetal—two different metals—in which the toasting is controlled by the expansion and contraction of the thermostat bimetal. When the thermostat has reached a preselected temperature, it operates a mechanical or electromechanical device to release the carriage, open the main power switch, and pop up the toast. Let us take a detailed look at the thermostatic control used with upright toasters.

There are three basic types of toaster controls: a clock-type timer, a single-stage control, and a two-stage control.

Clock-type timer. The standard clock timer was the first type of control device used in automatic pop-up toasters. This control arrangement used a spring-motored clock (Fig. 6-10) to time the toasting cycle. To start this type, the user, by pushing down the starting lever(s), wound the clock, lowered the carriage and latched it, and closed the switch. At the end of the time cycle the clock tripped the carriage latch, and the elevating spring lifted the carriage with the toasted bread. A toast color control

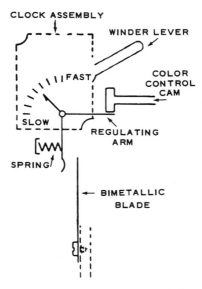

Figure 6-11. A spring-motored clock timer with thermostatic compensator: (A) clock assembly; (B) winder lever, which is linked to the starting handle; (C) spring for the regulating arm; (D) regulating arm; (E) color control cam; and (F) bimetallic blade.

(light to dark) enabled the user to vary the time cycle by altering the speed of the clock. To get uniform toast from both hot and cold starts without changing the color control on this model, it was necessary to preheat the toaster for the first slice of bread.

To make the system more "automatic," a compensator was added to the control arrangement (Fig. 6-11). This compensator comprised two parts: (1) a bimetallic blade which bent proportionately as the temperature increased inside the toaster, and (2) a spring-loaded regulating lever on the clock which was actuated by the bimetallic blade. (This regulating lever was usually independent of the toast color control and, of course, did not change its position.) As the toaster temperature rose, the bimetallic blade moved toward the clock-regulating lever, later striking it, and still later moving the lever to increase the speed of the clock. Thus the time cycle was varied automatically to suit the starting temperature of the toaster. The toast color control was the same as that used on the standard clock timer model, but with this later

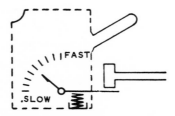

Figure 6-10. A simple spring-motored clock timer.

design it was possible to get uniform toast whether started hot or cold without changing the color control or preheating.

Few toasters manufactured today employ clock-type timing mechanisms, but it is important to know how they work since one of these "old-timers" may come into your shop at any time.

Single-stage time control. The single-stage, or single-cycle, time control utilizes the thermostatic principle that time is required for the bi-metal blade to respond to heat. Conveniently, still more time is added to the toasting cycle if fresh, moist bread is used, for heat—as we all know—is carried away as moisture evaporates. On the other hand, if dry bread is used, heat accumulates more rapidly inside the toaster, and the thermostat blade begins its movement toward the switch proportionately sooner. Thus, this type of timing adjusts itself automatically to accumulated heat and to almost any texture of bread.

Early applications of this principle included

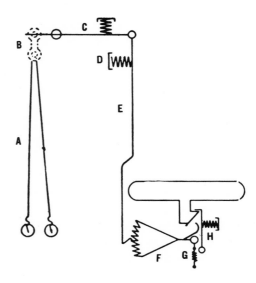

Figure 6-13. Hot-wire carriage-latch release. The hot wire (A) is shown at the left, fastened at the bottom to two terminals with its upper end passed through the insulator (B) on one end of the lever (C), whose fulcrum is at point 1. A spring keeps a downward pressure on the right-hand end of the lever (C). When the user depresses the carriage, it is held down by the spring-loaded latch (H). As the toaster heats, the hot wire (A), being in series with the elements, heats and expands, allowing the lever (C) to go down slightly on the right-hand end. This downward motion is transmitted to the pawl lever (E), which slips easily over the ratchet-edge of the sector (F). A soft spring (D) holds the pawl gently against the sector. Now when the thermostat shuts off, the hot wire cools and contracts, resulting in a slight upward motion on the right-hand end of the lever (C), which is transmitted through the pawl lever (E) to the sector (F), tripping the latch (H). As the carriage rises, it strikes the offset in the pawl lever (E), disengaging it from the sector (F) so that the centering spring (G) may return the sector to its proper position for the next cycle.

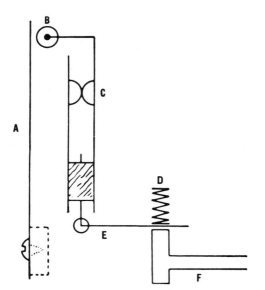

Figure 6-12. A single-stage thermostatic timer. The bimetallic blade (A), which is in direct contact with the bread, curles toward the right as the temperature rises until the end of the time cycle when it strikes the switch insulator (B), opening the switch (C). The switch tilting arm (E) is held against the color control cam (F) by the spring (D).

not only oven-type models but also flip-over toasters. When one of these models shuts off at the end of the time cycle, a signal bell informs the user that it is time to flip the toast or to trip the carriage latch, as the case may be. The signal bell is actuated by a low-resistance electromagnet.

The toast color control enables the user to vary the distance the bimetallic blade must travel to open the switch. Further development of the preceding models includes the pop-up feature to replace the signal bell.

The hot-wire principle is frequently employed in this type to draw the carriage-latch release (Fig. 6-13). A slender loop of special wire shaped like a hairpin with long legs has its upper portion—the loop—pulled tautly over the insulated end of a lever. The ends of the two legs are fastened to terminals which serve not only to anchor the ends of the hot wire but also to connect it in series with the heating elements. Then when the user closes the main switch by depressing the carriage, the current flows through the hot wire, thereby heating and expanding it. This small movement, which is transmitted to the lever at the point where the loop passes over, is multiplied at the other end of the lever by the position of the fulcrum. As the lever moves with this gradual expansion, it gathers a grip on the carriage-latch release. At the end of the time cycle the thermostat shuts off the current, the hot wire cools, contracts, and draws the carriage-latch release.

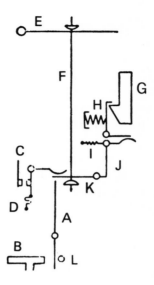

Figure 6-15. Two-stage thermostatic timing with auxiliary heater. This sketch shows a part of the mechanism at the beginning of a cycle with the carriage (G) already latched down. As the auxiliary element heats the bimetallic blade, it bends on both sides of the pivot toward the left until the free movement of the lower part of the blade is hindered by the color-control cam (B); this causes the upper end of the blade to exert pressure against and to close the auxiliary switch (C), short-circuiting the auxiliary heater. (An overcenter spring is shown as part 4, which gives the auxiliary switch a snap action.) The bimetallic blade now begins its second stage of timing—cooling. It can be seen that before the blade had reached its left extreme, the trip link (K) dropped slightly and into the path of the returning bimetallic blade. As the blade nears the starting point, it will strike the link (K) moving the tripping arm (J) toward the right, thereby tripping the latch (H). Then the rising carriage, on reaching the top, will lift the reset lever (E), transmitting an upward motion to the trace (F), which in turn lifts the link (K) above the blade and reopens the auxiliary switch to prepare the mechanism for the next cycle. A centering spring for the tripping arm (J) is shown as (I). (L) is the cold-position regulating stop.

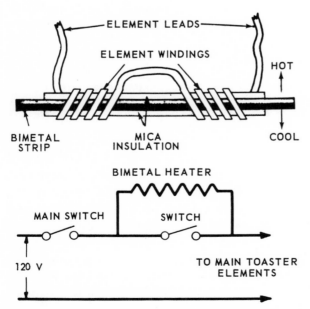

Figure 6-14. (Top) Bimetal strip is generally used to operate a mechanical timer. (Bottom) Schematic diagram of the element which heats the bimetal strip.

Some single-stage thermostatically timed control systems employ a small auxiliary heating element, situated directly adjacent to a thermostat blade somewhat heavier than those used in other types of timers, which snaps with sufficient force to trip the carriage latch and open the switch. Earlier models of this design did not have the pop-up feature, but were automatic in every other particular. A pilot lamp informed the user when the toast was done; then the carriage had to be raised manually.

Two-stage time control. This method is termed *two-stage* or *two-cycle* because the thermostatic timing cycle is in two parts, heating and cooling. A small auxiliary heating element is wrapped around the bimetallic blade and is connected in series with the main heating elements. There are two switches in this mechanism, a main switch and a short-circuiting switch for the auxiliary heater.

When the carriage is depressed, the main switch is closed, but the short-circuiting switch for the auxiliary is open, allowing the auxiliary to heat along with the main elements. As the auxiliary heats, the bimetallic blade begins its movement toward a regulating stop. When the blade strikes the stop, the short-circuiting switch is closed, thereby bypassing the auxiliary heating element, although the main elements continue heating. The cooling bimetallic blade now begins its second stage of timing—its return toward the starting point. As the blade reaches the end of this return movement, it trips the carriage latch, the carriage rises, and the toaster is ready for another cycle.

Automatic-carriage movement. In this type of upright toaster the carriage descends almost magically after the bread is dropped into the slot and a minute or so later the finished toast rises quietly. No levers are necessary—the only control is the color-selector button.

There are several ways of accomplishing this. For example, one design uses a single-stage timing system: The weight of a slice of bread, when dropped into the slot, depresses a lever which, through a linking mechanism, trips the switch to the ON position where it remains until it is tripped off by the thermostat. With the current on, the hot wire expands, allowing the carriage to descend. At the end of the time cycle the thermostat trips the switch to the OFF position, whereupon the hot wire cools and contracts, thereby lifting the carriage to its upper position. As the toast is lifted from the slot, the

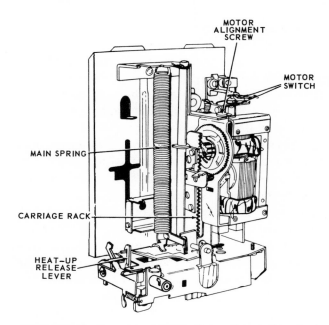

Figure 6-16. In some toasters the racks are lowered by a motor rather than by a handle.

starting lever in the carriage returns to its upper position, and the toaster is prepared for the next cycle.

In another design of the automatic-carriage movement, an electric motor (Fig. 6-16) is employed to pull down the carriage, but in this brand a two-stage thermostatic timing is used.

Keep-warm control. This simple mechanical adjunct with a selector knob indicating KEEP WARM in one position and POP UP in the other is incorporated into some models. If this control is set at KEEP WARM, the current will shut off as usual at the end of the cycle, but the carriage will not rise until the user trips the latch.

A special control knob or lever is sometimes incorporated into the design of the toaster which permits the user to change from high heat for toasting to low heat for warming snacks and pastries.

Timer adjustments. Most toaster timing devices are set at the factory to make medium toast on the MEDIUM setting of the color control. Erratic operation usually indicates a disjointing of the timer components or some other integral fault, but assuming that all other parts are operating properly, if the bread test reveals that the toast color control must be turned too close to either extreme for medium toast—and that the toast color can be varied through the control—the need for an adjustment is quite clear.

Each toaster has different limits of heat-up time, cool-down time (but remember that not all toasters have a cool-down time), and the amount of current drawn during cool-down time. The heat-up time is the elapsed time from the moment the cycle is started to the faint click indicating shunting of the control switch. Cool-down time is the time that elapses from shunting of the control switch allowing the bimetal to cool to the moment when the toast pops up. When checking the timing, be sure to cool the toaster to room temperature before making timer adjustments, and always begin the bread test from a cold start in order to prove the accuracy of the adjustment. Also be sure the toaster shell cover is in position when timing the cycle. Without the shell cover the heat is not distributed normally inside the toaster, and therefore the thermostat and timing cycles cannot be properly set. Consult the manufacturer's service manual for the make in hand for measured clearances of critical adjustments. Here are the adjustment instructions taken from the service manual for the timing of the toaster shown in Fig. 6-17. Remember that these are only typical instructions; exact data for a specific toaster should be taken from its service manual.

Adjusting the Timing

With the toaster completely assembled and cool, operate it at 120 V (regulated) and through an ammeter to observe cool-down current through one cycle at the DARK setting. Record the heat-up time, cool-down time, and the total elapsed time. Each toaster of this model must operate within the following limits.

Current: 7.2 to 7.5 A

Heat-up: 76 to 109 s
Cool-down: 18 to 32 s
Total time: 128 s (maximum)

Current: 7.5 to 7.8 A

Heat-up: 70 to 103 s
Cool-down: 18 to 32 s
Total time: 122 s (maximum)

While the toaster is still hot, shift color control knob to LIGHT setting and check for latch down and no buzz. If it fails to latch down, back out the cool-down screw one-half a turn; repeat the timing check. This latter procedure may be repeated as long as the heat-up and cool-down times remain within specification.

If adjustments are required, remove the carriage knob, control knob, and base end. The heat-up adjustment screw is the deeper of the two. Turning the screw in (clockwise) decreases the elapsed time.

Turning the screw out (counterclockwise) increases the elapsed time. One full turn of the heat-up adjustment screw changes the heat-up time about 18 s. The cool-down adjustment screw is the shallower of the two. Turning the screw in (clockwise) decreases the elapsed time. Turning the screw out (counterclockwise) increases the elapsed time. One full turn of the cool-down adjustment screw changes the cool-down time about 10 s. Note that a heat-up time adjustment will change the cool-down time but not necessarily equally. When the heat-up time conforms to the specification, correct the cool-down time if and as required. Check for latch-down by moving the control knob to LIGHT and push down the carriage. If the carriage fails to latch down, increase cool-down time but keep within specification.

In some models, the single-stage timer adjustment can be made from the bottom of the toaster by merely turning a regulating screw; in others, repositioning of the toast color-control button is all that is necessary. The point to remember, however, is that in this type of timer if the toast is too dark with the color control set at MEDIUM, the adjustment you make must shorten the distance the thermostat blade will have to travel to open the switch; if the toast is too light, you must increase the distance. Be careful not to bend the bimetallic blade.

Two-stage thermostatic timer adjustments are

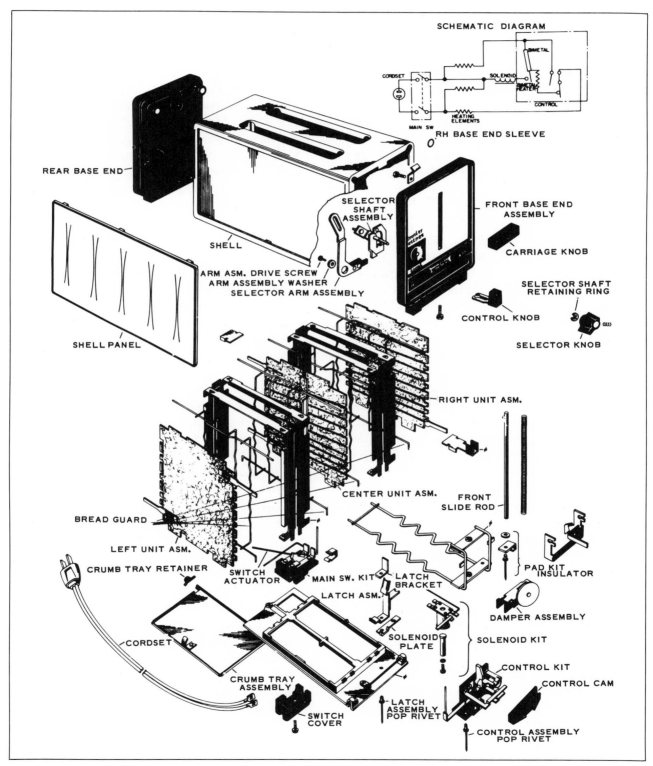

Figure 6-17. Major parts of two-slice automatic upright toaster.

somewhat similar to the one just described except that the adjustment must cause the bimetallic blade with its auxiliary heater to travel a shorter distance before it closes the auxiliary heater short-circuiting switch, if the toast is too dark, or a greater distance, if it is too light. Before attempting any adjustments, though, make certain that the auxiliary as well as all other heating elements are operating, as any variation in resistance will alter the time cycle.

Other Service Considerations

A very common problem with toasters is caused by food particles, which can affect both mechanical and electrical operations. Small crumbs, burnt raisins, and the like accumulating inside the toaster can readily contaminate thermostat or switch contacts, or jam up the release mechanism, the wire guides that hold the bread in place, etc. An air hose will do a good job of loosening these particles. You can usually scrape off food that has been burnt on the contacts and other parts. But when doing this, be very careful not to damage a heater element.

The base, the control knobs, the handles, and other plastic parts should be examined for damage, for chipped, cracked, or broken parts are an almost certain indication that the toaster has been dropped and the interior parts have sustained more serious damage. When appearances lead to the suspicion that a toaster has been dropped, scrutinize every part and subassembly with more than ordinary care so that you will overlook nothing that is required to restore every function to satisfactory condition.

Do not be too eager to dismantle an automatic toaster; first explore every possibility of making adjustments or learning the cause of failure from the outside. In some models numerous repairs and adjustments can be effected from the bottom after the removal of nothing more than the crumb tray. Furthermore, it is always desirable to learn the cause of failure quickly, especially if appearances indicate that repairs may be costly, in which case it is best to quote an estimate before going too far—customers do not like unpleasant surprises.

When you take apart a particular model for the first time, it is a good idea to spend an extra few minutes to study its method of operation and to learn the exact purpose of every part. Such an examination not only makes servicing a great deal easier the next time you handle the same model, but it also enables you to know just how many adjustments and services may be performed without dismantling.

Always remember to pad the workbench with several thicknesses of soft cloth to prevent scratching the toaster before laying it on its side or turning it upside down, and be sure to keep tools and loose parts away from the padded area.

With the outer shell removed from some models, the bread guide wires are free to drop out when the chassis is inverted. Such a helter-skelter spilling of these slender rods can do serious damage to other parts of the toaster, for almost invariably some of the guide wires become entangled in one or more heating elements. Therefore, always remove the shell with the toaster right side up. Then, if the work necessitates removal of the guide wires, lift them out one or two at a time. On the other hand, if you do not have to remove guide wires, fasten them in place by pulling a piece of cellophane tape tautly over the upper ends of each row of guide wires, continuing the tape down the ends of the toaster's inner frame about 2 in—but be sure to remove the tape before putting back the outer shell.

With some models it is possible to put the toaster through its cycle with the shell removed so that the various parts can be observed in actual operation. And there are models whose timing and carriage-latch tripping mechanisms are situated beneath the toaster, directly above the crumb tray. To watch this type of mechanism in operation, first expose the toaster's underside; then elevate the toaster 3 or 4 in with wooden blocks under the corners; next place a small mirror on the workbench below the mechanism; finally aim a flashlight about in line with your vision toward the mirror to enable you to see everything that is going on.

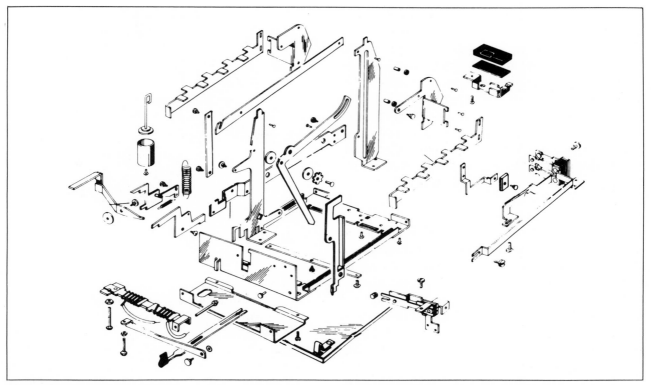

Figure 6-18. The many mechanical parts of a latch assembly are shown in this exploded drawing. All these parts must be clean and free to move.

The designs of the latching and pop-up mechanisms vary greatly from one model or make to another. Three of the more popular latching and pop-up arrangements are described and illustrated here. In the model shown in Fig. 6-19A, the latch knob, which is manually depressed to start the toasting cycle, is mounted on the leaf which, in turn, is mounted on the operating lever. The operating lever, therefore, pivots down with the latch knob. As the carriage lever is joined to the operating lever by the connecting link, this part also pivots down, one arm lowering the carriage mechanism to the bottom of the toasting well and the other arm swinging down and forward to be trapped and held by the trip latch. At the back of the operating lever a toggle arm and toggle are mounted. The movement of the lever when the latch knob is depressed throws the toggle against the blades of the main switch and thus closes both sides of the circuit. The heating units and the bimetal heater are thereby energized, and the toasting cycle is started. The bimetal heater is mounted on a pivot pin at the right hand end and is supported by the heat-up adjustment screw at

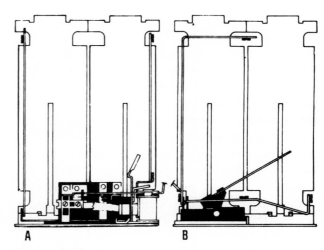

Figure 6-19. Two common pop-up arrangements.

the left. As it heats, it arches in the middle, and the edge slides up the face of the switch arm. The switch arm is spring-loaded against the bimetal by the spring contact of the shorting switch. When the bimetal clears the step in the switch arm, the latter part swings forward with the step under the bimetal and thus permits the spring contact to move forward and close the contacts in the shorting switch. This shunts the current past the bimetal heater, and the bimetal begins to cool. The bimetal is now supported at the right-hand end by the pivot pin and in the middle by the step in the switch arm. It straightens as it cools, and the left-hand end moves up against the compensator assembly. As the left side of the compensator assembly rises, the right-hand end moves down against the trip latch. When the trip latch is pushed down, the trapped tip of the carriage lever is released, and, under the pull of the carriage spring, all parts are returned to their starting positions. The cam-like form of the carriage lever causes the switch arm to pivot back out from under the bimetal, permitting this part to fall and opening the shorting switch in preparation for a new toasting cycle. The foregoing toasting cycle may be interrupted at any time by lifting the latch knob. This causes the leaf to pivot and push the trip latch down, thereby releasing the carriage lever exactly as it is released by the compensator assembly at the completion of the toasting cycle.

When the handle is depressed in the model shown in Fig. 6-19B, the two sets of contacts on the main switch are closed, energizing the heaters, which are connected in a parallel circuit. This also energizes the bimetal heater assembly, causing the bimetal to arch, and in turn raises the bimetal yoke high enough for the bell crank to drop under the yoke. This causes the shorting switch to close, shunting out the bimetal heater assembly, allowing it to cool. The bimetal heater assembly will continue to raise until the adjustment screw hits the compensator assembly which stops the upward motion of the bimetal assembly, forcing the latch to release the carriage. The carriage will return to the UP position under

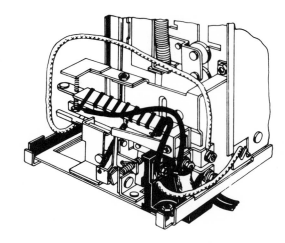

Figure 6-20. Solenoid-type pop-up arrangement.

spring tension and release the main switch to the OPEN position.

When the carriage knob is depressed in the model shown in Fig. 6-20, which is frequently called a *relay* or *solenoid* type of toaster, the carriage is latched down, the main switch is closed, and the timer is actuated. A bimetal heater in series with the outer heaters causes the bimetal to arch. When it has arched sufficiently, a contact in the control closes, shunting the bimetal heater, allowing the bimetal to cool. When the bimetal makes contact with the cool-down adjustment screw, the solenoid (relay) is energized. The solenoid core pulls the latch, releasing the carriage, which will then raise and open the main switch.

Most toasters have a dashpot or flywheel assembly to prevent the racks from returning to the up position too rapidly when the toasting cycle has been completed. There are several types of dashpots; some work with liquid, while others use suction, friction, or air pressure. If the dashpot is not working or the flywheel assembly becomes loose on its shaft, the racks will move up too fast and throw the toast out of the toaster.

Analysis of complaints. Here is an analysis of complaints that are most frquently encountered by a service technician when servicing upright toasters.

Toaster will not heat.

1. Check the cordset, main switch (in closed position), welds, bus bars, wiring, and all the heater elements for continuity. (Remember that in most models continuity can be obtained only when the rack is pushed down to close the main switch.) Repair or replace the inoperative components.
2. Check the main switch for condition and action. Contacts must be clean and touch with adequate pressure. Contact points can be cleaned with a very fine ignition file or fine crocus cloth. If faulty, replace the switch.
3. Check for loose connections. If any are found, clean and tighten them.
4. Check the latch and lifter arm assembly to make sure the switch contacts are closed and stay closed when the operating lever is depressed. Adjust the lever to make good contact or replace the assembly.

Carriage will not stay down.

1. See that the arm on the latch keeper is held down in place by the latch release assembly or catch. Check the movement of all parts. They should be free with no stickiness from foreign materials (dirt, grease, and grime), burrs, or bends. Clean and lubricate with heat-resistant lubricant. Repair or replace any defective parts.
2. In a solenoid-type (relay) toaster, check the clearance between the solenoid coil latch release and tip of the latch keeper. Adjust by bending the latch release so a gap of $\frac{1}{64}$ to $\frac{3}{64}$ in is obtained. Check to see that the carriage is being fully depressed and is not binding against the grill wire frame. Check the lifter arm for a bent or binding condition. Replace any defective parts.
3. Check for a burned solenoid coil or for damage to the latch. (Some latches use a small coil spring that may break, fall out of the toaster, and be lost, and unless you know that the spring was supposed to have been there you may be at a loss to determine

why the catch is not working properly.) Replace any faulty components.
4. Check the thermostat or timer cool-down adjustment screw for clearance from contact. Readjust if necessary.

Toast will not pop up.

1. In a solenoid-type toaster, check for an open solenoid and proper core and latch action. (When the bimetal touches and cools down the adjustment screw, the solenoid attracts the release latch, causing the carriage to raise and the main switch to open.) Check for a bent or broken latch. Replace defective parts.
2. Check the timing mechanism or thermostat for proper action. Readjust or replace, if faulty.
3. Check the various latch assemblies and carriage for free and proper action. Clean and lubricate, if necessary.
4. Check the action of dashpot or flywheel assembly for proper pushup. Repair or replace if either the dashpot or flywheel assembly is defective.
5. Check the return or carriage-elevator spring to be sure that it is not broken, jammed, distorted, or disconnected. Replace if it is broken or has no tension.
6. Check for contamination of the contacts in the timer or thermostat. Clean, if necessary.
7. Check for a shorted cool-down or heat-up switch. Replace, if necessary.
8. Check for a distorted bimetal strip. Replace, if necessary.

Carriage rises too slowly.

1. Check for bent, binding, fouled, or corroded front and rear slide rods and carriage assemblies. Clean and lubricate rods, or straighten if possible. If they are defective and cannot be corrected, replace them.
2. Check the return or carriage-elevator spring tension. Adjust or replace as required.
3. Check for bent or binding lifter arm. If the resistance is due to the dashpot, grasp the rod to the dashpot piston and rotate the

piston inside the dashpot cylinder forcibly to ease the friction exerted by the asbestos washer of the dashpot piston. If the dashpot is worn, replace it. In some models check for loose flywheel assembly. Tighten if necessary.

Carriage rises too quickly. This complaint can usually be corrected by tamping down the washer on the dashpot piston. Also clean the slide rod and damper sleeve (if one is employed) with alcohol. If the dashpot is worn, replace it. In some models, be certain that the flywheel assembly is free to ride on the slide rod and is not hung up. Also check to be sure that the flywheel is not loose; tighten if necessary.

Toaster is noisy or throws toast out. The dashpot is inoperative. Replace the plunger and/or the dashpot. If a flywheel assembly is used, check it for looseness; tighten if necessary. Also check the return spring for excessive tension; adjust tension on lift spring.

Toast burns.

1. Check the switch. Clean or, if faulty, replace.
2. Check for a shorted cool-down or heat-up switch. Replace, if necessary.
3. Check the timing or thermostat calibration to be sure that it is properly set. If not, adjust, or if the unit is defective, replace.
4. Check for a distorted bimetal strip. Replace, if necessary.

Toast too light or too dark.

1. Check the timing for heat-up, cool-down, and total time. Adjust according to specifications in the service manual.
2. Check for a defective timer assembly or thermostat. If faulty, replace.
3. With some toaster models, if the toast is too light at all settings of the selector control, the slide button may be in PASTRIES position. Move to TOAST position and check again.

One side is untoasted. Check for a defective heater element. If found, replace.

Servicing Horizontal Toasters

Most horizontal toasters are very similar in operation. The typical model shown here employs a single-element heater wound around a ceramic core which is positioned along the rear of the toaster in direct line with the height of the bread rack. A series of reflectors directs the heat to both the top and bottom of the bread on the rack simultaneously. As shown in Fig. 6-21, the operating lever assembly is mounted to the timer frame by means of a pivot pin. One end of the operating lever extends through the frame and is positioned between the arms of the double-pole switch. The push button is mounted on the other end of the operating lever. When the push button is depressed, the main switch contacts close as a result of the movement of the operating lever assembly. At the same time, the operating lever pushes down on the bell crank so its tab clears the arm of the bimetal assembly. The latch spring pulls the bell crank to the right and causes the latch to rotate and move forward until it catches the pin on the operating lever assembly. This movement of the bell crank pushes down on the spring contact of the shunting switch and holds the switch open. With the main switch closed and the shunting switch open, the heating element and bimetal heater are now energized in series.

The right-hand end of the bimetal assembly is supported by the point of the heat-up adjustment screw, and the left-hand end is mounted on a pivot pin. As the bimetal heats up, it arches in the middle, and since both ends are supported, a creeping counterclockwise rotation of the bimetal arm results. This causes the end of the arm to move to the right across the tab of the bell crank, and when it clears the tab, the bell crank is free to rotate and the force of the spring contact closes the shunting switch. The current is now shunted past the bimetal heater, and the bimetal begins to cool. As it cools, it straightens, and, with the bimetal arm trapped by the tab of the bell crank, the right-hand end of the bimetal rises until the point of the cool-down adjustment screw strikes the compensator.

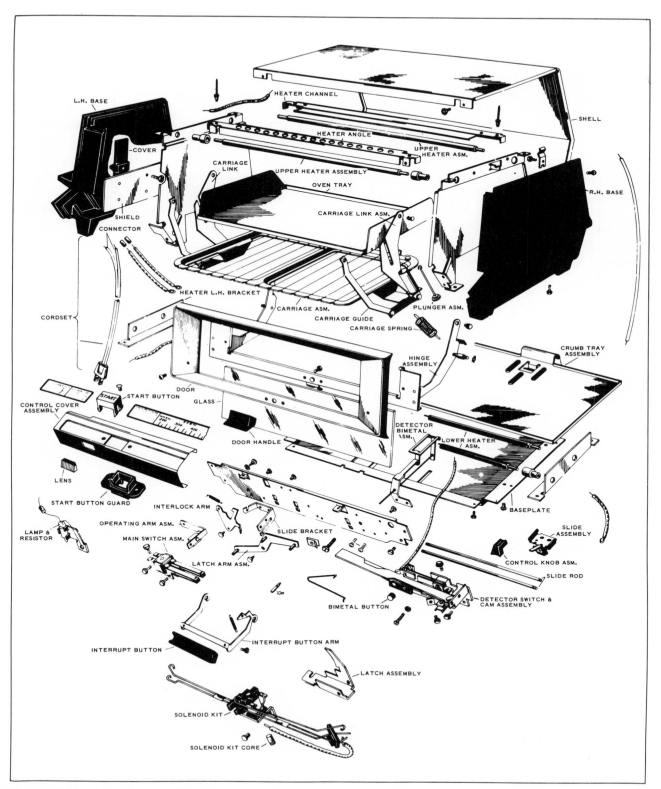

Figure 6-21. Schematic diagram and major parts of a typical horizontal toaster.

Since it can rise no further, continued cooling pushes the bimetal arm against the tab on the bell crank and drives it to the left. This causes the latch to rotate and free the pin on the operating lever assembly. The tension of the operating lever spring causes the operating lever to spring up. This movement automatically opens the main switch contacts, ending the toasting cycle. This is indicated by a slight click and the immediate fading of the bright glow from the element. The toasting cycle may be cut off at any time by lifting the push button.

Analysis of complaints. Here is an analysis of complaints that are most frequently encountered by service technicians when servicing horizontal toasters.

Toaster fails to heat.

1. Check cordset for continuity. If defective, replace.
2. Check for an open heater. If defective, replace heater.
3. In some models, check for an open bimetal heater. If open, replace the bimetal heater assembly.
4. Check each lead and terminal for an open condition. Take proper corrective action, if necessary.
5. Check the main switch contacts. They may be dirty or low in contact pressure. Increase the contact pressure by slightly bending down the upper stationary contact arms. However, care must exercised to be sure both contacts are open in the OFF position; otherwise a hazardous condition may exist. Clean the contacts with a contact file.
6. On some models, check for a damaged or missing button on the end of the operating lever assembly. Take proper corrective action, if necessary.

Toaster will not shut off.

1. Check for fused main switch contacts. If fused, replace switch.
2. Make certain there is enough gap between the contacts in the OFF position. Make proper adjustments, if necessary.

3. Check the condition of the operating lever spring. Make certain it is properly positioned.
4. Check for a loose or improperly positioned compensator bimetal. Take necessary corrective action.
5. Check the timing against the manufacturer's specifications. Adjust controls, if necessary.
6. In some models, check the shunting switch to make certain the contacts close properly for the cool-down part of the cycle. If necessary, increase the contact pressure by bending the upper stationary contact arm down and cleaning it with a contact file. If the contacts are badly damaged from overheating, replace the assembly.
7. In some models, check for a damaged or missing button on the bell crank and for a bent or binding operating-lever pivot. Take necessary corrective action.

Push button will not hold down.

1. Check the latch spring. Make certain it is not rubbing against the frame. If the spring tension is doubtful, replace the spring.
2. Check for a worn or loose roller pin. If necessary, replace the operating-lever assembly.

Toast is too light or too dark.

1. With the toaster completely assembled, check the timing against specifications in the service manual. Adjust as needed.
2. Check the timer for binding parts. Remove bind, if necessary.

Uneven toasting.

1. Excessive discoloration of the reflector or dirt and crumbs accumulated on the crumb tray could be the cause of this problem. If necessary, either clean or replace the reflectors.
2. Be sure the heating elements are operating properly. A difference in the reflection of heat may be caused by aging heating elements. Replace any faulty elements.
3. In some models, bread improperly posi-

tioned on the bread rack will not toast evenly. Make certain the customer is aware of the importance of positioning the bread on the rack of such models and emphasize to the customer the importance of the instruction booklet that comes with the toaster.

4. The type of bread used will quite often be a basis of this complaint. All breads do not toast alike. Those having more moisture, uneven texture, or less milk or sugar content may toast unevenly. Inform the customer of this fact after checking the toaster for other possible troubles.

AUTOMATIC GRILLS

Modern automatic grills are usually combination sandwich grills and waffle bakers; most units have flat grids which are interchangeable with waffle grids. These grids can be used for frying, too.

Servicing of Automatic Grills

The automatic grill is simple both electrically and mechanically. The two plates or grids which bake the waffles or toast the sandwiches (and can be used to fry on) are made of aluminum which heats up fast. Under the bottom grid, and above the top one, are heating elements, usually electrically connected in series. There are also a cordset and a thermostat which are connected in series with the elements; the latter regulates the operating temperature of the grill. A control knob on the thermostat permits the user to select the desired operating temperature or turn off the appliance. In addition to the foregoing parts, some automatic models feature a neon lamp to indicate when the desired temperature has been reached. This indicator lamp along with its series-connected carbon resistor is connected in parallel with the heater elements. (The purpose of the resistor is to drop the line voltage to the proper operating voltage

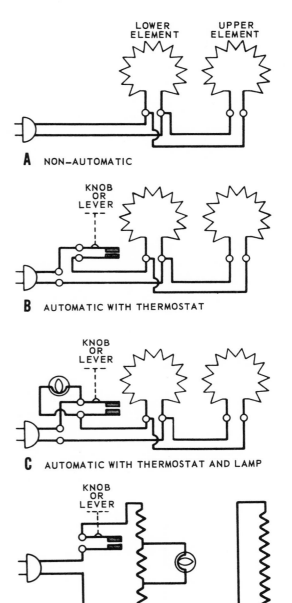

A NON-AUTOMATIC

B AUTOMATIC WITH THERMOSTAT

C AUTOMATIC WITH THERMOSTAT AND LAMP

D AUTOMATIC WITH SERIES ELEMENTS

Figure 6-22. Major parts of a typical automatic grill.

for neon lamps, about 70 V.)

When the control knob is advanced from the OFF position, the contacts of the bimetal thermostat close and the full line voltage is present across the heater elements and the indicator lamp assembly. This is the start of the heating cycle, and it will continue as long as the thermostat contacts remain closed; the elements will produce heat and the neon lamp will stay lit. But when the grill reaches the temperature determined by the control knob setting, the thermostat contacts open the circuit. The indicating lamp will then go out, signaling that the heating cycle is completed. The automatic grill is now ready for more batter and the next heating cycle.

As for mechanical parts, the upper-grid-assembly leveling hinge, though it has a round hinge pin fitted into a round hole in one hinge member as in any other hinge, has an elongated hole in the other member so that the upper grid, when it is closed, will automatically adjust itself as the waffle batter rises. When used as a sandwich toaster, this self-leveling hinge allows the upper grid to rest squarely on the upper side of a sandwich of almost any thickness. The hinge wires which connect the upper and lower grid elements are in some models concealed in the hinge, in which case the hinge is entirely enclosed to protect the wires from mechanical injury. Additional protection for these wires is provided in some makes by the armoring of each insulated conductor with a closed steel spring. Some makes have been designed with the leads outside the hinge, in which case these wires are nearly always armored. When replacement is required, be sure to replace both wires with precisely the right kind, use extreme care to ensure proper placement and thus avoid pinching, and allow enough length that the grids will not be hinge-bound. And, inasmuch as a broken or kinked spring armor is certain to cause a ground in time, it is also a good idea to replace both the armor and the wires whenever either is damaged. Other mechanical parts usually include the heating-unit reflector or baffle, the outer shell, the base, the handles, and the feet which obviously require no detailed description.

Probably the most universal complaint concerning the waffle-maker portion of the automatic grill is not that it will not operate but simply that the waffles stick to the grids. This usually happens because users do not read their instruction booklets, or because these booklets have long since been lost.

Before the waffle maker is used, it should be preheated to a medium temperature and then seasoned as directed in the instruction booklet. Generally the seasoning process involves coating the grids with a thin film of vegetable oil. Then one waffle should be cooked and discarded, as it has absorbed the excess oil. The grids are now conditioned and need no further application of oil. After use, the grids should be wiped clean with a damp cloth but should not be washed in soapy water, as this will remove the oil film. If for any reason the grids must be washed, they must be reseasoned as before.

While most manufacturers employ a form of sheathed heater element, a few still use the common open-coil type stretched taught through nonflammable bushings or around insulating supports. Before replacing an open-coil element, make sure that every insulating support is in good condition and is securely fastened, for a chipped, broken, loose, or missing support will allow the element to sag and later result in a ground. Remember, too, when stretching the new coil to somewhat less than the required length, to stretch it evenly throughout so that no hot spots will be formed. As the new element is threaded through the bushings or around the supports, keep a uniform tension on the coil, also to avoid the forming of hot spots and to prevent sagging. Use the same method of connecting the new element leads as the original unless the manufacturer's service manual offers an optional method for field service. No trimming of element leads for reconnecting is recommended because with two elements connected in series, any shortening of one or the other will result in unequal heat intensity.

If the grids become warped, they must be replaced. Grids which have become severely blackened may be cleaned on the cooking side only with a wire brush, after which they must be reseasoned.

The thermostat keeps the waffle maker at a

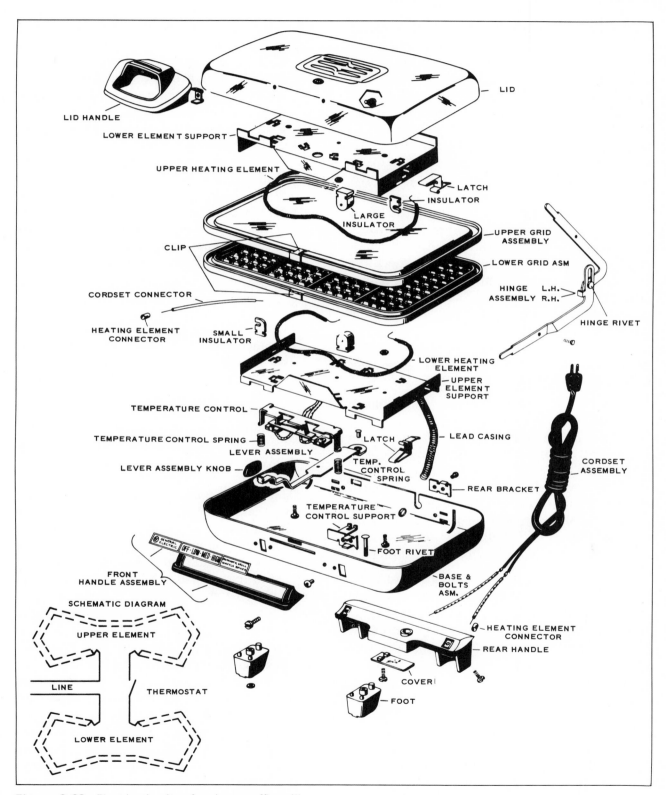

LID HANDLE

LID

LOWER ELEMENT SUPPORT

UPPER HEATING ELEMENT

LATCH
INSULATOR

LARGE
INSULATOR

UPPER GRID
ASSEMBLY

CLIP

LOWER GRID ASM

CORDSET CONNECTOR

HINGE   L.H.
ASSEMBLY   R.H.

HEATING ELEMENT
CONNECTOR

SMALL
INSULATOR

HINGE RIVET

LOWER HEATING
ELEMENT

UPPER
ELEMENT
SUPPORT

TEMPERATURE CONTROL

TEMPERATURE CONTROL SPRING

LATCH

LEAD CASING

CORDSET
ASSEMBLY

LEVER ASSEMBLY

LEVER ASSEMBLY KNOB

TEMP.
CONTROL
SPRING

REAR BRACKET

TEMPERATURE
CONTROL SUPPORT

FOOT RIVET

FRONT
HANDLE ASSEMBLY

GENERAL ELECTRIC   OFF·LOW·MED·HIGH

BASE &
BOLTS
ASM.

HEATING ELEMENT
CONNECTOR

REAR HANDLE

SCHEMATIC DIAGRAM

UPPER ELEMENT

COVER

LINE

THERMOSTAT

FOOT

LOWER ELEMENT

Figure 6-23. Electric circuits of various waffle grills.

predetermined temperature during the heating cycle. The control knob or lever, located at the thermostat, permits various degrees of brownness by settings of LIGHT, MEDIUM, and DARK. (These settings differ with the various models.)

The thermostat can cause problems. It can be stuck open, which can make the appliance inoperative. Loose connections and/or shorts at the thermostat terminals are also frequent sources of trouble. If the thermostat is defective, it should be replaced rather than repaired. But if the thermostat gets out of adjustment or is replaced by a new unit, the temperature settings must be readjusted to ensure the desired temperature. Run a temperature test and check the readings with those recommended in the manufacturer's service manual. The following is a *typical* automatic adjustment test; for an exact test, check the service manual.

Connect the grill through a wattmeter to a controlled 120-V ac source. The grill must conform to the following.

1. The wattage should be 0 with the control knob in the OFF position.
2. The wattage in the HIGH operating setting should be within 5 percent of the nameplate rating.
3. The temperature at the third or fourth cut-in with grids in place and the thermocouple placed in the center of the grids with the top closed should be $500°F \pm 15°F$ when adjusted at HIGH or DARK setting.

The thermostat in most models can be adjusted by turning the calibrating screw accessible through a hole in the bottom of the unit.

Analysis of complaints. While the automatic electrical grill is a simple device, things can go wrong. Here is an analysis of the most common customer complaints.

No heat.

1. Check for defective wall plug, cordset, and for loose connections at heating-element terminals. Repair or replace any faulty components.
2. Check for open heating elements. When ordering a new element, be sure to give make, model, and serial numbers and state whether an upper or lower element is desired. Following this precaution will ensure proper temperatures in the two grids after element replacement.
3. Check the thermostat or temperature control. Replace, if faulty.
4. Check for a defective lamp shunt. In a few models, the signal lamp shunt wire is part of the heating-element circuit. If the shunt wire is open or if its terminal connections are loose, the heating element will not heat, and the signal lamp will be burned out. Repair or replace as the case warrants.

Grill heats up too slowly or is insufficient for satisfactory service.

1. Check for loose terminals and switch-contact points. Replace and repair any faulty components.
2. Check the thermostat and heating elements. Replace, if defective.
3. Be sure the grill is drawing its rated power (usually approximately 1,000 to 1,200 W) from the voltage source. House voltage may be low.

Signal lamp does not light. This generally means that the bulb must be replaced. But remember that in certain models, heat indication is obtained by employing a part of the heating element as a signal lamp. Because the circuit is connected in series, any failure (opening) in the heating elements immediately causes a signal breakdown, in which case a new element may be required.

Grill is too hot. This complaint is usually due to either an incorrectly adjusted thermostat or a defective thermostat or temperature control. Adjust or replace the thermostat or temperature control as needed.

Handles get too hot. Examine the handles and the brackets on which they are mounted to be sure that they are firmly in place and properly offset from the base. Operate the grill at the HIGH setting for half an hour. The handles may be expected to be hot but should not burn the fingers when the grill is lifted. If heat is excessive, check the temperature against the limits specified in the service manual. If the feet burn a table or table covers, check as for excessively hot handles.

Waffles do not brown on top. Check the cold resistance of both heating elements individually. If the elements are within the tolerance specified by the manufacturer in the model's service manual (usually 5 Ω for most grills), examine lead crimp connectors carefully for bad contact. The problem may also be caused by the customer's technique in cooking waffles. Insufficient or too-thin batter or keeping the top open too long after batter is poured will cause the bottom to cook longer than the top. Instructing the customer, as with all appliances, is most important and is a part of the service technician's job.

Waffles brown unevenly. This can be caused by allowing the grids to accumulate too much browned grease in some areas or by batter being too thick to spread evenly over the entire grid.

Waffles stick. Check the grill for operation within the temperature limits specified in the service manual. If the grill does not conform to these limits, replace the thermostat. If the temperature is within the limits, this complaint may also be caused by any of the following:

1. Improperly seasoned grids
2. Grids scoured and not reseasoned
3. Grill opened before waffles are cooked
4. Not enough shortening in the waffle mix
5. Too much sugar in the mix
6. Insufficient preheat time (batter poured before light goes out)

These problems can be cleared up by training the customer, who should be urged to read the instruction booklet.

The so-called *doughnut baker* operates in the same manner as a waffle maker except that there is usually no thermostat. It may, however, have a heat indicator. The doughnut baker has a doughnut-shaped mold or molds, half in the hinged top and half in the base. There are

two heating elements—one in the upper portion and one in the base. Servicing the doughnut baker is the same as for grills except for the omission of the thermostat.

ROASTERS, TABLE OVENS, AND ROTISSERIES

Roasters, rotisseries, and table ovens are electrically operated, portable cooking devices, each having somewhat different characteristics and each suited for a special use. But they are alike in that they operate on the same basic principle.

Servicing Roasters

The electrical parts of a roaster include a group of heating elements, one encircling the sides of the liner, and one or two beneath it; a thermostat—to maintain an even temperature—which may be varied by the user through the control knob to suit the cooking operation; a pilot lamp on most makes which indicates whether the power is on or off; a cordset; and, in some models, a timer. A broiler unit is available for several makes and is used with the roaster body, but may not be connected simultaneously with the roaster elements because of

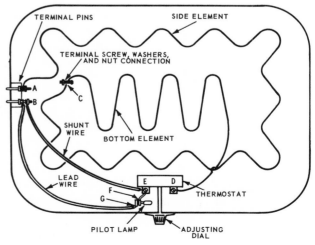

Figure 6-24. The circuit diagram of a typical roaster-broiler.

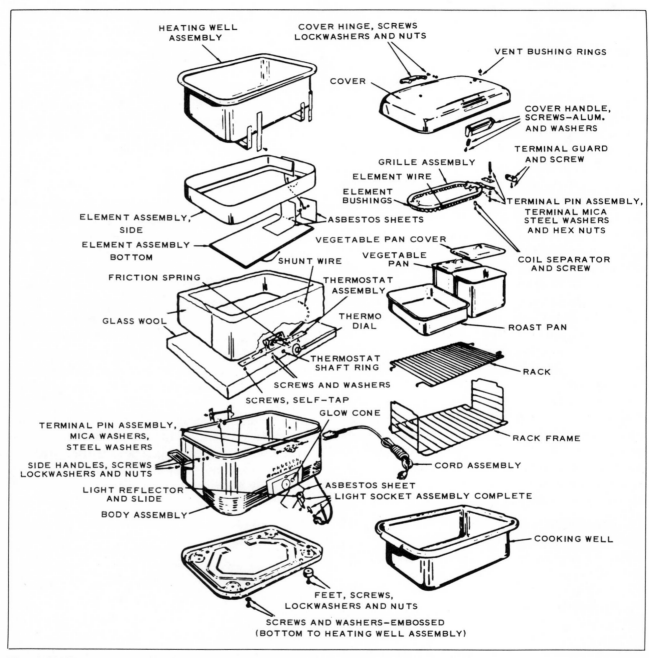

Figure 6-25. The disassembled view of a typical automatic roaster-broiler combination.

the severe overload that the two devices would impose upon an ordinary residential convenience outlet circuit. This accessory is used whenever it is desirable to have the heat source above the food.

The other principal parts of an electrical roaster are the liner, the insulation, a top cover, and an outer jacket. The outer jacket is enough larger than the liner to allow space for fiberglass insulation between the sides of the jacket and the liner and also beneath it. The top cover in some models is unattached; in others, hinged.

Most hinged-cover models feature a linking mechanism and operating knob for opening and closing the cover, making handling of the hot cover unnecessary. The large roasting pan, which fits loosely into the liner, can be easily lifted out for cleaning. The rest of the utensils and racks need no description.

The most common complaints about a roaster, as well as other cooking appliances, is that the unit will not heat, that the user receives a shock when the appliance is touched, and, in certain cases, that the thermostat is inoperative or causes underheating or overheating because of faulty adjustment. In addition, indicator lamps may burn out and require replacement.

With all heavy-current appliances like roasters, it is most important that plugs of cordsets be kept clean and bright at all times. Once they are permitted to get dirty or allowed to corrode, they become very hot and corrode more from arcing. This makes them get hotter and arc still more, until they become defective and the cordsets must be replaced.

Open elements are one of the most common causes of insufficient heat. To learn which element is faulty, try a continuity test. In the case of the side element, it may be fastened to the liner by several methods; in some makes, it is tied in place with asbestos tape; in some others, it is cemented to the liner; another manufacturer provides supporting clips on the liner for this purpose; still another may employ some combination of the foregoing. A few manufacturers recommend that the liner be sent —some, even the entire roaster body—to the factory service station for side-element replacement. Others supply this element separately together with the related parts and materials required for this operation. When confronted with a side-element burnout, therefore, consult the manufacturer's service manual before removing the faulty element to find out what service procedure applies to that make and model. Before replacing a side element, be sure that everything you will need for the job is at hand—such as asbestos tape and waterglass—in order to avoid unnecessary handling of the element, the connecting wires, and the insulation. The specific service manual for the make in hand will tell what miscellaneous supplies are needed for side-element replacement.

The bottom element(s) are relatively simple to replace, for it is laid flat against the bottom of the liner and usually can be reached by merely removing the bottom cover and the lower insulation bat. Be sure to use the installation process recommended by the manufacturer.

All automatic roasters are provided with some type of thermostats or clock-timing devices to control the cooking process. Thermostat-equipped roasters, in most cases, allow the user to choose the desired cooking temperature by turning a dial or sliding a lever on the appliance. This setting provides various spring tensions on the bimetallic blade of the thermostat and in this way prevents the bimetallic blade from going into action until the predetermined temperature has been reached. Timer-equipped roasters employ devices which work in the same manner as range-oven timers (see Chap. 9 in *Refrigeration, Air Conditioning, Range and Oven Servicing*) and permit the user to select a specific temperature and time limit.

While most roaster thermostats are provided with a calibrating device, a few are not. Lacking specific information, the following technique may be helpful as a general procedure for ascertaining the need for adjustment or replacement after a temperature test is performed. If a customer has not complained of inaccurate temperatures, it is most unlikely that it is possible to improve the calibration of the thermostat unless the mean temperature is more than 25 to 35°F above or below that which is called for on the dial. For example, if testing with the dial set at 400°F and the thermocouple meter shows that the shut-off point is actually 460°F and the turn-on point is 380°F, by averaging these two figures, you arrive at a mean temperature of 420°F. As pointed out above, if the customer has not complained of temperature inaccuracy, it is generally safe to presume that the thermostat is all right, for this is not a hair-splitting adjustment. But if the customer had complained—in this instance of overheating —make the necessary adjustment or replace the thermostat, as the case may be. Except for calibrating, do not attempt any other service on thermostats.

Here is an analysis of the most common complaints about roasters.

Roaster fails to heat.

1. Check for defective cordset or plug. Replace, if necessary.
2. Check for loose connections. If any are found, clean and tighten.
3. Check the thermostat for condition and action. If faulty, replace.
4. Check the timer (if used) for condition and action. If faulty, replace.
5. Check the heater elements. Replace if necessary.
6. Check the switch. Replace if necessary.

Roaster gives insufficient heat.

1. Check the fit of the lid. Take necessary corrective action.
2. Check the heating elements. Replace if necessary.
3. Check the thermostat or timer (if used). Replace either or both, if faulty.
4. Check adjustment of thermostat. Readjust if needed.

Roaster fails to broil (in some models).

1. Check for open broiler element. Replace if necessary.
2. Check the temperature control. Adjust, repair, or replace as is needed.

Indicator will not light. Test the lamp and lamp circuit for continuity. Repair or replace as required.

Shocks user. Check for electrical ground. When found, take necessary corrective action.

Servicing Table Ovens

While sold under several names, most people call these appliances table ovens or toaster ovens. These appliances will bake or broil, make toast, thaw and heat frozen foods, and provide many other useful purposes—right on the dining table, if desired.

Some table-oven units follow the same basic principle as those of the horizontal toaster described earlier in this chapter. Let us take a look at a typical model (shown in Fig. 6-21) and see how it operates. For it to act as a toaster, the operating knob, when depressed with the door closed, moves elements of the latch assembly which simultaneously engage a pawl lever "cocking" the toast-timer control; actuate the function switch cutting a resistance shunt out of the bimetal heater circuit; withdraw and hold the bell clapper; rotate a cam which actuates the main switch and depress a pin and spring which provide force upon cutoff. Current flows through the heating elements to initiate toasting and through a resistance ribbon wrapped around the bimetal, heating it gradually to the point where it switches to a "cool-down" contact. The bimetal heater is then deenergized and the cool-down period begins. When cool-down is complete, a contact energizes a solenoid which releases the latch and pulls the bell clapper, ringing the bell. The pawl, being released, allows the force stored in the spring to return all parts to the OFF position.

The operating knob, when lifted to an oven setting with the door closed, moves elements of the latch assembly which simultaneously actuate the main switch and rotate an arm of the oven temperature control to determine cutoff temperature. A small heater in series with the heating elements and mounted on the bimetal bracket transmits heat directly to the bimetal to anticipate the heat received from the oven cavity. This minimizes initial heat-up overshoot when in the oven mode. During oven operation, the temperature control cycles the elements on and off to maintain the selected temperature. Also, the function switch contacts are closed, allowing current to flow through the shunt and the toast timer control bimetal heater, keeping it energized at a low level. This is required to produce normal toast color if the unit is operated for toasting immediately after being used for oven operation.

The door is linked to a slide assembly which

supports the toast rack, moving it forward when open and back when closed. A pin and roller on the slide assembly ensure that the main switch is ON only when the door is fully closed. In addition, when in the toast cycle, operation can be terminated by opening the door to full horizontal position. A projection on the slide triggers the latch and the bell clapper, initiating the shutoff sequence. The signal light is on whenever the main switch is closed.

Some of the newer table ovens work on a different method. Unlike the conventional timed toasters, operation is controlled by a sensing mechanism described as the *detector bimetal assembly.* This mechanism has two bimetals, a detector bimetal and a compensator bimetal, arranged as shown in Fig. 6-26. Since the bimetals are arranged to act in opposition to each other, heat from a specific source is self-canceling. Air temperature effect on the compensator bimetal cancels the air temperature effect on the detector bimetal. The heating element effect on the compensator bimetal cancels the heating element effect on the detector bimetal, leaving only the bread surface effect on the detector bimetal. The following steps are generally the basic operation of one of these units when set on TOAST, OVEN, and BROWN TOP SIDE.

TOAST setting. When the control is set on one of the TOAST settings, the following switch contacts close: top browning switch, oven switch, and detector switch. When the START bar is depressed, the main switch contacts close, and the four heaters are energized. In the meantime, the detector begins to bend in a downward motion and, depending on the TOAST setting, will open the detector switch contacts. When the detector switch opens, the current flows through the trip bimetal. The trip bimetal heats up rapidly because it resists the flow of current and, therefore, bends in an upward motion. This self-heating bimetal pushes upward against the adjusting screw in the latch assembly until it releases the latch arm opening the door and the main switch. The door swings upward exposing the toast.

OVEN setting. When the control is set on one of the OVEN settings, the top browning switch and detector switch are closed. The oven switch remains open, and, therefore, the trip bimetal stays out of the circuit. When the START arm is depressed, the main switch contacts close and the four heaters are energized. As the unit heats up, the detector bimetal bends in a downward motion and will, depending on the oven setting, open the detector switch contacts. The circuit is then open, and the heaters begin to cool down. The detector bimetal next continues to operate as a thermostat, cycling off and on, maintaining a predetermined oven temperature.

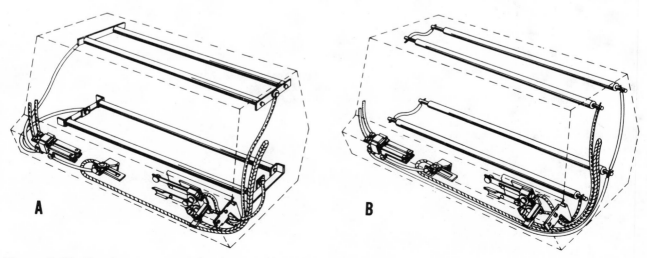

Figure 6-26. Two arrangements of a detector bimetal assemblies.

The unit will continue to cycle until the door is opened manually. When the door is opened, the main switch is also opened.

BROWN TOP SIDE setting. When the control lever is set at BROWN TOP SIDE, the top browning switch is open and the detector switch contacts remain closed. When the START arm is depressed, the main switch contacts close and the two upper heaters only are energized. The two upper heaters will remain energized until the door is opened manually.

Analysis of complaints. While the problems discussed previously for the various toaster and grill appliances are usually encountered, here are some additional complaints often received about table ovens or toaster ovens.

Oven does not heat, light out. Check the cordset and main switch for continuity. Examine the bus wires for open welds. Be sure that the actuator spring from the latch assembly to the main switch is in place and is in good condition. Be certain that the main switch is firmly supported in its bracket. Also make sure that the main switch contacts are not contaminated or open because of low contact pressure. If contaminated, clean the contacts with a contact file. Check the contact pressure with a gram testing gauge. Replace any faulty parts.

Oven does not heat, light on. Check the welds. Check the heating elements and the oven temperature control for continuity. Repair or replace any faulty parts.

Top or bottom element does not heat. Check the welds. Check the elements for continuity. If top elements are out, check the anticipator heater (if used). Repair or replace any defective parts.

Oven temperature too high or too low. Check the adjustment of the heat control or thermostat. Readjust or replace as is necessary.

Unit operates correctly on TOAST *but not on* OVEN. Check the oven temperature control for operation and calibration. Be sure that the adjusting screw has not been turned by the customer. Readjust or replace as is necessary.

Unit operates correctly on OVEN *but not on*

TOAST. Check the welds. Check the toast-timer control for operation and calibration. Be sure that the function switch is open in the TOAST position only; clean the contacts and adjust if required.

Unit operates correctly, but light is out. Check the lamp and resistor. Check the welds. Repair or replace as is necessary.

Operating lever will not latch down. Check the latch mechanism, pawl, bell clapper, solenoid, springs, and pivots for free and proper action. Preferably compare them with an assembly known to be good. Look for missing or broken pins, springs, and C or hairpin clips, and for rusted or bent parts. If you hear the solenoid buzz, check the calibration of the toast color control. Check for a missing spring inside the solenoid. Take whatever corrective measures are necessary.

Toast too light or too dark. Check the adjustment according to the service manual. Readjust or replace the temperature control or thermostat as needed.

Bread toasts on top side only on TOAST *setting.*

1. The top browning switch contacts may be contaminated or open because of poor contact pressure. If contaminated, clean the contacts with a contact file. Check the contact pressure with a gram gauge.
2. Check the top browning switch for damage. Replace if necessary.
3. Check the lower heater elements for continuity. They may be open because of damage. Replace if necessary.

Bread toasts unevenly.

1. Check each heater element for continuity. Replace if necessary.
2. While the unit is operating on the TOAST or OVEN setting, visually check each heater element for even glow. If an element(s) is defective, replace.

Door will not latch.

1. Check the latch spring; it may be disengaged or broken.
2. In some models, check the trip bimetal adjustment. The screw may be turned in too far. Adjust as necessary according to the service manual.
3. Check for a bent door catch. Correct the problem.
4. Check for a faulty lead dressing around the latch mechanism. Take proper corrective action.
5. There is a remote possibility in some models that the latch arm assembly may be engaged by the latch assembly while the door is open. If this is so, the door will not close. To correct, simply trip the OPEN bar (interrupt arm) before closing the door.

Door swings open almost immediately on TOAST *setting.*

1. In some models check the trip bimetal adjustment; the adjusting screw may be turned in too far.
2. In some models check the detector switch contacts for contamination or low contact pressure. If contaminated, clean the contacts with a contact file. Check the contact pressure with a gram gauge. If the detector contact pressure is not within limits, replace the detector switch assembly.

Door will not swing open on TOAST *setting.*

1. The oven switch contacts may be contaminated or they may remain open because of poor contact pressure. If contaminated, clean the contacts with a contact file. Check the contact pressure with a gram gauge.
2. In some models, check the trip bimetal adjustment. The adjusting screw may be turned out too far.
3. In some models check the detector switch contacts. They may be fused. If so, replace the detector switch assembly.
4. Check for binding parts such as the door

hinges, top of door rubbing against shell, detector push rod, and latch arm. Take proper corrective action.

Door swings open on OVEN *setting.* Check the oven switch. The contacts may be closed on the OVEN setting causing the door to swing open because the trip bimetal is in the circuit.

Servicing Rotisseries

The portable electrical rotisseries on the market today differ widely in the number of optional cooking features they offer. Many basic models provide for only rotary broiling and contain just a heating element and a motor for driving a spit assembly. The more sophisticated models, on the other hand, feature several cooking methods in one appliance. For example, a whole chicken—or any other piece of meat which will fit into the cooking compartment—may be impaled on the spit where it will be rotated automatically under the broiler heating unit. On several models the top is so designed that it can be utilized for surface cooking or as a warming compartment. Or, if the rotisserie is so equipped, it can be converted into a roaster by inserting an auxiliary bake unit in the bottom of the cooking compartment.

All rotisseries have two basic electric operating parts: (1) the broiler heat unit which, being in the ceiling of the cooking compartment of some

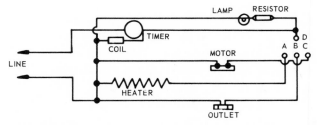

Figure 6-27. Typical push-button type of rotisserie-broiler circuit.

Push button	Terminals connected	Elements in circuit
Broil	D to A	Heater
Rotisserie	D to A and C	Heater and Motor
Outlet	D to B	Outlet

oven-type units, serves also to heat the pan or the warming compartment on the top side of the appliance, and (2) the motor which turns the spit.

The type and quantity of control and signaling devices used with a rotisserie depend upon the make and model. They include a pilot lamp to indicate when the rotisserie is operating, a buzzer to inform the user when the cooking cycle has been completed, a heat control, a spit-motor switch, and a timer. A receptacle to which a bake unit can be connected is provided on some models. As a safeguard against overloading the wiring of the rotisserie as well as of the house circuit, which would occur if the bake and the broiler units were connected simultaneously, the bake-unit receptacle is usually controlled by a double-throw switch or push-button arrangement which will energize the broiler unit in one position and the bake unit in the other.

If the rotisserie is equipped with this adjunct, the bake unit may be tested separately with the prods, after which it can be connected to the bake-unit receptacle and the unit tested for actual heating through the controls. Make sure that the plug fits snugly into the receptacle to ensure a good electric connection. Also, check the double-throw switch to be certain that its contacts close tightly in both positions.

The heat control used in some models is a type of infinite-control switch. This device controls the temperature of the cooking compartment by periodically interrupting the flow of electric energy to the heating element. The duration of these interruptions may be varied by the user through the control dial. If the dial is set at one-quarter of full heat, the current will flow to the element 15 s out of every minute; at one-half heat, 30 s out of each minute; at full heat, the current will flow uninterrupted. Hence, an infinite number of variations in heat intensity are available between the high and the low extremes.

Oversimplified for the sake of clarity, here is how this type of switch operates. A cam within the switch mechanism is rotated at a constant speed by a motor. (In at least one make the spit motor is used for this purpose.) Built into the switch is a set of normally closed contacts which may be shifted toward or away from the cam by turning the control dial. Now when the dial is set at the highest heat, the contacts will be just out of reach of the cam so that current will flow uninterruptedly. But if the dial is turned to, say, a medium heat, the contacts will be thereby moved into a position where the cam, as it rotates, will open them and hold them open for about half a revolution. If the dial is set at one-quarter heat, the cam will open the contacts and hold them open for three-quarters of a revolution, and so on. But unless the manufacturer's service manual gives specific instructions for adjusting any type of infinite-control switch, do not attempt service on this control. With this exception, therefore, it is more economical in the long run to replace the control if it is faulty than to repair it.

Simple two- or three-heat control in uniform steps is accomplished with a common two- or three-heat switch used in conjunction with a two-element broiler unit. In a broiler unit consisting of two elements of equal wattage, up to three heats may be obtained by switching. High heat is delivered when both elements receive their full-line voltage; medium (or one-half) heat when one of the elements receives its full-line voltage; and low (or one-quarter) heat when the two elements are connected in series.

The simple two- or three-heat manual-control switch used in conjunction with two 120-V elements presents no problem. If the switch is faulty, renew it.

The purpose of the timer assembly, of course, is to turn automatically all parts of the rotisserie at the end of a predetermined time. Because this subassembly is more often replaced than repaired, no detailed discussion of its working parts is called for here. That is, whether electric, spring-motored, or powered by the spit motor, the timer should be replaced if it is faulty unless the manufacturer's manual gives directions for making adjustments. When the timer must be replaced, consult first the service manual or the jobber to find out whether an exchange plan is offered. A point of interest, however, is that the timer in some models is powered by the spit motor, some of which are controlled by an ON-OFF switch. Remember, therefore, when servicing a rotisserie of this type that, if the spit-motor switch is turned off, neither the spit not the timer will operate. Other manufacturers employ an independent timer.

The principal mechanical parts of a rotisserie include the connecting shafts, the couplings, and the gear train through which motion is conveyed from the motor to the spit (and also to the controls in some models) at a suitable speed. As a rule, the gear train reduces the armature speed to 2 to 8 r/min.

The motor employed to drive the spit assembly is usually of the shaded-pole type. Except for minor adjustments, it is usually more economical to renew a faulty motor than to repair it.

When servicing the motor and gear assembly, keep in mind that free rotation of all parts is essential and that any linking or connecting shafts must be precisely aligned with their couplings without abnormal end play. Excessive end play, particularly in a "floating" shaft, will allow it to work away from its coupling socket at one end or the other, with the result that the coupling's indexing member will be "chewed" away gradually until the coupling and/or the shaft end is ruined.

In replacing any of these mechanical parts, therefore, look for the original cause of the trouble and be sure to eliminate that at the same time. For example, if a damaged coupling or shaft is discovered, try to find out why the part failed. It is possible that the shaft and coupling engagement is too shallow, in which case the other parts must be realigned to effect a deeper engagement of the indexing members. In the same inquiring manner, look for the cause of gear damage before installing new gears.

Sheathed heating elements in a rotisserie (the most commonly used) are relatively simple to replace; unless the manufacturer recommends a revised procedure for replacement, merely observe how the original unit was installed and

follow the same method. Open-coil elements, however, require more care in handling and installation. First of all, make sure that the insulating supports for the element are intact and are securely fastened. Then stretch the coil evenly throughout its entire length to somewhat less than the required measure so that as it is threaded into place the coil can be kept under slight tension. When properly installed, the coil should be uniformly stretched and taut from end to end. If the turns of the coil are more closely gathered in one place than in another, hot spots will occur at these points and uneven heating will result. It is worth a little extra time, therefore, to work with extreme patience in order to do a good job. No attempt, as previously mentioned, should be made to splice a heating element.

Analysis of complaints. In addition to the servicing information given previously for roasters and table ovens, here are some specific complaints often received about rotisserie units.

No heat, indicator lamp does not light, motor(s) does not run.

1. Check for a defective cordset. If a continuity test shows the cordset to be faulty, replace it.
2. Be sure that the timer switch contacts are not stuck in an OPEN position.

No heat at any setting but indicator lamp lights.

1. Check for an open heating element. If the element is bad, replace it.
2. Check for a defective selector switch. If faulty, replace it.

No heat at any setting, indicator lamp does not light, but timer motor runs. Check the thermostat contact for an open contact. If found, replace the thermostat.

Insufficient heat.

1. Check the fit of the lid or cover on the deck. There should be no noticeable gap.
2. Check the wattage on BAKE-ROAST and on BROIL. If in either setting it is more than 5 percent below the nameplate rating, a high

resistance in one or more of the connections is indicated.

3. Check the thermostat as directed in the service manual and correct the setting if necessary.
4. Make sure that the customer is not using the meat thermometer temperature given in the user's booklet for the thermostat settings.
5. If nothing is wrong with the appliance or the manner in which it is used, the trouble may be very low voltage due to low line voltage in the customer's area, inadequate wiring in the customer's home, or a defective outlet (hot cord plug). A comparison of heat-up time in the shop and in the customer's home should reveal any of these conditions.

Indicator lamp and heat remain on, regardless of thermostat setting. Check the calibration of the thermostat. Take corrective action as suggested in the manufacturer's service manual.

Unit gets too hot. Check the adjustment of the thermostat. Readjust as directed in the service manual or, if the thermostat is faulty, replace.

Spit motor does not run, indicator lamp lights.

1. Check the spit motor. If defective, replace.
2. Check the selector switch. If defective, replace.

Broiling complaints. Check the wattage of the upper heater circuit. If it is within tolerance and low voltage at the customer's home is not indicated, make sure the customer is using the correct broiling techniques including the following:

1. Shelf in top position with offset up
2. Temperature control set at BROIL
3. Proper button depressed
4. Timer set at STAYS ON
5. Lid supported on the clip at the front
6. The reflector and glass door in place
7. The broiler rack positioned on the recommended shelf
8. The automatic timer set at STAYS ON
9. The BROIL button depressed

10. The unit preheated for 10 min

Unit is noisy. Any geared motor produces noise, but when the load on the spit is in reasonable balance, the noise level should not be objectionable. Although there is no satisfactory method of defining an acceptable noise limit, experience with the appliance will soon enable you to exercise sound judgment. Use petroleum jelly to eliminate any squeak in the split slot. If the window is loose, putting a spring clip—usually available from the manufacturer—between the window and door frame will generally eliminate the rattle.

OTHER COOKING APPLIANCES

Electrical fry pans, warming trays, sauce pans, egg cookers, potato bakers, woks, bean pots, casseroles, deep- or French-fryers, snack keepers, skillets, trivets, bun warmers, pizza bakers, gourmet chef pans, chafing dishes, fondue pots, kabob grills, and similar appliances have a heating element permanently built into the body of the unit. The temperature of the unit is controlled by a thermostat which may be built into the pot or body, or it may be a part of the removable plug or handle. In the latter case, a watertight seal around the electric connections on the appliance makes it possible to immerse the unit completely in water once the plug or handle has been removed. Always check the user's instruction booklet to be sure that the appliance can be immersed.

The wiring diagram in Fig. 6-28 is typical for many of the cooking appliances mentioned previously at the beginning of this section. With the unit plugged into the 120-V ac outlet and the control turned on, the thermostat contacts close and the appliance starts to heat. When the appliance reaches the temperature indicated on the control setting, the thermostat contacts open. Up until this time the neon lamp has been on. When the thermostat contacts open, the neon

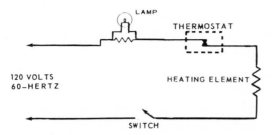

Figure 6-28. Typical schematic diagram for many cooking appliances.

lamp also goes out and will so remain until the contacts are closed again. In other words, the lamp indicates when current is flowing through the heating element.

In almost all these cooking appliances, the heating element, as previously stated, is permanently built into the cooking portion of the unit. The element is not accessible for servicing, and the entire body or at least its base portion must be replaced when any kind of element trouble occurs. Thus, the only repairs that can be done to most of these appliances are to the cordset/plug, control unit, and thermostat (in some appliances this cannot be serviced, either).

If the temperature of the cooking appliance is too high or too low for proper food preparation at the recommended thermostat or heating-probe setting, check the tmperature at that setting with a thermometer. Either a thermocouple or a less expensive liquid heat thermometer can be employed. When using a thermocouple, place its junction, the sensing element, on the bottom of the pan, running the wires over the rim of the pan to an indicating meter (see p. 138). The liquid- or oven-type thermometer is simply placed on the bottom of the pan with the indicator up. If a lid is available, it should be kept on the appliance except when reading the thermometer.

To set the temperature control adjustment on most cooking type appliances that have built-in control units, turn the control knob to the OFF position, loosen the set screw, and pull the control knob off the shaft. Turn the control shaft until the pilot lamp lights. With the control shaft in this position, replace the control knob on the shaft with the pointer at the lowest

mark on the scale. Place the thermometer in the center of the appliance, close the lid, and turn the control knob to the highest point on the scale. This point varies with make and model of the cooking appliance and may be 400, 420, or 450°F. Watch for the pilot lamp to go out; when it does, read the thermometer immediately. The thermometer should indicate a temperature about 20°F higher than the temperature indicated on the control dial. As an example, if the dial is set at 400°F, the thermometer should indicate about 420°F when the pilot lamp goes out to indicate that the thermostat has shut off the current. If the temperature control dial reading is not 20°F less than the thermometer reading, pull the knob off the shaft and reset it on the shaft so that it will read 20°F less than the thermometer. Then turn the control to the highest position again.

Let the appliance cool off until the pilot lamp lights again to indicate the thermostat has closed. Now read the thermometer again. This cut-in temperature should be about the same as that indicated by the control knob. If it is not, reset the control knob on the shaft to agree with the thermometer reading. Allow the appliance to cycle several times, checking the thermometer each time the current goes on or off. The average cut-in temperature should be the temperature indicated on the control dial. The cut-out temperature should average about 20°F higher.

If the appliance does not have a pilot lamp, you can use the test aid and an ammeter. The ammeter is plugged into one receptacle section and the appliance into the other. The ammeter needle will indicate when the unit cuts in and draws current.

Analysis of complaints. Because of similarity of the servicing problems in the appliances mentioned, a general analysis of complaints received by service technicians is given first, and the problems of specific appliances follow this.

No heat, indicator lamp does not light.

1. Check for open circuit in the cordset. If test shows no continuity, replace cordset.

2. Check the thermostat for contacts stuck open. This can be accomplished by connecting a jumper wire across the thermostat contacts and applying full power to the appliance. If the thermostat is defective, the indicator lamp will light and the appliance will begin to warm up. In most instances of a defective thermostat, it is necessary to replace the entire probe or handle unit.

No heat, indicator lamp lights. This is a good indication that the heating element is open. To confirm this, check the continuity of the element. If the heating element is open, it seldom can be repaired in cooking appliances of this type. If the probe and cord are good but the appliance still will not heat, more than likely the heating element is open. But should continuity be obtained through the heating element, it is possible that the connecting terminals between the heating element and the probe are bad, and these can be changed on some units by screwing off the old ones and replacing with new ones. In other units this is impossible. In a few models, the base contains the heating element and can be removed from the body of the unit and replaced or exchanged.

Temperature Incorrect. Complaints concerning temperature such as the heat and indicator lamp not cycling off at the selected operating temperatures usually involve the probe or thermostat. This can be checked as previously described, or here is another method of testing a probe or the thermostat temperature.

1. Fill the appliance half full of water.
2. Set the control at full ON (420°F).
3. Wait 5 min after water comes to a rolling boil.
4. Turn the control knob down until the light goes out.
5. If indicated temperature at this point is between 180 and 220°F, the control is satisfactory. If not, replace.

For a more accurate check, appliances with probe controls can be tested dry by the use of a thermocouple test set. Here is a typical electric cooking appliance temperature check using a thermocouple.

1. Insert the probe and connect to a regulated 120-V ac supply.
2. Place the thermocouple in the center of the appliance pan or container in a normal position.
3. The wattmeter should read 0 when the control dial is at OFF and the unit is at room temperature.
4. The wattmeter should read the rated wattage ±5 percent on the operating area of the dial.
5. Turn the control dial to exactly 420°F.
6. Indicated temperature should be from 400 to 440°F, taking the average of the third, fourth, and fifth cut-in temperatures.
7. If not within limits, replace or exchange the probe or thermostat.

Unit overheats. See if the thermostat contacts are fused together. Also check the adjustment of the thermostat. While most thermostats cannot be recalibrated, a few can be; check the service manual. If the thermostat is faulty, replace it.

Insufficient or low heat. Be sure there are no loose terminal connections. If the terminals are dirty, pitted, or worn, they should be cleaned or replaced as the situation warrants. Also check the contact points on the thermostat. If they are not making proper connection, they can act as a high resistance which will reduce the current flow through the heating element. If the thermostat is faulty, it should be replaced.

Neon bulb does not light, but appliance heats. Check the resistor in series with the neon bulb with an ohmmeter. With most units, it should read between 20,000 and 85,000 Ω. If the resistor is good, check the lamp. Also make sure that the lamp leads are not touching, which would short the circuit and keep the neon lamp from operating. If the bulb or resistor is faulty, replace it. On appliances with temperature controls and a thermostat in the handle, a broken seal will permit moisture that could cause a short to enter the handle. Replace the seal if damaged.

Utensil body is dented. Small dents in an aluminum utensil body can frequently be hammered out with a leather or plastic mallet, with extreme care. Since aluminum bends fairly easily, do not hit it too hard.

Food Cookers

These units are usually employed both to cook food and to keep it warm. Their operation is simple. After a specific amount of water is put into the base pan, the control lever is pushed to the ON position. This closes the control contacts which are kept closed by the magnet holding to the keeper plate which is a special alloy. When the water is boiled out or evaporated, the keeper plate loses its magnetic qualities and releases the magnet, causing the control to break contact and snap back into the OFF position, striking the bell. Here are some of the complaints most often encountered.

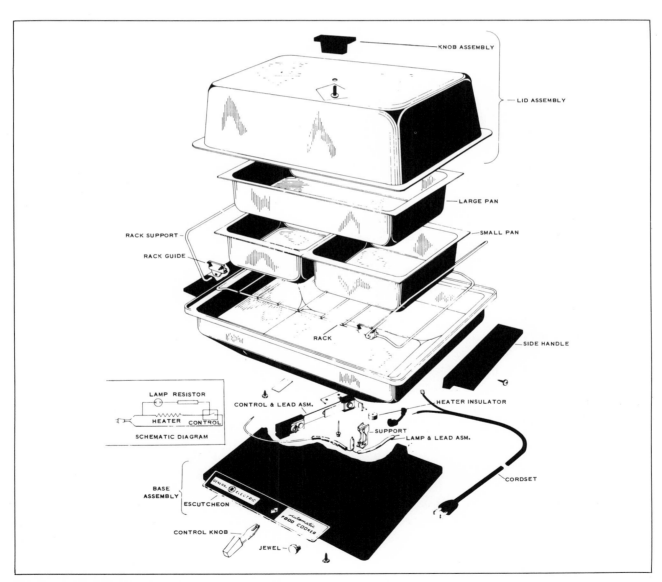

Figure 6-29. Disassembled view of a typical food cooker.

Appliance will not stay on.

1. Be sure the magnet is clean and check for metal chips on the magnet. Clean as necessary.
2. Check for nonmagnetized magnet. Magnet should hold to metal scale or contact file. Replace if necessary.
3. Be certain that the lead dress is proper so as not to impede movement of the control. Take necessary corrective action.

No contact.

1. Be sure that the blades in the control are not bent. Replace control, if required.
2. Be sure that there is no film on the contacts of the control. Clean with a contact file, if necessary.

Control binds. Check for a bent baseplate or improper lead dress. Straighten the baseplate or dress the leads according to the wiring layout in the service manual.

Slow Cookers

Of all food type cookers, the slow cooking types are presently the most popular. These unique cookers, in their insulated cases, use very low wattages (70 to 80 W on the LOW switch position—140 to 160 W on HIGH). Also, they are easy to service since about all that can go wrong with them are a burned-out heating element, a faulty control switch, or a defective cordset. That is, test the appliance for opens, shorts, and ground and when the trouble is found, take the appropriate action. Generally, if the heating elements are faulty—either the HIGH, LOW, or both elements—the entire base assembly (the elements are usually built into it) must be replaced.

Skillets or Frypans

Skillets, or frypans as they are often called, are subject to the same troubles as already described on p. 181. However, because they receive more constant use and are exposed to more moisture than other appliances, we will look at them in more detail. The most likely

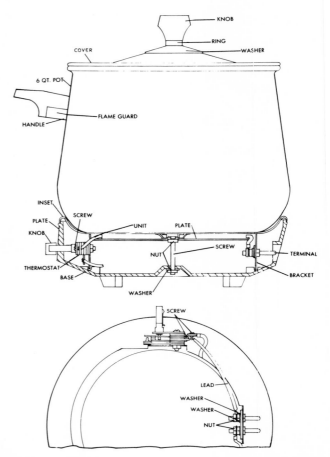

Figure 6-30. Major parts of a typical "slow" food cooker.

parts to cause problems are those that have mechanical action (thermostat and control mechanism) and those that moisture will affect.

The following are the most common complaints received on the operation of a skillet or frypan:

Appliance does not heat or does not get hot enough.

1. Check heating element. If faulty, replace the entire pan body and element assembly. This is rather costly, and often the customer will prefer to purchase a new skillet, since it would cost only slightly more than the repairs on the old one.
2. Check for an open circuit in cordset. If faulty, replace or repair.

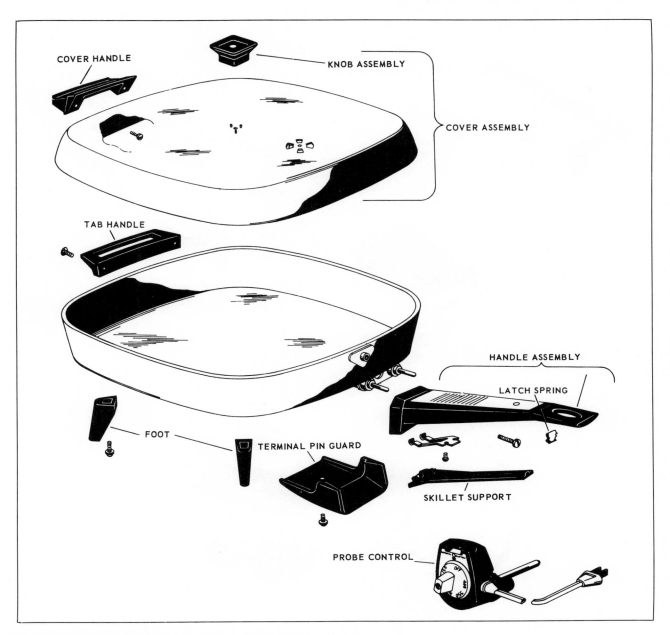

Figure 6-31. Disassembled view of a modern skillet or frypan.

3. Check the terminals for loose connections. Tighten, if necessary.
4. Check the condition of the contact points in the thermostatic control unit. Clean the points and adjust their lineup if necessary.
5. Check the thermostatic control unit. Adjust if necessary or replace if faulty.

Skillet will not shut off automatically or overheats.

1. Check to determine if the contact points in the thermostatic control are stuck together. If the points are fused, replace thermostat assembly.

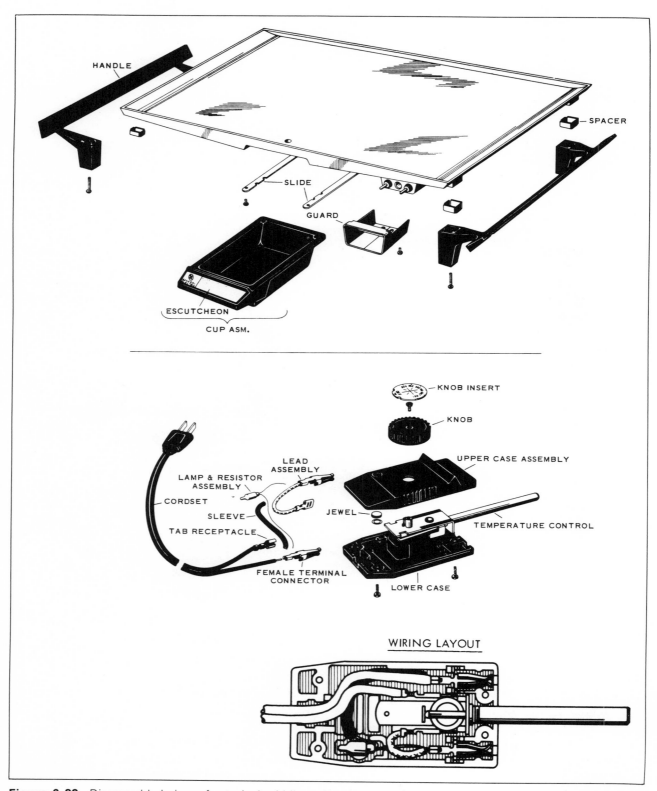

Figure 6-32. Disassembled view of a typical griddle and probe control.

2. Check thermostatic adjustment. Reset heat adjustment if necessary.

Temperature is too high or too low. Check the temperature at the various settings with a thermometer as described on p. 181. Readjust the control or replace as necessary.

Neon lamp does not light; skillet heats. Check for burned-out bulb or a shorted resistor. Replace either unit if required. When replacing the resistor in some units, it is necessary to coat the resistor and its connections with special insulating varnish (available at any wholesale appliance supply house) to make the joints waterproof. Check the service manual for full details.

Electric saucepans are somewhat similar in construction and operation to electric skillets. Both are thermostatically controlled appliances, and both have heating elements cast or welded in the base. Actually, the only major difference

is the shape of the pan. Electrically and mechanically, most modern saucepan appliances are subject to the same troubles as the skillet.

Deep-Fat Fryers

Deep-fat fryers, which are used to cook chicken, fish, French fried potatoes, onion rings, doughnuts, and other fried foods, are similar to skillets and saucepans except that the heat is usually supplied to the sides instead of the bottom.

Complaints from the customer on this appliance will be that the unit does not operate, or the temperature is too high or too low. For nonoperation, make the usual checks on the detachable cord before attempting to disassemble the fryer. Since most units have a pilot light, check lights when the fryer is plugged in. If the lamp does not light, there is an open circuit in the fryer or the lamp itself is burned out. In any

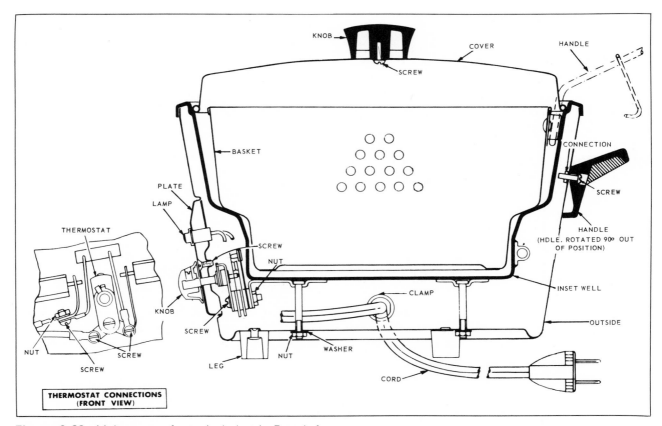

Figure 6-33. Major parts of a typical electric French fryer.

case, the fryer must be disassembled to find the fault.

Incorrect operating temperatures will be caused by either a defective or improperly adjusted thermostat. Readjust the control or replace if faulty.

Popcorn Poppers

Most popcorn poppers have an open-wire heating element mounted on ceramic insulation blocks and a standard cordset to supply power. If defective, either can easily be replaced. While many of the older popcorn poppers did not have a thermostat, most modern ones do. That is, their heat is either regulated or set so that just the correct amount is obtained for popping the corn without having to shake or stir it. In most electric popcorn poppers, the thermostat remains closed until the temperature of the pan reaches about 460°F. When this temperature is reached, the thermostat opens to prevent the corn from burning.

Kettles

Most electrical kettles have a sheathed heating element in series with an automatic cutoff switch. When the cordset is attached to the kettle and plugged into a 120-V ac outlet, the water heats up and commences to boil. If the kettle boils dry or is plugged in without water, the bimetal in the cutoff mechanism is activated because of the excessive heat and releases the cutoff switch, opening the contacts. If the cutoff switch opens, the kettle must be cooled off and the reset lever in the base must be pushed in the direction of the arrow to close the switch.

Some kettles are fully automatic; in them an automatic control switch is activated by steam temperature inside the dome. A neon lamp, usually located in the top of the handle, glows as soon as the kettle automatically cuts back from about 1,500 to 300 W. A slide switch is conveniently located in the top of the handle so that the kettle may be operated on continuous fast boil. With the kettle on the automatic setting, when the water reaches a boil, the switch bimetal opens the switch controls so that

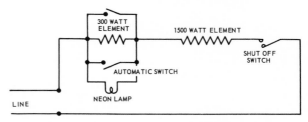

Figure 6-34. Schematic diagram of an automatic kettle. When the automatic vapor control switch is open, the two elements are in series and the combined wattage of the circuit is 300 W at 120 V.

the two elements are in series and the combined wattage of about 300 W maintains an even, slow boil. Should the kettle boil dry or be energized without water, it is protected by a cutoff switch in the base which disconnects both elements from the circuit. A slide button is provided in the base to reset this switch after the kettle has cooled or water has been added.

Here are the major complaints received and how to proceed to analyze them.

Water does not heat up.

1. Make sure the reset lever is in the CLOSED position.
2. Check the cutoff switch controls for contamination and/or lack of contact pressure. Take whatever corrective action is necessary.
3. Check the continuity of the cordset and heating elements. Replace if faulty.
4. Check the operation of the cutoff switch. Replace if faulty.
5. In some models, check continuity of manual slide switch and control switch to ensure that contacts are closed. Replace, if faulty.

Kettle cuts off before water boils or during boiling. Check operation of the cutoff switch. Replace, if faulty.

Kettle heats slowly. This is usually caused by a heavy accumulation of scale around the heater element. Excessive scale reduces the efficiency of the element because it acts as a heat insulator; therefore, more time and more power are re-

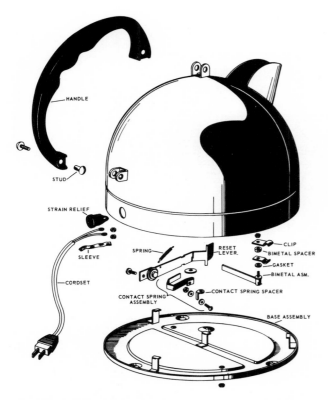

Figure 6-35. Disassembled view of an automatic kettle.

quired to boil water. Remove scale with a recommended scale remover.

Kettle leaks. Most minor leaks can be corrected by soldering. If the leak is difficult to locate, remove the base cover. Put a little water in the kettle, hold it over a clean surface, and plug the unit into a 120-V ac outlet. Do not rest the bottom of the kettle on the surface. Water on the clean surface will indicate the location of the leak. To repair a leak, use 50/50 solder and a proper flux. Use a suitable source of heat—propane, gas, or portable acetylene torch. Silver solder must be used to attach the bus wires to the heater element units. Protect the element end seal and bottom of the kettle from excessive heat by means of a brazing shield.

Kettle does not cutoff (damaged by overheat). Each kettle is tested for cutoff before it leaves the factory. Therefore, failure of the cut off mechanism is extremely rare unless it has been

tampered with inadvertently by the user or someone not familiar with the proper service techniques, or placed over direct heat. When servicing a kettle that has been damaged by overheating and the complaint is that it fails to cut off, turn the unit over and visually check the base cover for discoloration. If discoloration is evident in the form of a circular pattern, it usually indicates that the kettle was left standing too long on a lighted gas or electric range element. The resultant heat given off is enough to destroy or change the characteristics of the bimetal in the cutoff mechanism. In many cases, the product will operate normally for a long period until the kettle is either plugged in dry or is allowed to boil dry. The damaged bimetal then fails to function.

No pilot light. Check operation of kettle by boiling 1 qt water with the automatic control switch in the AUTO position. The wattage should drop from 1,500 to 300 W when the boiling point is reached. If there is no wattage drop, check the automatic control switch boil adjustment. If kettle operates normally, check the pilot lamp connection in the handle. If connections are proper, replace harness assembly.

To perform the cutoff test, check the specification given in the manufacturer's service manual. A typical set of specifications reads as follows.

Specification: At 90 V alternating current the kettle must cut off in 17 to 27 s from the time it is emptied of boiling water.

Procedure

1. Pour approximately 1 qt cold water into the kettle.
2. Plug the unit into a voltage-controlled ac outlet with an ammeter in the circuit.
3. Allow the water to boil for 2 min. *Note:* The kettle must not cut off before or while the water is boiling.
4. With the voltage set at 90 V, pour the boiling water from the kettle. Note the exact time the kettle is empty.
5. Check the time the kettle cuts off as indicated when the ammeter needle drops to zero. Do not leave the kettle plugged in dry longer than 50 s.

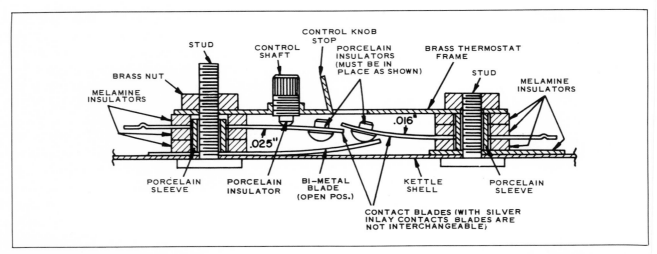

Figure 6-36. A typical kettle automatic control switch assembly.   *Note:* Correctly position bimetal blade with brass side down.

Note: If the kettle has cut off within the proper time, i.e., 17 to 27 s, reset it before returning it to the customer. If the kettle does not meet these specifications, adjust the cutoff switch.

Table Ranges

The tabletop electrical hot plate, now usually called a table range, has been a popular kitchen appliance for a long time. Today, most hot plates are used as an auxiliary to the kitchen range or as heating source of such appliances as fondue cookers, oriental woks, omelette cookers, chafing dishes, bean pots, and so forth.

The basic table range consists of one or two heater elements, a neon indicator lamp, a cordset, and usually a thermostat. While most table units have thermostat controls—some with as many as 7 to 10 positions—a few models have a single three-position switch reading OFF, MEDIUM, HIGH or two separate switches marked MEDIUM and HIGH. There are two separate heating elements in these appliances; one is brought into play at MEDIUM and both at HIGH. It is entirely possible for a table range to operate at one heat but not the other, since the elements can burn out independently. In a thermostat-controlled unit an open element means no heat regardless of the setting of the control.

While at one time open-coil elements were

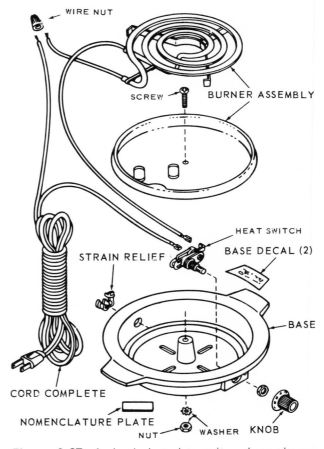

Figure 6-37. A simple hot plate unit such as shown here may be used in conjunction with such appliances as fondue cookers, oriental woks, omelette cookers, and chafing dishes.

most widely used, today the sheathed or rod type is popular. As previously mentioned, the open-coil-type elements are usually wound on porcelain forms called bricks. These bricks are molded with circular grooves in them to hold the coiled resistance wire and support it so it cannot sag when heated. The bricks are rather fragile and are often cracked or chipped if liquids spill on them when they are hot, or if something is dropped on them. There is a cement that can be used for brick repair, but it is usually much better to replace the brick than to try to repair it. Bricks are inexpensive and the job looks much more professional when you return it to your customer with a new brick in it rather than a patched-up repair job.

In the sheathed-type element, the Nichrome resistance wire is first imbedded in a refractory material, such as magnesium oxide, and then encased in a metal tube. The refractory material is a good electric insulator and a good conductor of heat. Thus, it prevents the resistance wire from shorting and readily conducts the heat to the metal tube. As with the other appliances using sheathed elements, they cannot be repaired and the entire element must be replaced if it is faulty.

The servicing of table ranges is simple. Since there are only three basic parts in this appliance, check for opens and shorts in these three areas and the trouble will be quickly found. As in other appliances, the most likely places for opens in table ranges are at or near the connections. A visual examination will usually help in finding the exact location because the open-circuited part will either be loose or badly discolored due to overheating. Another place where trouble can develop, if the table range is moved around a great deal, is in the cord—either where it enters the table range or at the plug. If the trouble is not immediately apparent, check the continuity of the table range with an ohmmeter. First, with the switch turned on, test across the prongs of the plug. The meter will indicate infinite resistance in case of an open circuit whereas a short would cause a zero reading. Next, test for continuity from each prong to the case and frame of

the table range to make sure no grounds are present. If there is no continuity here, there are no short circuits to ground. If an open circuit is indicated, test the continuity of each part.

The most common trouble is a burned-out heating element. Next in the order of frequency come defective cords, bad connections, and defective switches. The only satisfactory cure for a burned-out heating element is replacement with a new coil of the same size. A sheathed-type element must always be replaced as a unit. You can sometimes use a mending sleeve, of the type shown in Fig. 5-4, for splicing a broken coil, or you can stretch the old coil enough to make a new connection at the end. To do this requires at least as much time as to install a new element and you will not make as much money as you would by doing the job right. Unless you replace the element, your customer will have no assurance of satisfactory use, because an element that has burned out once usually will not last long.

Heat is responsible for most bad connections in the table range. Expansion caused by heat loosens the connections, thereby greatly increasing their resistance, which in turn causes more heat and arcing. Heat also causes oxidation of parts, making them brittle and easy to break. Again, replacement is the only permanent repair. Screws and nuts used for connections inside the table range should be nickel plated. Brass or steel nuts and bolts will weld together or burn and should not be used. Extreme care must be taken when attempting to tighten or loosen connections to avoid breaking the porcelain.

Switches as a rule do not give very much trouble. When they do, something is usually worn out or burned inside the switch so repair is impractical. Replace the switch with a new one of the correct size and type. Tag the wires as you disconnect them from the old switch so you will know how to wire the new one.

In general, whenever you have a table range to repair, replace the defective part with a new one. While most table ranges are simple in construction, they get very hard service because of the intense heat, so only a new replacement part will give a satisfactory result.

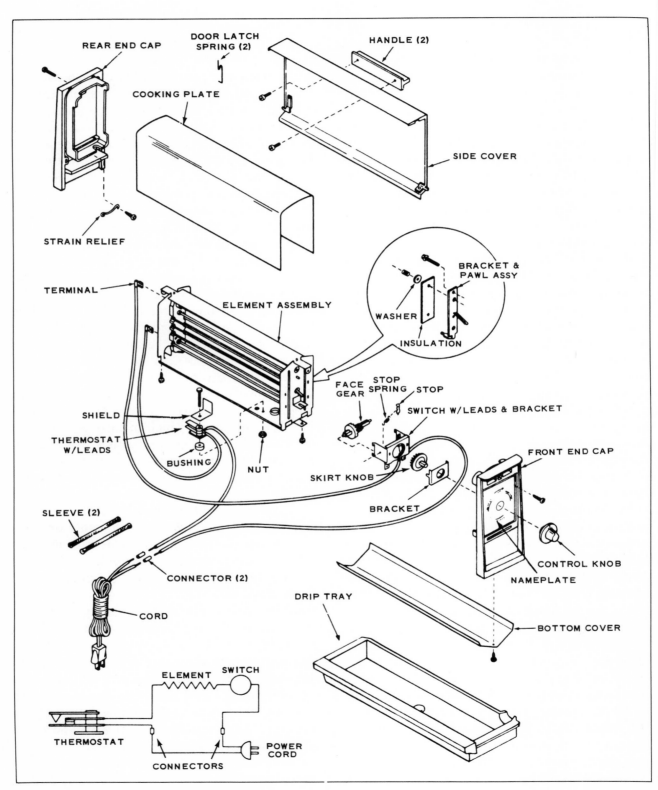

Figure 6-38. Major parts of a bacon grill.

Bacon Grill

As shown in Fig. 6-38, the bacon grill—electrically—consists of a heating element, switch, cordset, and thermostat. The thermostat is used to control the bacon cooking time, as well as the cooking temperature. If the temperature at the thermostat goes above 410°F, the thermostat opens; if it goes below 340°F, it closes. Thus it sets up a "cycling" of heating and cooling of the heating element. The heating and cooling causes the expansion and contraction of the heating element; this allows the end of the element assembly, with the "adjustment bracket and pawl" attached, to move a short distance. This movement provides the turning of the gear, the setting of which determines the cooking time. The pawl moves over an 18-tooth gear and operates as a ratchet. Each cycle causes the pawl to turn the gear approximately one-eighteenth of its rotation. A spring and stop is used in addition to the pawl to control the increment of turning, each cycle. An adjustment screw is provided on the element assembly for adjusting the bracket so the pawl operates smoothly each cycle, and does not "hang-up" on top of the teeth. When the gear reaches the OFF position, the ON-OFF switch opens because of a groove provided in the gear for this purpose. At this time the control knob should indicate OFF. At all settings of the control knob other than OFF, the ON-OFF switch will be closed until the OFF position is reached.

To adjust the timer pawl-bracket adjustment in the illustrated typical bacon grill, remove the cooking plate and perform the following:

1. Turn the adjustment screw clockwise until the tongue of the spring pawl rests on the topmost tooth of the ratchet part of the gear. This will draw the plate approximately parallel to the ceramic support. This should be accomplished without forcing the adjustment screw or bending the adjustment plate.
2. While observing the spring pawl, turn the adjustment screw counterclockwise until the spring pawl moves *forward* and the tongue drops into the groove between the teeth of the ratchet gear.

3. Applying a slight counterclockwise pressure on the control knob, to take up any backlash, turn the adjustment screw $2\frac{1}{2}$ turns clockwise for the final setting. Remember that most timer controls such as this should always be turned in a clockwise direction. That is, never try to turn the knob backward; damage to the control mechanism can result. If you want to turn the knob to OFF, turn it in the clockwise direction to the OFF position.

To perform an operational check on the timer pawl-bracket assembly, proceed as follows:

1. Connect power to the unit and observe the travel of the spring pawl while manually cycling the thermostat. This can be done by connecting and disconnecting the service cord plug from the ac outlet. The power should be kept on long enough to allow full expansion of the element (6 to 15 s, per specifications). *Caution:* Do not attempt adjustment while the unit is connected to the ac line. Remember that the heater wires are also "hot" electrically.
2. Observe the travel of the spring pawl during the cycling. It should travel far enough to pick up the next tooth and complete a full position on the dial indicator.
3. In normal operation with the cycling completed automatically, the control knob will stop at the OFF position. The next position clockwise will also be an OFF position.
4. When an operating unit is turned off manually, the contraction of the cooling element will occasionally pick up this second OFF position.

Warming Trays

Like most of the other cooking appliances described in this section, the electric circuit of a warming tray is simple. It consists of a heating element, thermostat, and cordset all connected in series. In most units, a neon indicator light assembly is connected in parallel with the heater element and remains lit as long as current flows through the element. The thermostat permits the user to select a range of tempera-

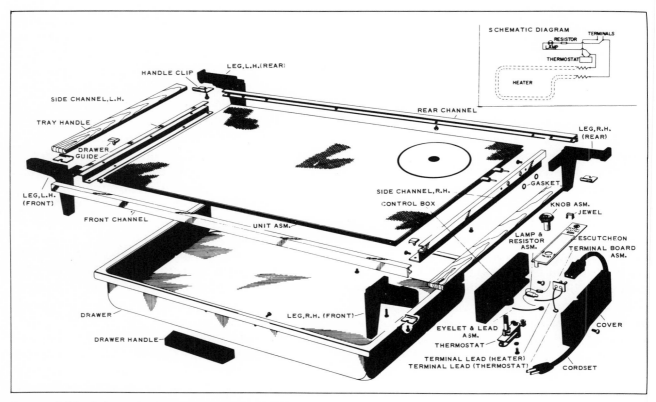

Figure 6-39. Schematic diagram and disassembled view of a typical warming tray.

tures. The flat heating surface—made of metal or a special heat-resistant glass—has the heating element built into it. Some units have a warming drawer suspended below the heating element.

Analysis of complaints. Here are the complaints encountered most often when servicing warming trays and what to do about them.

No heat at any temperature setting, indicator lamp does not light.

1. Check for a defective cordset. If no continuity is found, replace cordset.
2. Check for a defective thermostat. If found, replace.

No heat, but indicator lamp lights. This problem shows that there is an open heater element. If continuity is not obtained across the heater terminals, its element is open. As a rule, the heater element is sealed into the tray assembly, and the entire unit must be replaced.

Appliance blows fuses or draws excessive current.

1. Check for shorted cordset. Replace, if faulty.
2. Check for shorted heating element. Replace entire tray assembly if necessary.

Heat does not cycle off, and indicator light remains on at all times. See if the thermostat is shorted or its contacts are stuck in a closed position. Replace the thermostat, if faulty.

Tray is not level. On buffet serving appliances such as warming trays, the frame may not always be level. First check for loose or missing screws in the frame assembly. Take the proper corrective action. If the frame appears to be warped, it usually can be corrected as follows.

1. Place the tray on a level surface and determine which legs are short.
2. Place a cardboard or wood spacer approxi-

mately $\frac{1}{2}$ in thick under the legs touching the surface.

3. Press firmly down on the top of the tray on the corners directly over the legs to be aligned.
4. Remove the spacers; test for evenness. If warpage still exists, repeat precedure until the tray is level.

Baby-Food Warmers

Most baby dishes have a heater circuit which consists of a heater board immersed in oil, a temperature-control thermostat, and a safety fuse. When the cordset is connected to the dish and plugged into an electrical outlet, current flows through the thermostat, heater element, and safety fuse. The current passing through the heater element produces heat which is transferred to the food. When the heat in the food reaches a point slightly higher than body temperature, the thermostat opens. The dish can be disconnected and the food will be kept warm for approximately 30 min by the heat stored in the oil.

Some baby dishes have a signal light in series with the thermostat, heater element, and safety fuse to indicate when current is flowing through the circuit. When the food reaches its proper temperature and the thermostat opens, the signal light goes off.

Analysis of complaints. Here are the major complaints that a technician faces when servicing a baby food warming dish.

Oil leaks. Check the complete dish for any evidence of oil, especially around the terminals. If you detect any indication of an oil leak, the complete dish should be replaced.

Temperature is out of specifications. Check the dish temperature as follows.

1. Place $4\frac{3}{4}$ oz water (amount contained in a small baby food jar) in each food section.
2. Plug the unit into a controlled 120-V ac outlet and allow it to heat up for about 30 min.
3. Check the water temperature in the center tray with a maximum-reading thermometer

(0 to 150°F) by holding it in the water for about 1 min. The temperature should be 120°F $\pm$ 10°F.

4. If the temperature is not within these limits, the dish should be replaced.

Suction cup will not hold. Check the suction cup for holding power by pressing the dish on a flat, smooth surface. If it will not hold, replace the suction cup.

No heat. Check the dish and cordset for continuity and replace either if defective. Test the continuity of the dish between the two large outside pins.

COFFEE MAKERS

While electrical coffee makers come in various sizes and shapes and operate on different principles, they all are basically the same electrically. All have an electric heating element, almost all are thermostatically controlled, and all are timed so that heat is applied for just the right amount of time to brew perfect coffee. Many are also equipped to keep the coffee warm afterward. Some have adjustable thermostats so that the coffee can be varied in strength by permitting it to brew for different lengths of time.

While automatic coffee makers are essentially the same electrically, they make the coffee in several ways. The three most popular types are the percolator, brewer, and drip types.

In the percolator type of coffee maker the heated water is forced upward repeatedly through a percolating tube which extends from the center of the base into a basket containing the coffee grounds located in the upper portion of the coffee-making unit.

In the brewer-type coffee maker (often called the *vacuum* coffee maker) all the heated water is forced into the upper bowl at one time where it is kept with the coffee grounds until it drains down into the lower bowl to complete the coffee-making process.

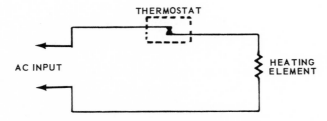

Figure 6-40. Schematic diagram for a coffee maker using a "bump" thermostat and single heating element.

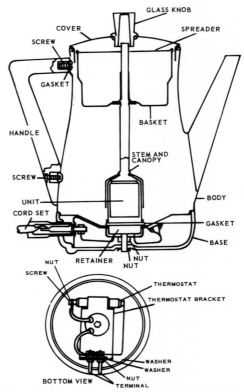

Automatic drip coffee makers are fairly new to the field. In this type, heated water is permitted to drip through the coffee grounds into a carafe or container below. A filter prevents the grounds from going into the coffee.

Servicing the Percolator Type

The simplest automatic percolator-type coffee maker comprises (electrically) a heating element, a cordset, and a control thermostat. A few models rely on a fixed-heat thermostat, while most provide the user with an adjustable one (weak to strong). Some have rather sensitive

Figure 6-41. Cutaway view of percolator above with fixed thermostat.

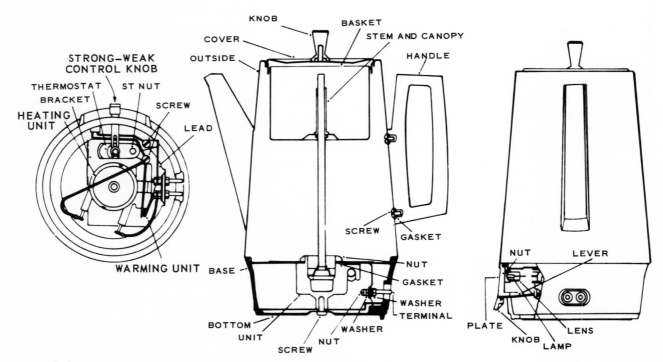

Figure 6-42. Cutaway view of percolator with adjustable thermostat.

thermostats that "bump" off and on within close limits to keep the coffee hot after it has percolated. Others, especially adjustable percolators, employ a warming element which turns on to keep the coffee hot after the main heating element has turned off.

The operating principle of the percolator is a simple one. At the start of the coffee-brewing cycle, the coffee grounds are placed in the basket in the top of the coffee maker. The proper amount of *cold* water is then added to the reservoir portion of the pot and the control dial or lever is set at the desired position between MILD and STRONG. The percolator's cordset is then plugged into a 120-V ac source.

In virtually every electrical percolator, a small well or pump chamber is provided at the bottom of the pot into which the valve (of the valve and pump stem assembly) is fitted. In operation, the small quantity of water in the well or pump chamber boils almost immediately because the heat is concentrated directly in, under, or around the well. The pressure thus created by the boiling water in the well rapidly increases until it closes the valve so that for the moment no more cold water will enter this small chamber. With the valve closed and the only outlet being through the pump stem, the rising pressure forces the small amount of water up through the stem and into the basket, where it seeps down (percolates) through the coffee grounds and returns to the reservoir. Thus each time the pump stem is emptied, the pressure recedes and allows the valve to open again and admit another small quantity of water, and the perking cycle is repeated. This continues until the water in the reservoir becomes hot enough to open the thermostat contacts. When these contacts open, current will not flow through the heater element and the percolating comes to an end. Most percolators are designed in such a manner that the coffee does not repercolate after the initial perking cycle has been completed.

The percolator just described is of the so-called single-element type. Most percolators today are of the two-element variety. As shown in Fig. 6-43, when the percolator is cool, the tempera-

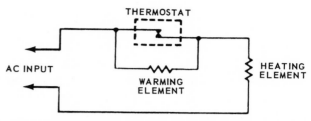

Figure 6-43. Schematic diagram for a coffee maker using a warming element in addition to the heating element.

ture control assembly (thermostat) switch is closed and the "keep-hot" or warming heater assembly is shorted out of the circuit. When the coffee maker is energized, the pump heater (high-heat or perking element) operates at its normal capacity; percolation takes place, and the temperature of the liquid gradually rises. At a predetermined brew temperature, or strength, which may be changed by means of the control knob, the bimetal blade opens the temperature control (thermostat) switch. This throws the warming assembly in series with the pump heater. The combined resistance of the pump heater and warming heater reduces the current flow to a point at which the total wattage is sufficient to maintain the brew at drinking temperature. In other words, for all practical purposes, the pump heater or high-heat element is out of operation, and the warming heater operates continuously to keep the coffee hot. Incidentally, if a coffee maker has a neon indicator lamp, it will light when the warming heater comes into operation or when the thermostat opens.

The percolating cycle can restart only after the temperature in the reservoir drops low enough to allow the thermostat contacts to close. When this occurs, full current flows through the pump heater element once again. But in a two-element percolator, this will not occur under normal circumstances since the thermostat closing temperature is set far below the "keep-hot" temperature.

When servicing coffee makers, be sure to check the service manual very carefully since manufacturers do use special features. For example, in some cases, the particular electric circuit consists of a booster unit controlled by a thermostat switch, a pump unit in parallel controlled by another thermostat switch, and a "keep-hot" unit in series with a pilot light and in parallel with the thermostat switch which controls the pump unit. At any setting of the control lever other than REHEAT, both thermostat switches are closed when the percolator is cold. When it is connected to power, both units are energized, but the "keep-hot" unit and the pilot light are shorted out. Under the influence of the two units, the temperature of the brew rises rapidly, and at an appropriate point the booster thermostat opens, cutting that unit out of the circuit. At a slightly higher temperature, the pump thermostat opens and throws the "keep-hot" element and pilot light in series with the pump unit. The additional resistance which this puts into the circuit effectively reduces the flow of current so that the total heat generated by the two units amounts to only enough to keep the brew hot, and percolation stops at this point.

The temperatures at which the booster and pump circuits open are controlled by the position of the cam that is operated by the control lever. The cam is designed with a track, or slope, for each of the two thermostat switches. The one associated with the switch which controls the booster unit progressively lowers the temperature at which the switch opens as the lever is moved from REHEAT to MILD to MEDIUM to STRONG. The one associated with the switch which controls the pump unit raises the cutoff temperature very abruptly when the lever is moved from REHEAT to MILD and continues to raise the cutoff point progressively but moderately as the lever is moved from MILD to MEDIUM to STRONG. The pump thermostat should be open when the lever is in the REHEAT position.

The theory responsible for this cam design is relatively simple. Mild coffee requires a short percolation period, and strong coffee requires a longer period of percolation. Since percolation ends when the pump circuit opens in response to the heat of the brew, it follows that the more rapidly the brew is heated, the milder the resultant coffee. For mild coffee, therefore, the booster unit is retained in the circuit until the brew closely approaches the cutoff point of the pump unit, and the period of percolation is thereby shortened. For stronger coffee, the booster unit is cut out at a lower temperature and the pump unit at a higher temperature to extend the period of percolation.

Analysis of complaints. If the customer does not give specific information on the difficulty being encountered, make the following inspection of the percolator. Visually inspect for signs of breakdown, such as burned-out unit, defective cords, and broken leads. If the defective part is readily distinguishable and beyond repair, replace it.

If no obvious defects are visible, check the electric circuit for continuity as follows. Check for a short with an ohmmeter or continuity checker of some type. If a short is evident, trace out and correct or replace the defective component. If this does not solve the problem, make the following operation test of the percolator.

1. Set the flavor selector to STRONG.
2. Fill the percolator to its capacity with cool tap water.
3. Insert the basket and pump stem assembly. Put on the cover.
4. Insert the stem of a thermometer or thermocouple into the water in a manner suggested in the service manual. Be sure the temperature-measuring device does not touch the metal of the pot.
5. Insert a wattmeter into the circuit and read wattage. It must be within the tolerance (usually ± 5 percent) specified in the service manual or on the nameplate. If it is outside these limits, the interpretation of problems is as follows.
   a. Absence of wattage indicates an open circuit in the cordset, the pump heater (high-heat element), or the pump-heater leads.
   b. A very low reading indicates an open

Figure 6-44. This illustrates a typical 12-in thermometer with a clip for testing the heat of liquids.

circuit in the control assembly or the control assembly lead.

c. Any other wattage indicates a pump heater which is out of tolerance, corroded control contacts, or high resistance at one of the terminal connections.

6. Watch the wattmeter for the drop in wattage which indicates the opening of the control contacts. This should occur between 180 and 195°F on the thermometer, and the wattage should then drop below LOW (see the service manual for exact amount). With most models, during the "keep-hot" cycle the wattage should be about 60 W ± 10 percent.

a. In some models, absence of wattage indicates an open circuit in the warming-heater assembly or its connections.

b. Any other wattage indicates a warming-heater assembly which is out of tolerance or has a high resistance at one of its terminal connections.

c. A temperature reading which is out of the acceptable range requires readjustment of the control. This is done as directed in the service manual. Most models have a calibration adjustment screw that is accessible through a small hole in the bottom. As a rule, turning the screw counterclockwise raises the temperature, while turning it clockwise lowers it. In some models, a one-quarter turn will change the temperature as much as 20°F. As previously stated, this adjustment should be made so that the warming-heater element cuts into the circuit between 180 and 195°F. If this cannot be accomplished, the thermostat should be replaced.

If a customer has a specific complaint, follow it out. Here are some of the more common customer complaints and how to remedy them.

Appliance does not heat at all. Heating element (high heat or pump heater) burned out because of one or more of the following.

1. Corrosion and/or coffee stain allowed to accumulate contrary to operating instructions. See the user's instruction booklet.
2. Appliance was plugged in when dry.
3. Less than 2½ cups of water used (in some models).
4. Water leak shorted the connections.
5. Manufacturing defect in the element.

The heating element in all cases should be replaced, and the customer informed of the cause of the failure. Other causes of this problem could include a loose connection and a defective cordset, or the thermostat's contacts set in an OPEN position. Some older percolators had a fuse—either screw-type or strip-type—in the ac line. Check for it in older models.

Appliance does not stop percolating. A typical median percolating time varies from about 14 to 18 min for an eight-cup percolator and 7 to 10 min for a four-cup unit. If the coffee maker does not operate within these general limits, the thermostat could be defective and should be replaced. In some models, the warming element may be shorted. Check and take necessary corrective action.

Coffee maker shuts off too soon or repercolates.

1. In some models, check the warming heater wattage after the cut off temperature is reached to determine whether the wattage is within tolerance (refer to service manual).
2. In some models, if the warming heater is

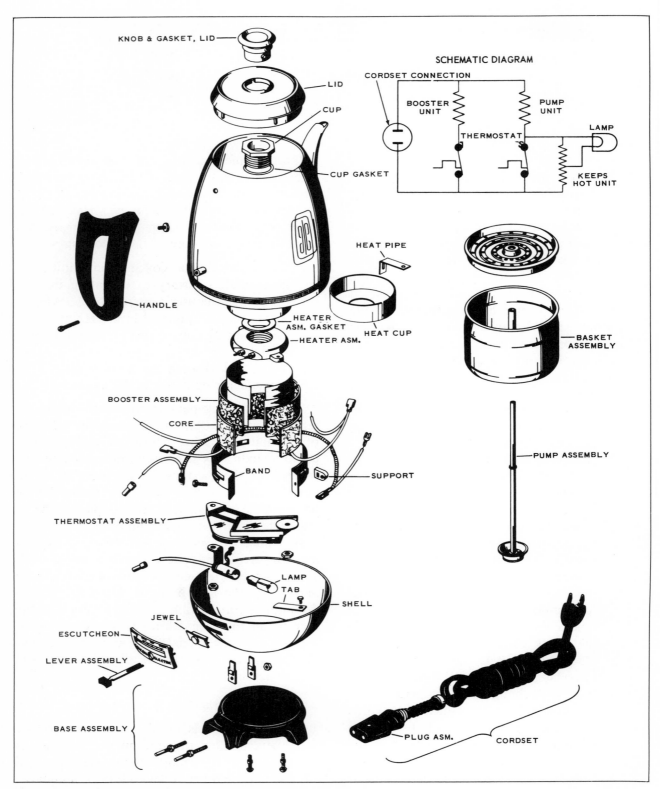

Figure 6-45. Disassembled view of a typical percolator coffee maker.

within tolerance, the customer may be using the unit in a cold or drafty location.

3. In some models, shifting the brew control to MILD after the cutoff temperature is reached will tend to prevent reperking.
4. Check the pump stem assembly. If faulty, replace.
5. Check the thermostat and the continuity of the warming or "keep-hot" element if one is employed. If either is faulty, replace.

Slow to perk. (Remember that 7 to 18 min of percolation is essential for satisfactory flavor extraction.)

1. Check for low voltage in the customer's home. Advise customer.
2. Check the wattage during the percolating cycle. If defective, replace the thermostat.
3. Check for stem partially clogged with corrosion. Clean or replace.
4. Check the heating element for surface coated with lime and coffee stains. Clean or replace.
5. Check the pump jacket for looseness on the stem. Some manufacturers have a feeler gauge available for the checking of the pump clearance. When using one of these feeler gauges, if a pump easily accepts the "no-go" end of the gauge, or if a pump refuses to accept the "go" end, replace the pump.

Coffee weak or not hot enough.

1. Check the pump clearance. A feeler gauge, if available, should be used. If pump fails clearance test, replace.
2. Check the percolator cutoff temperature (see the service manual). If it is below the tolerance or at the low end of the acceptable range, readjust the control for a higher cutoff within the test limits. Replace thermostat, if defective.
3. Check for coffee grounds in or under the pump. Clean, if necessary.
4. This complaint could also be due to the customer starting the coffee cycle with hot water instead of the recommended cold water or not using enough coffee. Customer education.
5. See if the pump may not fit tightly or seat properly in the cup of the pump heater. Check for rough spots on the rim of the pump heater cup or on that part of the pump which contacts the pump cup. Repair or replace the pump.

Water warms but does not perk.

1. In some models, be sure that the pump heater or high-heat element is not coated with coffee stains or lime deposits. If it is clean it.
2. Check for a defective pump. Replace the pump assembly if faulty.
3. Check for open thermostat contacts. If coffee maker has an indicator lamp, it will light instantly when plugged in if the thermostat contacts are stuck open. If the thermostat is faulty, replace it.
4. In some models, check the linkage for contact between the brew selection knob and thermostat. On many models when the base and the percolator are assembled, the brew selection offset arm on the thermostat must fall into the slot on the control shaft. This can be checked by placing the top of the percolator to the ear and flipping the control knob back and forth. A metallic sound will be heard on both sides of the rotation if the unit is assembled properly.

Coffee boils.

1. Check for proper cutoff temperature and test for open warming element. If components are defective, replace.
2. Be sure that the warming element is flush and tight against the bottom of the body. If not, repair it.
3. Check the pump assembly for blockage or other defect. Clean or replace, as necessary.
4. Check the thermostat for looseness or damage. If faulty, replace.
5. Check for improperly dressed lead interfering with the thermostat's action. Repair, if necessary.

Coffee will not keep warm. Check warming heater and if found defective replace.

Reperks intermittently after coffee is made. Check warming heater and if found defective replace.

Lid falls off. Raise the bumps in the lid using a rounded punch against a soft wood block. If the knobs break twice in the same lid, replace the lid.

Tight lid. Flatten the bumps in the lid.

Customer has difficulty putting lid on. Check for the basket being too low in relation to the body rim. Repair or replace the defective part.

Lid knob is loose. Check for a missing or worn gasket. Replace, if necessary.

Light fails to come on. Be sure that the lamp is correctly positioned in its bracket; otherwise replace the lamp, resistor, lead, and boot assembly.

Pot leaks around element. Replace the asbestos washer between the element and the pot. If the leaks appear to have damaged the thermostat, replace it also. Be sure to tighten the element securely, using a extra deep socket or special wrench provided by the manufacturer. Be very careful when doing this not to dent or mar the rim; this can cause leaks. A genuine body leak cannot usually be repaired.

Coffee tastes bitter. Complaints that the coffee tastes bitter usually mean that the customer has not followed cleaning instructions in the user's booklet with sufficient care. Check the percolator visually for signs of accumulated residue and stains. Clean if necessary, using a commercial stain remover which may be purchased locally.

Before returning a coffee maker to a customer, be sure to check it for current leakage. True, the possibility of excessive current leakage developing in even an immersible coffee maker is most unlikely. Yet thorough servicing requires that a current leakage check be made. With the ac cord plugged into a 120-V ac outlet, measure the leakage between the vessel and a known earth ground (water pipe, conduit, etc.), using a leakage current tester. The current must not exceed 0.5 mA.

If the manufacturer recommends it, a hi-pot test can be used to check current leakage. A typical test is the hi-pot at 1200-V 60-Hz ac for 1 s, or 1000 V for 1 min with the control level at the MILD or STRONG position.

An alternate method to measure leakage is by using a 1500-Ω 10-W resistor, paralleled by an 0.15-μf capacitor. Measure the voltage across the resistor with an ac voltmeter of 1000 Ω/V or more. The reading should not exceed 0.75 V rms $\left(\frac{0.75V}{1500\Omega} = 0.5\ mA\right)$. If the leakage is greater than 0.5 mA, remove the base vent screw and then fill the coffee maker to the top cup mark with cool water. Plug the percolator into an ac outlet and complete the perking cycle. If there is moisture trapped in the base, this operation will usually dry out the moisture and eliminate the current leakage. After the drying out process, replace the base vent screw and gasket. Recheck for current leakage.

Any leakage not within the limits stated by the manufacturer (usually 0.5 mA) will be considered a shock hazard and must be corrected before returning the appliance to the customer.

Servicing Coffee Urns

Most electrical coffee urns—the type used at parties, for meetings, in offices, etc.—are just large coffee percolators and, for the most part, operate in the same way as their smaller counterparts. Figure 6-46 shows a cutaway view of a 12- to 30-cup coffee urn. While the coffee urn is cold, the thermostat contact remains closed. The lamp is shorted out of the circuit. When the unit is connected to a 120-V ac source, the current flows through the pump heater and thermostat in series and in parallel with the "keep-hot" unit. At a predetermined brew temperature, the thermostat opens causing the lamp to glow and the pump heater cuts off. The "keep-hot" heater remains in the circuit.

The troubleshooting information given previously for smaller percolator units holds good for urns. About the only additional trouble that may occur is with the faucet. Leaks at the faucet or pump well can be corrected by tighten-

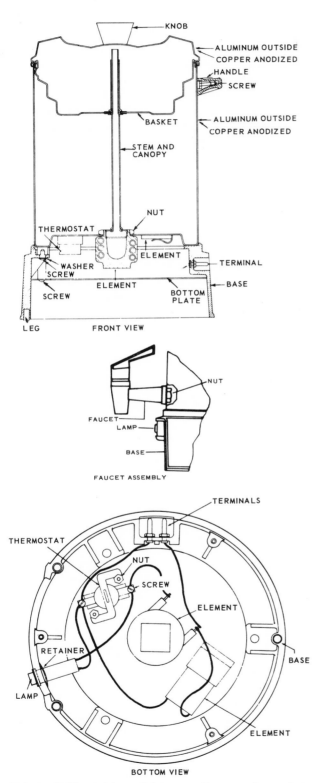

Figure 6-46. A 12- to 30-cup coffee urn: front view, faucet assembly, and bottom view.

ing the faucet nut or heater nut. If tightening either of these parts fails to correct the complaint, replace the appropriate gasket. Should the faucet drip, it is wise not to use pliers to remove or tighten the faucet top as they will mar the finish. Remove it by hand, replace the seat cup, and reassemble the faucet top by hand pressure.

Servicing Brewer-type Coffee Makers

The brewer or vacuum type of coffee maker was once extremely popular, but in recent years its use has been on a steady wane. However, some are still in use and may come into your service shop for repair.

The usual electric components of this type of coffee maker consist of a low- and high-heat element, a switch, a thermostat, an indicator light, and a cordset. To start the brewing cycle, water is placed in the lower bowl and the proper amount of dry coffee is placed in the upper bowl. When the switch is turned and current flows through both heater elements, a large rubber seal on the bottom of the upper bowl

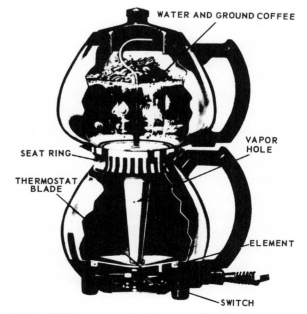

Figure 6-49. Schematic diagram of a typical brewer-type coffee maker.

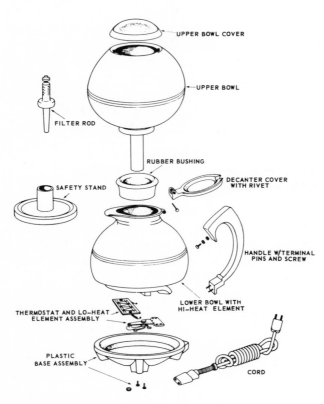

Figure 6-48. Exploded view of a typical vacuum-type coffee maker.

holds the pressure inside the lower bowl as the water heats up. When the water boils in the sealed lower bowl, the pressure thus generated forces all but a small quantity of the water up through the upper bowl spout and thence into the upper bowl with its coffee. This small amount of water boils away and the resulting steam agitates the coffee in the upper bowl. When most of or all the water is out of the lower bowl, the heat increases rapidly; the thermostat then shuts off the power and shorts out the high-heat element. A partial vacuum is thereby drawn in the lower bowl, for water and steam have been expelled but no air has entered. This vacuum is further intensified when the heat is turned off, for as the lower bowl cools, the remaining vapor condenses, the residue of water —if any—cools and shrinks in volume, and finally the brewed coffee is pulled through the filter and into the lower bowl. The coffee is

automatically kept at temperatures of between 160 and 180°F by the low-heat setting of the thermostat.

Although several manufacturers have developed a number of refinements in the brewer-type coffee-making cycle, the basic theory just described is all the service technician needs to know to make most repairs. Details of special modifications can be found in the service manuals.

Analysis of complaints. The brewer- or vacuum-type coffee maker develops the same heating troubles as the percolator. Check the cord and plug, the thermostat, and the heating element. In addition, brewer-type coffee makers are subject to the following special problems.

Coffee does not return to lower bowl, or cycles back and forth between the upper and lower bowls.

1. Determine whether the thermostat is out of adjustment. Readjust or replace it.
2. Check the operation of the thermostat. Replace if faulty.
3. Check for a faulty seat ring or gasket on the bottom of the upper bowl. If the ring has cracks or breaks in it when flexed, replace it with a new one.
4. If the bowls do not fit snugly because of distortion of the top bowl, use the special tool (available from the manufacturer) to reform the top so that the gasket or ring fits properly.

Coffee boils over.

1. Check for a faulty seat ring or gasket. If defective, replace.

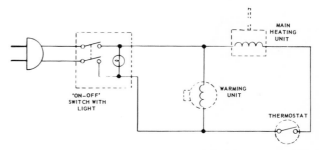

Figure 6-49. Schematic diagram of a typical brewer-type coffee maker.

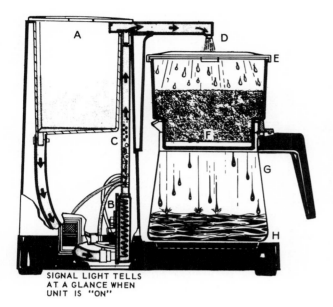

SIGNAL LIGHT TELLS
AT A GLANCE WHEN
UNIT IS "ON"

Figure 6-50. Operation of a typical drip coffee maker.

2. Check for an excessively high temperature setting. Adjust the thermostat or replace it.
3. Check for an incorrect fitting of filter cloth. Instruct the customer on the correct technique of using it.
4. Check for a hole in the filter cloth. Replace the filter cloth, if necessary.
5. Check for a vacuum leak around the handle-holding screw. Repair or replace defective part as necessary.

Coffee does not remain in upper bowl long enough. The thermostat is out of adjustment. Readjust or replace. Check the service manual for the specific length of time of this period since it varies with different models.

Servicing Automatic Drip Coffee Makers

In the past few years, automatic drip coffee makers have become fairly popular for home use. Figure 6-50 shows the operation of a typical drip coffee maker. As detailed, gravity feeds water from the reservoir (A) to the internal tank heater (B). With the touch of a switch, water begins the heating, and within seconds is heated, forced up the tube (C) and out the dripper

spout (D). The spreader (E) evenly distributes water over the ground coffee, through the permanent filter (F), and into the glass carafe (G). Warming plate (H) keeps coffee hot to the last drop. While many manufacturers have developed slight variations, this is basically how an automatic drip coffee maker operates.

Analysis of complaints. As with other coffee makers, troubles do arise with automatic drip coffee makers. Here are the most common problems and how they can be serviced.

Unit does not pump all the water out of reservoir.

1. Check the hose, and if pinched, replace it.
2. See whether the unit is dirty. If it is, clean according to special cleaning instructions below.

Unit leaks water down the front.

1. Check the unit and spout adjustment. Readjust as necessary.
2. Replace the spout if it still leaks.
3. Replace unit if it still leaks.

Unit does not pump at all.

1. Check the continuity of cordset, thermostat, and switch; any defective part should be replaced. (After replacing parts, make sure the hose is not pinched.)
2. If unit is burned out, instruct customer to make sure water is in reservoir before plugging in and turning on the coffee maker.

Unit does not restart on second brew cycle. Check for a correct warming bracket. If parts are correct, change thermostat. Check for a pinched hose; check the assembly of the warming bracket and make sure it is down tight.

Unit restarts while warming.

1. Check warming unit. If it shows no continuity or if the plate does not get warm, replace it.
2. Check for a correct warming bracket. If parts are correct, change the thermostat. If the unit has been starting to pump on the "keep-warm" cycle, give it the following test after taking corrective action. Adjust volt-

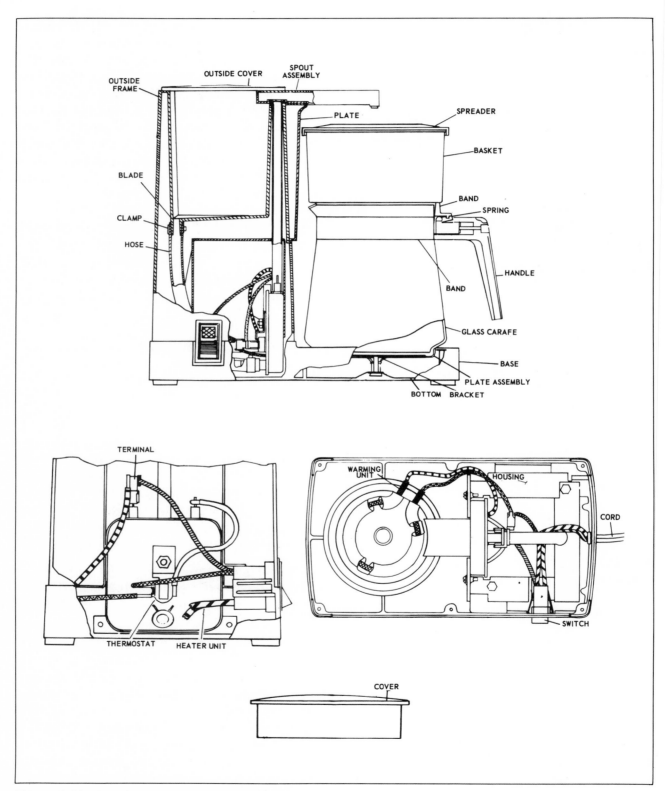

Figure 6-51. Parts of a typical drip coffee maker.

age to 110 V if available (otherwise use standard line voltage) and pump 8 cups of water and let stand on "keep warm" for 2 h. Put the cover on the carafe with a piece of paper on the top of the cover. If the unit starts pumping, it will put a water spot on the paper indicating a failure of the test.

Warming plate does not warm at all. Check the continuity of warming unit and replace, if faulty.

Coffee does not stay hot. Check the assembly of the warming bracket and make sure it is tight so that it conducts heat to the plate.

Special cleaning instructions. After continual use, a mineral deposit may form in the tubes and spout. The rate at which it forms will be determined by the hardness of water used and the frequency with which the coffee maker is used. Inform your customer that a special cleaning with brush and vinegar is necessary every 60 brewing cycles, i.e., once a month if the coffee maker is used twice daily. If the mineral deposit is not removed, excessive steaming will occur, and unpumped water will remain in the reservoir when the unit is making coffee. Here are the general steps in a typical special cleaning.

1. Unplug the cord from the wall outlet. Detach the spout from the coffee maker. Brush the spout out under running water. Shake dry.
2. Brush out the vertical transfer tube to which the spout connects, then invert the coffee maker (without carafe) to remove the mineral deposit.
3. Pour one cup of water into the reservoir, then empty into a sink, shaking slightly to remove the mineral deposit.
4. Replace the spout. Pour household vinegar (5 percent acetic acid) into the reservoir to the four-cup level. Let it stand for 30 min.
5. Place the carafe-basket-spreader assembly under the spout on the warming unit. Plug the cord into the wall outlet; turn the

switch on. Let one cup of the vinegar "pump" into the carafe. Turn the switch off. Let it stand for 30 min. Repeat the procedure about three times or until all the vinegar has been pumped into the carafe.
6. Unplug the cord from the wall outlet. Discard the vinegar. Rinse the reservoir with hot tap water. Remove any remaining mineral residue from the spreader with a dishcloth.
7. For final rinsing, fill the reservoir with cold water to the eight-cup level. Plug in the cord and turn the switch on to pump water through the basket-spreader assembly into the carafe. (The vinegar should not remain in the unit beyond the cleaning time.)

For specific instructions on cleaning a given automatic drip coffee maker, check the manufacturer's service manual for that model.

BLANKETS AND HEATING PADS

While electrical blankets and heating pads have much the same general appearance and similar circuitry, they serve two completely different functions. The electrical blanket is not supposed to warm the body, but rather replace body heat that is lost to the cooler air of the room. The heating pad, however, is supposed to give heat to the body in a concentrated form.

Servicing Electrical Blankets

In the simpler electrical blankets, the electrical portion comprises a flexible heating element, an adjustable thermostat, an indicator lamp, and an ON-OFF switch which sometimes is incorporated in the temperature-control knob of the thermostat. The heating element is sewn into the blanket fabric in a pattern that allows a uniform distribution of heat over almost the entire area of the blanket. The thermostat usually operates on the difference between the temperature for which it is set and the temperature of the room to keep the blanket temperature constant at a preselected level. The thermostat, the ON-OFF switch, and neon indicator lamp with its resistor are usually contained in a control box.

In a simple electrical blanket circuit, the line current flows through the heating element inside the blanket when the ON-OFF switch and the thermostat contacts are closed. The indicator lamp lights to show this current flow. When the preselected temperature in the blanket is reached, the thermostat contacts open, the current stops flowing through the heating element, and the indicator lamp, of course, goes out. Some blankets have dual controls and two separate heating elements, one for each half of the bed so that the temperature can be individually regulated for each side. Incidentally, electrical blankets are available in four basic sizes: twin, double, queen, and king.

Other control arrangements are used with electrical blankets. For instance, one design does not employ room temperature as the determining factor in the heat control. Rather, a small heater is wrapped around the thermostat control and is connected in series with the blanket heating element. The purpose of the thermostat heater is to simulate the actual blanket temperature.

As shown in Fig. 6-52, the line current will flow through both the thermostat heater and the blanket heating element. When the temperature has reached its preselected setting, the thermostat contacts open and the current is cut off to both the thermostat heater and blanket heating element. While both elements will tend to cool because of loss of current, the thermostat heater in the control box, being the smaller of the two,

will cool much quicker than the blanket element, and it will make the thermostat contacts close before the blanket temperature drops any great amount. To prevent the thermostat heater from overcontrolling the circuit, a small permanent magnet is attached to the arm of the thermostat. This magnet slows down effectively the turn-on action of the thermostat and tends to keep the two heaters in line with one another. In fact, small magnets (which do not make any electric contact whatsoever) are used quite frequently on the contact arms of electric blanket thermostats. As the thermostat bends the arm, the magnet gets closer and closer to the fixed contact which has a small iron washer attached to it. Without the magnets, the contact points would move rather slowly until they made contact; this could result in a certain degree of sparking, and in some instances, the contact would tend not to be firm. By use of the magnets, however, the contacts, when within a certain distance of each other, snap together and make a quick, firm contact. The same action occurs, only in reverse, on a break—the magnets hold the contacts together until the pull of the heating thermostat overcomes the magnetic attraction, and the contacts snap apart. This snap action tends to overcome any overcontrol or other similar problem.

Many blankets use small thermostats inside the blanket itself. These can be felt as tiny lumps. The safety thermostats have nothing to do with the operation of the blanket thermostat, or with its temperature control under normal conditions. They are merely a safety device and normally are closed. If anything should occur in the control box that would permit the blanket to stay on for too long a period of time, these safety thermostats will open and keep the blanket from becoming overly hot.

In some of the more elaborate electrical blankets using what is called "electronic" control, sensing elements are placed inside the blanket itself. In fact, many of the newer blankets utilize two separate and distinct electric circuits, a heater circuit and a sensor circuit. A bimetal thermal switch is used in place of

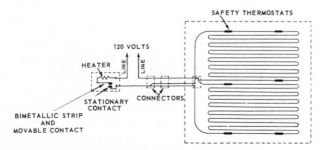

Figure 6-52. A simple electric blanket circuit.

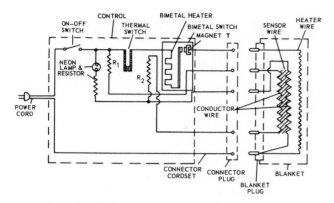

Figure 6-53. An electric blanket employing a sensor system.

the magnet relay in the sensor circuit to create the "recycling" feature. The heater circuit (Fig. 6-53) provides warmth to the blanket and consists of the following.

1. A heater element to create heat in the blanket.
2. A bimetal switch to control the heat of the heater element in the blanket.
3. A bimetal heater in the control box to simulate the temperature of the heater wire in the blanket.
4. A magnet to prevent the bimetal switch from cycling too frequently.
5. An ON-OFF switch to turn the control on and off manually.

The sensor circuit between the bimetal switches and the ON-OFF switch is necessary to turn the blanket off in the event of an overheat. A typical sensor circuit consists of the following.

1. A sensor wire made up of two conductor wires separated by a temperature-sensitive plastic material which decreases in electrical impedance as the temperature increases. This wire is distributed within the blanket.
2. A normally open double bimetal thermal switch with both bimetals moving in the same direction to compensate for variations in room temperature.
3. Two added resistors, one on each bimetal of the thermal switch.

This typical sensor circuit operates as follows. One resistor R_1 is connected to one of the conductor wires of the sensor wire, and the other resistor R_2 is connected in series with R_1 and the two conductor wires. When current passes through both conductors of the sensor wire, unequal amounts of heat are developed in R_1 and R_2. Therefore, R_2 bimetal deflects more because of the unequal heat and closes the thermal switch contacts.

If the conductors of the sensor wire short, no heat will be supplied by R_2 resistor and the contacts will not close. If an overheat occurs, the decreasing sensor impedance will tend to bypass resistor R_2 and allow less heat to be developed in it. When resistor R_2 cools down to approximately the heat in resistor R_1, the thermal switch contacts will open.

If the overheat is removed, the sensor impedance increases, and resistor R_2 again supplies more heat to its bimetal than R_1 resistor, causing the bimetal contacts to reclose.

Analysis of complaints. Customer training in the proper use of an electrical blanket is most important. It can prevent many unnecessary service calls. Two major complaints are the ones described below.

Blanket gives insufficient warmth. As stated earlier in the chapter, an automatic blanket should not be compared with a heating pad in which relatively high heat is concentrated in a small area and maintained regardless of room temperature. The blanket control system is designed to respond to changes in room temperature to compensate for the heat loss from the human body. Most people do not require any additional heat if the ambient temperature is 75°F or higher. At decreasing temperatures down to 50°F, the blanket control will cycle the heating circuit more frequently so as to equalize the heat loss and provide constant comfort. Personal preference as to operating temperature is provided by the setting of the comfort-selector dial. The exact setting can best be determined by the user after a short trial period. Thereafter, this setting can be maintained, and constant comfort will be assured. It will be unnecessary to constantly change the setting once it has been arrived at by trial. If the control is located near a radiator, cold-air return, or some other source of heat or cold directly affecting the control operation, the control should be moved to a location where the air around it is about the same temperature as that near the blanket. The control should not be put under a pillow or in an enclosed space.

Each person should find his or her own particular comfort setting on the control by setting the selector to HIGH and then adjusting to the preferred comfort point on the dial. Do not attempt to compare the dial setting on one control with that on another control. The control is operating correctly if your comfort is obtained at any setting between HIGH and LOW.

Customer receives shock from blanket. Explain to the customer that this is not a hazard by assuring him or her that a slight "tingle" or "shock" can be felt when contacting a sensitive portion of the body to a grounded object or another person beneath a blanket. This is due to a capacity effect between the body and the blanket wire. The "tingle" effect is extremely small and is just barely above the threshold of sensibility. Where two twin-bed blankets (or a king size blanket) are involved, a simple reversing of the wall plug of one of the blankets may correct the condition. One of the three following arrangements should correct the condition:

Beginning with the plugs in their usual position, reverse one plug. If the tingle still is felt, leave the first plug reversed and reverse the second plug. If the tingle still is felt, return the first plug to its original position and leave the second plug reversed.

The most common problems with electrical blankets are open blanket heating elements, dirty thermostat contacts, control-box defects, and bad cordsets. Open blanket heating elements or defective safety thermostats cannot be repaired and the entire blanket unit must be replaced or exchanged. Some electronic control units and the controls for blankets with internal sensing also cannot be serviced in the average service shop. (Some large-volume repair shops have

special blanket control testers made by the manufacturers of the blankets, but the average service shop cannot usually justify the cost of these special testers.) Dirty thermostat contacts can be cleaned as described in Chap. 5, while a continuity test will show up opens or shorts in cordsets or control boxes. If any parts are found defective, repair or replace them.

If the customer complains of too much or insufficient heat at a normal comfort-selector knob setting, calibration of a typical control may usually be changed by proceeding as follows.

1. Set knob to Position 5. [While this procedure is based on a typical 10-position switch (OFF and nine steps), the same technique can be used for any number of steps.]
2. With a knife blade or a small screwdriver, pry off the plug button in the center of the control knob.
3. Squeeze the projections together inside the knob with long-nose pliers and carefully lift the knob straight off.
4. Replace the knob on the shaft as follows.
   a. For a low-heat complaint, reposition so that Position 3 is in the center.
   b. For a high-heat complaint, reposition so that Position 7 is in the center.
5. Replace the plug button.
6. Recheck the calibration point by turning it slowly through the range from LOW to HIGH, listening particularly for any abnormal sound such as loud "trip" or double-click condition. If such is heard, replace the thermostat.

Servicing Heating Pads

Electrically, a heating pad is nothing but a small electrical blanket. As shown in Fig. 6-54, a typical simple heating circuit comprises a pad heating element, safety thermostat, cordset, and control box with a thermostat indicator lamp assembly and heat control switches with bias heaters. Some of the older style heating pads do not have a control box. They employ a dual heating element arrangement, one element with

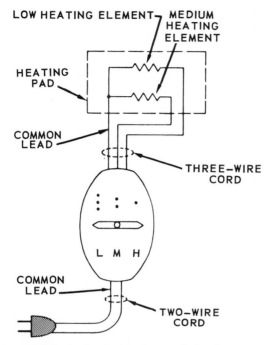

Figure 6-54. A simple heating-pad circuit.

about 20 W, the other about 40 W of heat. A special selector switch allowed the user to select either of these or both together, giving a choice of low, medium, and high heat. Thus when the switch was in LOW position, the 20-W element was in the circuit; when in the MEDIUM position, the 40-W element was used; and when in the HIGH position, both elements in a series arrangement were employed.

When a safety thermostat in the pad is used, it is normally closed and will not open unless a problem arises to cause an excessive amount of current to flow. Since the neon lamp assembly is usually connected to the line before the thermostat control, it remains on all the time the unit is plugged into an outlet; it does not cycle ON and OFF with the pad heater current. While heating pad circuitry may vary slightly from model to model, the same general concept is used by most makers.

The servicing techniques involved with heating pads are much the same as for electrical blankets. To check the operating temperatures, place a thermocouple between the pad the the protecting cloth cover. Connect the heating pad to a power source through a wattage testing circuit. Set the heat control switch on LOW and check the power consumption. It should be ± 10 percent of the nameplate rating. The temperature ranges for the three settings should be as specified in the service manual. If they are not, replace the control unit.

SPACE HEATERS

Portable electrical heaters are commonly employed in and around the home and are usually called *space* heaters, since their major function is to heat the air space in the room or area in which they are placed. They are made in many different sizes and shapes to suit various conditions of service and heat requirements, but, depending on the method of air circulation, they are of two general types, natural draft and forced draft. In the former type, the air rises by natural draft over electrically heated elements and is heated by direct contact with the heating element and by ascending through the natural draft. This air is distributed throughout the room. In forced-draft heaters, the air is blown by an electrical fan over an electrically heated element, which heats the air by contact. This heated air is then distributed throughout the area to be heated by the draft of the fan unit.

Servicing Natural-Draft Heaters

For many years, the so-called bowl-type heater, which was based on natural drafts, was the most popular space heater in home use. This heater obtained its name from the bowllike shape of its metallic reflector, which was mounted on a sturdy base and was provided with wire guards to prevent accidental contact with the cone-shaped heating element. The heating element consisted of coiled resistance wire wound on a ceramic form. Cordsets were connected directly to screw terminals at the base. The replacement coils were readily available and were simply screwed into place. The heater's effectiveness depended mostly on a shiny reflector. All that was required was to unplug the cordset, remove the protective grill, and clean the reflector with a soft cloth.

The modern heater of this type consists of perforated sheetmetal cases through which air can circulate over the surface of the heating element. The warmed air is caused to rise, thereby providing circulation of warm air in the room. The heating elements may consist of resistance wire wound on cylindrical insulators or resistance wire mounted on special heater strips or bars of suitable wattage.

The natural-draft heaters can be as simple as a cordset and a heating element. Most have an additional feature: a tipover switch. If the heater is tipped over, the spring-loaded tipover switch automatically opens and, since the switch is in series, shuts off the heater.

Analysis of complaints. Here are the major complaints that service technicians face with natural-drift heaters.

No heat.

1. Check the heating element and if faulty, replace. After ascertaining that the element is defective, carefully disconnect the wires from their terminals; be careful not to damage the asbestos or mica insulating washers when removing the element mounting insulators from the heating frame. If any part of the insulating material is damaged, replace it during the reassembly process. Check the rating plate of the heater; it will give the information needed to get a duplicate.
2. Check the tipover switch. If the contacts of the switch are stuck open or are broken, replace the switch.

Heater blows fuses, draws excessive current.

1. Check the cordset for a short. Replace if faulty.
2. Check the heating element for a short. If found, replace.
3. Check for a short circuit to the metal case. If found, remove the cause of the short. The hi-pot test, if recommended by the manufacturer, should be taken between one plug of the cordset and the metal case or grill.

A 1,100-V hi-pot test for 1 s is usually recommended to show up a short.

Unit does not produce sufficient heat. The best way to check a heater's performance is to measure its power consumption. The wattmeter should read the nameplate wattage ± 5 percent.

Unit does not turn off when tipped over. This is an indication that the tipover switch is shorted. Clean the switch or replace as is needed.

Servicing Fan-type, or Forced-Draft, Heaters

While bowl-type and convection heaters are still available, the vast majority of space heaters sold today are of the forced-draft type. Heaters of this type consist essentially of one or more heating units and an electrical fan, which blows the heated air through the heating units and circulates it in a given area. Forced-draft heaters are available with or without thermostats for room temperature control. Some deluxe heaters have switch-selected degrees of heat intensity or fully automatic thermostat control and also cut-off devices that turn off the current if the heater is upset accidentally. In a unit of this type, the thermostat, motor, heater, and tipover switch are in series; therefore, if the thermostat contact is closed, the element heats and the fan rotates, forcing air over the heater and out of the unit. When the room temperature reaches the thermostat setting, the thermostat contacts open and the heater is shut off. Some heaters also have an overheat protector which cuts off the heating element circuit if an overheat condition exists.

With some units, when the thermostat knob is turned to the extreme counterclockwise position, the unit is off. As the knob is turned clockwise, the points in the thermostat make contact and the heater is energized. The unit will continue to heat until the temperature of the room rises causing the bimetal to bend and the contacts to open.

Analysis of complaints. While forced-draft heaters are simple, they do break down. Here are the typical complaints and what should be done about them.

Unit gives no heat and fan does not operate.

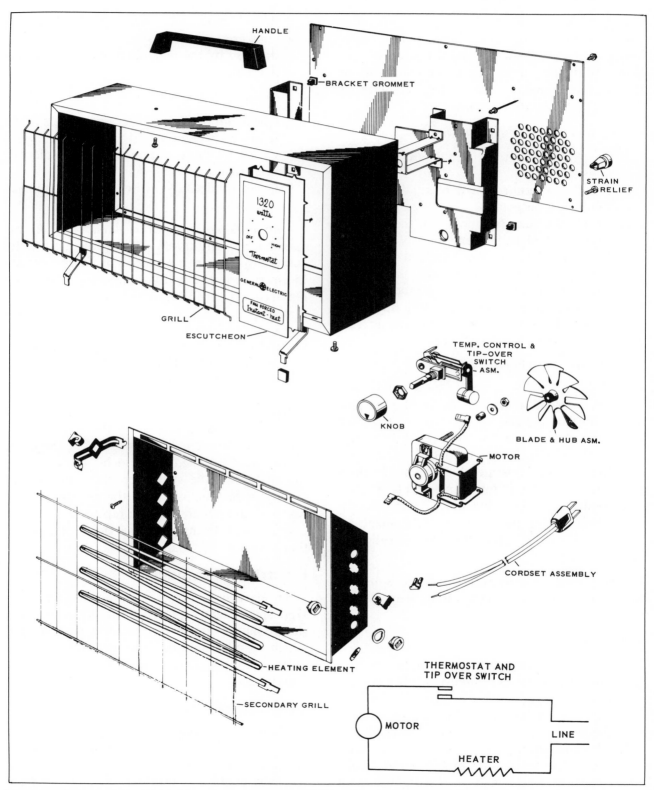

Figure 6-55. Schematic diagram and disassembled view of typical forced-draft heater.

Check the continuity of the cordset thermostat (or temperature control), tipover switch (if one is used), overheat protector, and all lead connections. With some units also check continuity of the heating elements and motor. Replace, if necessary, any faulty components.

Fan works but unit gives no heat.

1. Check the heating elements. If faulty, replace.
2. Check the control switch or thermostat. Replace, if necessary.
3. Check the fuse (if one is in the heating circuit). Replace if open.
4. Check for continuity of overheat protector switch (if one is employed). Repair or replace as necessary. In some models, reset overheat protector by depressing button.
5. Check the tipover switch (if one is employed). Replace if necessary.

Unit heats but fan does not turn.

1. Check for a motor bind. Check the air gap for unevenness and shift in the frames. Adjust or replace, if necessary. Check for binds between rotor shaft and bearings. Replace or realign as needed.
2. If the motor operates but the blade does not turn, check for a worn blade hub. Replace the blade assembly, if necessary.
3. Check for a poor bearing alignment or insufficient end play. Replace or realign, if necessary.

Thermostat or temperature control is out of adjustment. Replace thermostat or temperature control, if necessary.

Unit is noisy.

1. Check for extraneous material in the fan housing. Remove or take any other corrective measures.
2. Check for a fan that is binding or striking the frame. Realign or replace, as needed.
3. Check for loose parts. Tighten as necessary.
4. See if the fan motor makes noise. Lubricate the rotor shaft with a light film of SAE-30 motor oil. If felt wicks or lubricating cups are used, they should be lubricated as directed by the manufacturer in the service manual.

Unit operates intermittently.

1. Check for loose connections. Tighten or repair as necessary.
2. Check for faulty thermostat and replace if necessary.
3. Check for defective switch. If faulty, replace the switch.

Unit does not heat at medium or high, but operates normally at low. This problem is usually caused by an open element or faulty switch. In either case the faulty component should be replaced. In problems where one circuit normally operates while the others do not, always check the schematic diagram in the service manual or figure out the wiring arrangement by tracing it out. This will usually tell what portion of the circuit is not performing as it should.

Unit does not turn off when heater is tipped over. Check the position of the switch with relation to the plunger. In most cases, the latter should be positioned so as to hit the center of the switch. If the switch contacts are welded or otherwise touching with the plunger all the way out, the switch must be replaced.

For more information on servicing of fans and their motors, see Chap. 3.

Electrical Fireplace

An electric fireplace is simply a forced-air-type space heater placed in a metal housing shaped like a fireplace. Servicing such a unit, which usually must be done in a home, is the same as for electric forced-air heaters just described.

Steam Heaters

While not as common as the resistance-wire heaters, portable electric steam heaters are available for use on ac or dc circuits and are thermostatically controlled. Standard ratings are from

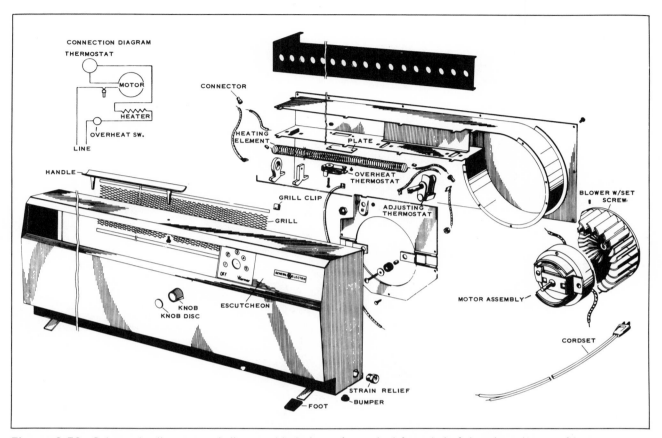

Figure 6-56. Schematic diagram and disassembled view of a typical forced-draft baseboard type of heater.

1000 to 3000 W for use on circuits of either 120 or 240 V.

Most steam heaters use heating elements of the immersion type. Heating elements of this construction are designed for operation when submerged in a liquid and must never be used in the open air. The liquid used in these electric steam heaters or radiators is permanent-type antifreeze. Be sure that no liquid is lost when servicing the heating unit. Any replacement must be made with the proper mixture of antifreeze as recommended by the manufacturers.

Electrically, the steam heater is the same as the natural-draft resistance-wire heater. That is, it has a cordset, a heating element, a thermostat, and usually a tipover switch. The servicing of these electrical components is the same as for the natural-draft heater.

PERSONAL CARE APPLIANCES

In recent years a complete series of hair drying, setting, detangling, and styling appliances have been introduced. These so-called personal care appliances are all of the heater type.

Servicing Hair Driers

There are two types of hair driers, salon and portable. In the portable type, a fan blows air through a soft plastic hose to a thin plastic bonnet. The bonnet is slipped over the head and tied with a drawstring. It is generally in two layers, with vent holes inside, so that the hot air is distributed evenly over the hair. An electrical heater in the fan housing warms the air to speed

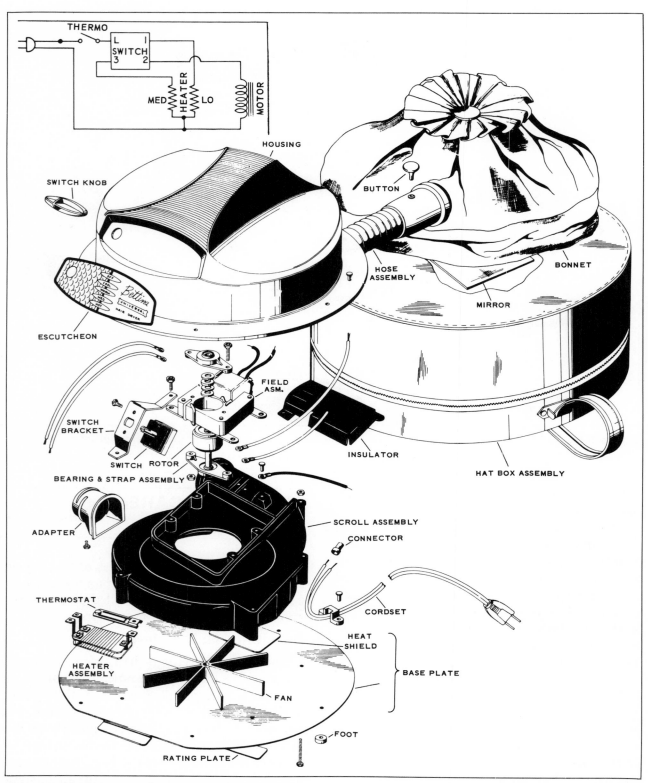

Figure 6-57. Schematic diagram and disassembled view of a typical portable hair dryer.

the drying. The salon type is basically the same except there is no hose and the user sits under a hood. The basic parts of both types of hair driers are a fan, a heater, and a bonnet or hood.

All but the very simplest have a selector switch so that the user, after turning the heater on, can choose cool, warm, or hot air. When a typical unit (as shown in Fig. 6-57) is plugged into a 120-V ac power source and the control knob turned to COOL, only the motor circuit is energized. With the control knob in the WARM position, in most units, one heating element and the motor circuit are energized. In the HOT position, both heating elements and motor are connected in parallel. The impeller or fan blade mounted on the rotor shaft draws air in through the grill and pushes it through the tunnel created by the orifice and case bottom. The air is heated by the heating element and passes through the ducts to the bonnet or hood. A temperature control or thermostat cuts out the heater circuit if an overheat condition exists and the motor continues to run. A fuse is often connected between the temperature control and heater as a secondary safety measure in case the temperature control fails to open.

Some salon-type hair driers feature a steam or moisture operation. In this style, with the steam unit properly filled with water, the attachment cord plugged in and installed into the sliding duct, the control knob is turned to the MIST setting. The steamer element and thermostat are usually in series, and the other circuits in the hair dryer are open. As the tank heats, the water boils, producing steam which goes through the duct to the hood and is expelled through the holes in the liner. When the tank boils dry, the heat rises causing the bimetal in the base of the tank to bend. The contact on the bimetal completes the circuit through the hair dryer, causing it to operate and at approximately the same time the bimetal opens the circuit to the heating element in the tank. The bimetal has a strip of Nichrome attached to it so that current passing through causes the bimetal to stay hot and keeps the hair dryer circuit energized and the tank circuit open. A fuse is often mounted

between the element and temperature control for safety in case the temperature control malfunctions or an overheat condition exists.

Most portable hair driers use the "open-wire" type element, mounted on a small mica card inside the air duct so that the air stream blows over the unit. Small metal eyelets are employed to make the connections, since the heater wire cannot be soldered. The wire leads are frequently fastened securely to the heater element itself, and all connections and disconnections must be made from the other end. If an element checks faulty, the entire element assembly must be replaced. This also holds true for the salon style, except that sheathed heating elements are usually used.

Portable hair driers generally use shaded-pole type motors, while most salon styles employ the universal type. The servicing of both is described in Chap. 2. The fans themselves are either a plastic disk with fins or a metal disk with blades. For solutions of fan problems, see Chap. 3.

If the plastic hose or bonnet should become damaged, replace it. In your work as a service technician, do not use any of the various do-it-yourself patching techniques for plastics.

Analysis of complaints. Here is a summary and analysis of complaints that customers usually give about their hair dryers.

Unit gives no heat at any position and/or motor does not operate.

1. Check all wiring connections and cordset for continuity. Take necessary corrective action.
2. Check the temperature control or thermostat for an open circuit. If defective, replace.
3. Check the switch for proper operation. If defective, replace.
4. Check the fuse. If blown, replace.

Unit operates intermittently.

1. Check the switch. If it is intermittent—that is, turns on one time, then refuses to operate the next—and is not very dirty, it may be all right after cleaning. The easiest way to clean it is to spray an aerosol cleaning compound into the body of the switch.

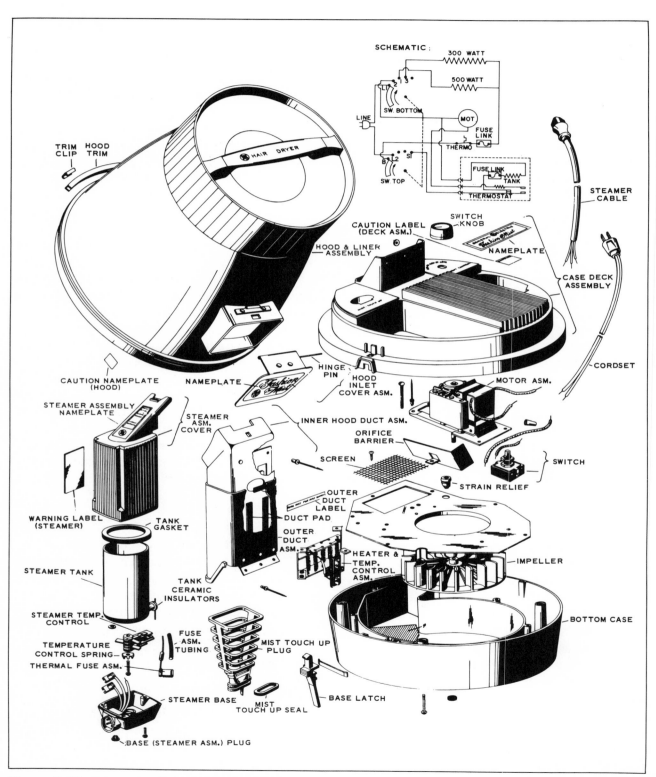

SCHEMATIC:

300 WATT

500 WATT

LINE

SW. BOTTOM

MOT

FUSE LINK

THERMO

SW. TOP

FUSE LINK

TANK

THERMOSTAT

STEAMER CABLE

TRIM CLIP

HOOD TRIM

HAIR DRYER

CAUTION LABEL (DECK ASM.)

SWITCH KNOB

NAMEPLATE

HOOD & LINER ASSEMBLY

CASE DECK ASSEMBLY

CAUTION NAMEPLATE (HOOD)

NAMEPLATE

HINGE PIN

HOOD INLET COVER ASM.

MOTOR ASM.

CORDSET

STEAMER ASSEMBLY NAMEPLATE

STEAMER ASM. COVER

INNER HOOD DUCT ASM.

ORIFICE BARRIER

SCREEN

SWITCH

STRAIN RELIEF

WARNING LABEL (STEAMER)

TANK GASKET

OUTER DUCT LABEL

DUCT PAD

OUTER DUCT ASM.

HEATER & TEMP. CONTROL ASM.

IMPELLER

STEAMER TANK

TANK CERAMIC INSULATORS

BOTTOM CASE

STEAMER TEMP. CONTROL

FUSE ASM. TUBING

MIST TOUCH UP PLUG

TEMPERATURE CONTROL SPRING

THERMAL FUSE ASM.

STEAMER BASE

MIST TOUCH UP SEAL

BASE LATCH

BASE (STEAMER ASM.) PLUG

Figure 6-58. Schematic diagram and disassembled view of a typical salon-style hair dryer with a moisture feature.

Spray it well, work the switch several times, then recheck the switch. If it still does not work properly, replace it.

2. Check for loose connections. If you find any refasten them. Replace the temperature control or thermostat if all connections are tight and intermittent cycling continues.

3. Check the fuse (if one is employed) between heater and temperature control or thermostat. If faulty or not set properly, take necessary action.

Unit gives no heat at any position, but motor runs.

1. Check the heater lead for proper connections. Correct as needed.

2. Check the heater contacts on the thermostat for contamination. Check the fuse link and heater element for continuity. The thermostat must be closed with a cool hair dryer. If you find any faulty components, replace them.

Motor will not run, but heater operates.

1. See that the leads and connections are in order. Take corrective action as needed.

2. Check switch and terminals. Also check the fuse link, if the unit has one. Take corrective action as needed.

3. Check the motor field for continuity. Replace the motor if faulty.

4. Check the bearings and fan or impeller for bind. Remove source of the bind. Lubricate the bearings if required with a *very* small amount of light oil. Wipe off any excess oil.

5. Check for a striking impeller (or fan blade) or foreign objects catching in the impeller or fan blade. In many instances, bobby pins and other objects have been dropped into the case and have jammed the fan. Spin the fan and see.

6. Check the thermostat for proper cutoff. Also check the thermostat or temperature control for continuity. Replace, if faulty.

Appliance is noisy.

1. Check the tightness of the motor mounting.

Tighten or replace as necessary.

2. Check whether the impeller or fan blade is secure on the shaft. If not, tighten.

3. Check for a warped impeller or fan blade. Replace. (The fan blade or impeller must be straight and balanced.)

4. Be sure that the motor bearing is free and that the fan (or impeller) is not striking or rubbing the housing. Take proper corrective action.

5. Check for loose objects in the blower compartments. If any are found, remove them.

Heat selection is improper.

1. Check wiring as shown in wiring layout or schematic. Take any corrective action that is necessary.

2. Check the heating element. If faulty, replace the complete heater.

3. Check the selector switch. If defective, replace it.

Air flow insufficient.

1. Check for foreign objects that are causing drag on the motor or impeller (or fan blade). If any are found, remove them.

2. Check for a binding motor or striking impeller (or fan blade). Take proper corrective action.

3. Check for an obstruction in air duct or air intake grill. If any is found, remove.

4. Check for an excessive air leakage around the air duct. Replace the duct seals, if necessary.

Thermostat cuts unit off repeatedly. This usually indicates a blockage of air flow resulting in overheating. Check the fan, hose, and bonnet opening. If full flow of air is felt, the thermostat may be defective. Replace any defective components.

Other complaints that are frequently lodged against hair dryers with moist or steam features are the following.

Unit will not steam.

1. Check continuity of steamer attachment

220

CHAPTER 6

cord and heating element. Replace any faulty components.

2. Check the steamer temperature control. Replace, if necessary.
3. Check the fuse in the steamer for continuity. Replace, if necessary.

Unit will not switch to hair drying from steam operation.

1. Check the temperature for proper operation. If faulty, replace.
2. Check the continuity of the attachment cord. If faulty, replace.
3. Check for a loose connection in the switch. If found, repair.

Steamer leaks.

1. Check the case for cracks. Replace the case if necessary.
2. Check the gasket. Replace the gasket if necessary.

Hair Setters and Curlers

Hair setters and curlers are designed for the quick setting of hair with mist, with conditioner, or dry. In 6 to 8 min after the typical unit shown here is plugged into a 120-V ac outlet and the knob turned to DRY, the temperature control will open and the indicator lamp will light, indicating that the rollers are ready for use. The temperature control or thermostat will continue to cycle on and off to maintain a satisfactory curler temperature.

Analysis of complaints. While most hair setters contain only one or two heater elements, an indicator light, a thermostat, a switch, and sometimes a fuse, troubles do arise. Here are the more common ones.

Unit will not turn on; no indicator light.

1. Check the switch for continuity and proper operation. Replace, if necessary.
2. Check continuity of the cordset. Replace, if faulty.
3. Check all wire connections for continuity and tightness. Take necessary corrective action.

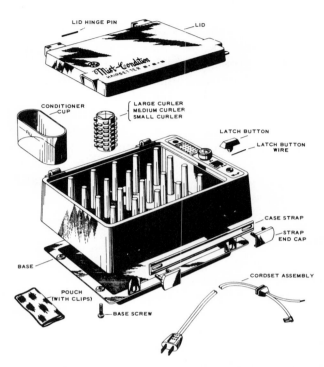

Figure 6-59. Schematic diagram and disassembled view of a typical hair-setter unit.

4. Check the fuse (if used). If blown, replace.

READY *light is on, no heat.* Check temperature control or heating unit for defective parts. Replace, if necessary.

Unit heats, no READY *light.* Check the neon assembly and replace necessary component.

Unit gets too hot or not hot enough. Make the following test:

1. Melt an indentation in the center of the top of a curler with a soldering iron, down to the metal. Place the junction of the thermocouple in the indentation next to the metal and melt the plastic around the thermocouple wire holding it firmly in place. (Do not wrap the wire around the curler.)
2. Connect the ends of the thermocouple wire to the test meter. Place all the curlers in their respective locations and substitute the test curler for the large one on the extreme right in the middle row.
3. Turn the unit on and allow it to heat for

10 min. The temperature should be between 150 and 180°F.

4. If the unit does not meet this specification, the temperature control or thermostat must be replaced. (Do not attempt to calibrate the temperature control or thermostat.)

Unit leaks. Replace the complete hair setter unit.

Detanglers

Most hair detanglers on the market today consist of a charger base and a power handle. The base incorporates an induction charging system in which a magnetic field in the well creates enough voltage in a coil contained in the handle to supply the recharging current required to charge the battery in the handle. The handle has an ON-OFF switch and a movable comb permanently attached to the motor output which goes in a back-and-forth motion next to the outer comb. The comb has two rows of teeth moulded in one piece and permanently attached to the handle.

The detangler charger base incorporates a sealed induction coil connected to the 120-V ac power outlet. This coil induces a small current in a mating coil contained in the handle. The current is rectified and serves to continuously recharge the battery also in the handle. The handle must be stored in the well of the charger and the charger must be connected to a continuously energized outlet to maintain a full battery charge. *Note*: Some fixtures are so wired that the outlet is controlled by a switch which, if not on, results in an inadequate battery charge and either poor or no operation. The motor in the handle drives the center comb in a back-and-forth movement. The power handle should be thoroughly rinsed under running water to remove any soil or accumulation of hair. The charger base should be disconnected from the power outlet and wiped clean with a damp cloth. Heavy accumulation of soil or hair on the handle or in the charger well will result in poor seating of the handle and prevent proper charging.

Analysis of complaints. Here are the complaints that are usually made in conjunction with hair detanglers.

Power handle will not run, runs slowly, or produces inadequate power. To test the recharger and handle, plug the charger to be tested into an energized 120-V outlet. Insert a steel screwdriver into the charger well and up against center metal post. A magnetic vibration should occur. If there is no vibration, the charger should be replaced. Check for excessive heat after the charger has been plugged in for at least $\frac{1}{2}$ h. Because of the induction system usually used in the manufacture of detanglers, the battery condition cannot be measured directly. However, the following procedures will enable you to determine whether the handle is defective and should be replaced or the battery merely needs to be charged.

1. If the handle does not run, operate the switch several times to be certain it is indexing properly.
2. Turn the switch off.
3. Place the handle in a known good recharger unit.
4. After 1 min, turn the switch on.
5. If you see no motion of the shaft, it is bad and should be replaced.
6. If motion is detected after Step 4 and is sufficient to operate the shaft at least once, turn the switch off. Leave the handle in the charger for an extended period before deciding on its condition. A 16- to 18-h charge should result in full capacity, power, and speed.

Handle is noisy. With the switch off, check the main shaft for looseness. Slight side-to-side movement is normal. Excessive movement indicates a defective assembly. Replace the handle.

Unit runs with normal speed but stalls in use. Check the gear mesh by grasping the center comb and stalling. If the motor also stalls, check for a poor battery charge as detailed previously. However, if the motor continues to run, the gear mesh is bad and the handle should be replaced.

Charger overheats. The center post in the well

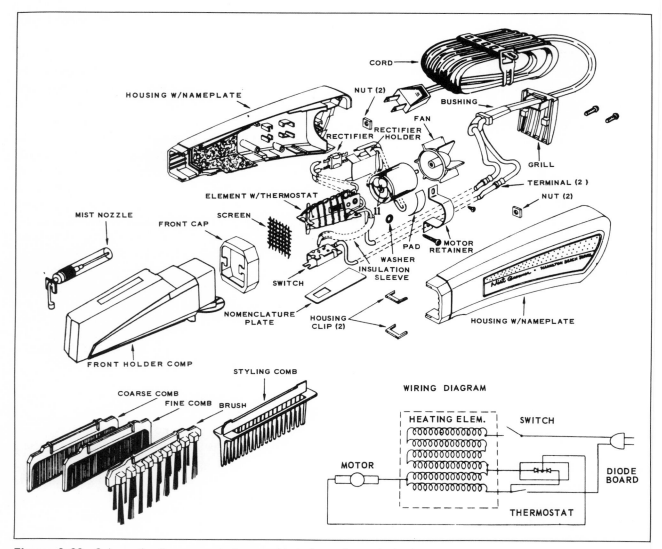

Figure 6-60. Schematic diagram and disassembled view of a typical mist groomer.

should be warm to the touch but not hot. If the heat is excessive, replace the charger since this indicates a defective coil.

Service on many detanglers is handled with the manufacturer on an exchange basis; that is, rather than providing any service, the manufacturer exchanges the handle and/or the charger for a new unit.

Styling-Grooming Dryers

While many types or styles are available, this hand-held appliance designed for drying and/or styling hair usually features a contoured handle which encases a switch, thermostat, fuse, motor, and heater. With the cordset connected to an appropriate power source and the switch at STYLE or DRY, air is drawn into the screened air intake, through the heating unit, and out the exit grill over the comb (or brush).

To test a styling dryer, plug the unit into a 120-V, 60-Hz ac outlet through a wattmeter. With the switch at STYLE the wattmeter should read between about 170 and 210 W. With the switch at DRY, the wattmeter should read between about 380 and 420 W.

To check the thermostat, remove comb (or

brush), then stand the dryer on the air outlet to block the circulation. With the switch at HIGH, the thermostat should cut out between 5 and 30 s. It may require a couple of minutes to cool down with the circulation *unrestricted* before the thermostat will cut in. Extended operation with the circulation blocked may cause the fuse to blow which will require a complete replacement of the appliance.

Some styling-grooming dryers, or styling combs as they are sometimes called, have a mist feature. To mist-style the hair, the user exchanges the drying nozzle for a mist nozzle while warm air is emitted from around the comb or brush. If such models will not supply mist, check the reservoir assembly as follows.

1. Make certain the reservoir contains water and is not too full.
2. Clean off the outside of the mist nozzle to remove any mineral residue left by water from previous use.
3. Replace the reservoir assembly.

Most manufacturers of styling-grooming dryers do not recommend repair of their units; service should be by replacement only.

Shave Cream Dispensers

Shave cream dispensers are designed to accept a standard 6- or 11-oz aerosol can of shaving cream. They supply hot, moist lather at a pre-regulated temperature and contain a heating system that shuts off automatically. With the aerosol can properly in place, the unit connected to a 120-V ac outlet, and the heat button pushed, contact is made internally so that within approximately 45 to 60 s the shave cream will have been heated and the dispenser will be ready for use. During this period a buzzer sounds; at the end of the heat-up period, the buzzer shuts off, as does the heating cycle.

Analysis of complaints. Here is an analysis of common customer complaints with heated shave cream dispensers.

Dispenser will not turn on; no buzzer. Plug the unit into a 120-V, ac regulated outlet to confirm the complaint.

Lather leaks. This will be evidence by lather oozing up around the button when checked with a full aerosol can of shave cream.

Lather is too hot or too cold. Accept the customer's complaint without attempting to verify it.

No lather. Check with a full aerosol can of shave cream.

With most manufacturer's models, because of the intricate *sealed* electric system, no repairs should be made or attempted. Service is by replacement only.

Massagers

Several types of electrical massage units are on the market today. Some fasten to the back of the user's hand with straps, while those with plastic cases have several attachments which are applied to the skin and scalp. Most massagers are similar in that they take electric energy and change it to mechanical vibrations at a fairly high speed to give a soothing massage when the hand or an attachment is rubbed over the skin. Some have built-in heating units.

The two most common methods of producing the vibrations are eccentric weight and coil action. The former method usually employs a standard series-wound motor driving an eccentric weight mounted on the front end of the armature shaft. The eccentric weight produces a strong vibratory movement. A soft-rubber pad covering the base transmits this vibration as a massage action to the body. A fan on the rear end of the armature shaft is often used to cool the motor in extended operation.

In the second method, a strong magnetic field produced by a coil wound on a special core and fed by a 60-Hz alternating current causes only the stud bracket to be attracted and repelled, setting up the desired vibration. The various attachments, when slipped onto the stud, transmit this vibration as a massage action to the body.

Some massagers, as previously stated, have an additional switch position which connects a small heater plate to a secondary winding on the coil providing low, gentle heat with soothing massage. The heater circuit is electrically isolated from the power line for safety reasons. A third switch

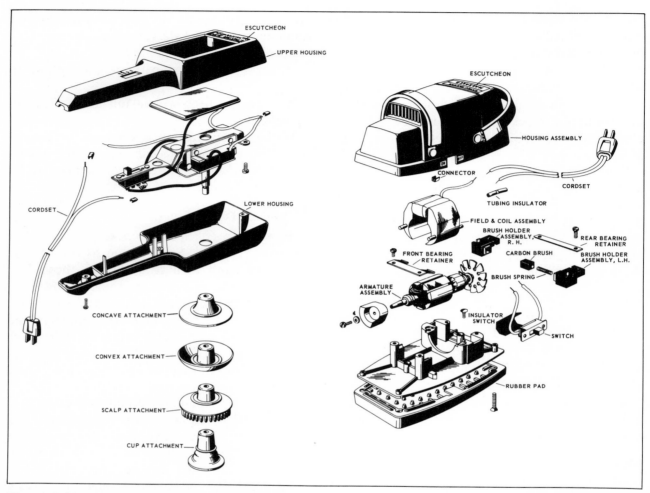

Figure 6-61. Disassembled views of the two basic types of massagers: (left) coil action and (right) eccentric weight.

position is sometimes connected to a series with the coil. This changes the waveform of the applied alternating current in such a way that the amplitude of vibration is increased, giving a "high" mode of operation for deep massage.

Analysis of complaints. Little can go wrong with a massager. Here is an analysis of complaints for a typical standard model.

Motor will not run. Check the circuit for continuity. Check for intermittent switch and cordset. Check the condition of the brushes and commutator. Spin the armature with a finger to check for a bent fan or a bearing misalignment causing bind. Take proper corrective action.

Massager is noisy. Since this is a vibratory device, it is essential that all parts be firmly secured and all screws be drawn up tight. If an unusual noise is encountered, check for a bad bearing alignment, an armature striking the field, a loose weight or fan, improper lead dress, or a foreign object inside the case. Take proper corrective action. Also check end play of the armature assembly; it should be a maximum of about 0.010 in. If excessive, replace the assembly.

Spring straps are defective. Replacement is the only solution, but it is wise to warn the customer that with many models, these straps are available only as part of the complete housing assembly.

With many coil-type massagers, service is limited to exchange only for any functional failure since none of the operating components is available for separate replacement. In fact, even when the components are available, it is usually necessary to replace the complete coil mechanism. In either event, the cordset, case, and appearance items may be replaced, if required.

Cord replacement in many models may be made by clipping the old cord $\frac{1}{2}$ in from the terminal connector and slipping shrinkable tubing over the cordset leads. Then make connections with crimp connectors. Pull the tubing down over the entire splice and heat with a match until shrinkage takes place and a well-insulated connection results. Redress as the original, making sure a strain-relief knot is in the recess of the housing.

The heating element used in many massagers is just a wirewound resistor covered with a special insulating compound. If it has been found to be defective, it should be replaced. As a rule, the resistor is soldered in place.

Wrinkle Removers

A wrinkle remover operates much like a steam iron; it uses steam to remove the wrinkles from clothing. Electrically, it consists of a thermostat, fuse, and heater which is located in the water tank. The heater converts the water in the tank to steam. Here are some of the more common problems that occur with a typical wrinkle remover and analyses of them.

Unit will not steam. Check the continuity of the cordset, fuse, temperature control, and heating element. Replace any defective components.

Unit leaks. Check tank top seal. If necessary, replace cover assembly.

Fuses blow.

1. Check wiring layout for proper lead dress. Take necessary corrective action.
2. Check the temperature control. If faulty, replace.

MISCELLANEOUS HEATER-TYPE APPLIANCES

Several other heater-type appliances on the market are usually classified as miscellaneous. Let us take a look at some of them.

Fire Starters

Fire starters of the type used for barbecues as well as fireplaces are common today. Most contain a sheathed heater element and a fan. There is little that can go wrong with the unit. However, here is a typical analysis of complaints usually received about fire starters.

Heating element will not work.

1. Check all connections for continuity. Take proper corrective action.
2. Check the heating element for continuity. Replace if faulty.

Fan motor will not run.

1. Check the field coil for continuity. Replace the motor if defective.
2. Check the motor for binding. Replace the motor, if defective.

Both fan motor and heater will not work.

1. Check the cordset for continuity. Replace if necessary.
2. Check the internal connection. Repair as necessary.

Motor is noisy.

1. Check for loose fan blade. Tighten the blade, if necessary.
2. Check for loose bearing-mounting screws. Tighten screws, if necessary.
3. Check for worn or defective bearings. Replace, if faulty.

If the connector or lead wire becomes loose from the heating unit, it should be reassembled with silver solder and a low flame at right angles to the connection and parallel with the vertical plane of the bracket. A replacement connector should be slipped over the wire and away from the soldered connection. If a bracket breaks loose from the heater tube, it can also be resoldered with silver solder.

Vaporizers

A vaporizer works on the same basic principle as the other appliances mentioned in this chapter; that is, when an electric current flows through a resistance, heat is generated. But with the heat-type appliances previously described, metallic heating elements are employed; with vaporizers, water is used as the conducting material. Like metal, water offers electrical resistance, thus forming heat.

The vaporizer consists electrically of only a cordset and two metal electrodes. (Some have a safety fuse in the circuit.) The current-carrying line is connected to the two electrodes in a housing which is located in either a plastic glass or ceramic container. When water is placed in the container and the cordset plugged into a power outlet, the circuit is completed and current flows through the water. As we know, water is a fair conducting substance, but it has electrical resistance and thus can generate heat. When the water starts to boil, the resulting steam is used to vaporize the medication placed in the medicine holder. The current will flow as long as there is water in the container. Once all the water has boiled away, there is no conducting substance left, so the vaporizer shuts itself off. Bottle sterilizers, bottle warmers, and some egg cookers also use water for its electrical resistance.

About the only problems that arise in a vaporizer are an open cordset or faulty electrodes. The electrodes require an occasional cleaning since many of the medications used in a vaporizer have a grease base, which can coat the electrodes and hinder the current-conducting process. Because of the chemicals in the water in some localities, mineral deposits must be scraped off the electrodes. Keep in mind that the closer the two electrodes are to each other, the greater the current flow and the quicker the water will boil. If the electrodes become defective, the entire housing unit should be replaced.

LIGHT-PRODUCING APPLIANCES

Make-up Mirrors

While not a heater-type appliance in the true sense of the word, they are included in this chapter. The typical make-up mirror (Fig. 6-62) is a free standing appliance featuring dual flip-over mirrors and two miniature fluorescent lamps. Included is a convenience outlet and adjustable filters to provide any one of four light shades, which may be obtained by adjusting the light selectors to desired settings.

Here is an analysis of complaints generally received from make-up mirror customers:

One light.
1. Check to be certain lamp pins are making full contact with each socket. Pins should be perpendicular to the slot in the socket and securely held by the socket contacts. If properly set, reverse lamps to confirm a possible bad lamp.
2. Check sockets and wiring. If defective, repair or replace as is needed.

No light.
1. Check wiring layout for loose connections or damaged sockets. Repair or replace as is necessary.
2. Check cordset. If defective, replace.
3. Check to see that both lamps are properly inserted in sockets. Correct if necessary.
4. Check switch. If faulty, replace.
5. Check the ballast output voltage. Plug the cordset into a 120-V ac source. Have switch ON and with a VOM set at 300 V ac test diagonally across each pair of sockets. The voltage should be about 120 V. If not, replace the ballasts.

No change in light colors or improper light colors.
1. Check to see that filters are properly aligned. If not, align so when slider is at full left, openings in the filters are up and in line with the reflectors.
2. Check spring and chain linkage for breakage

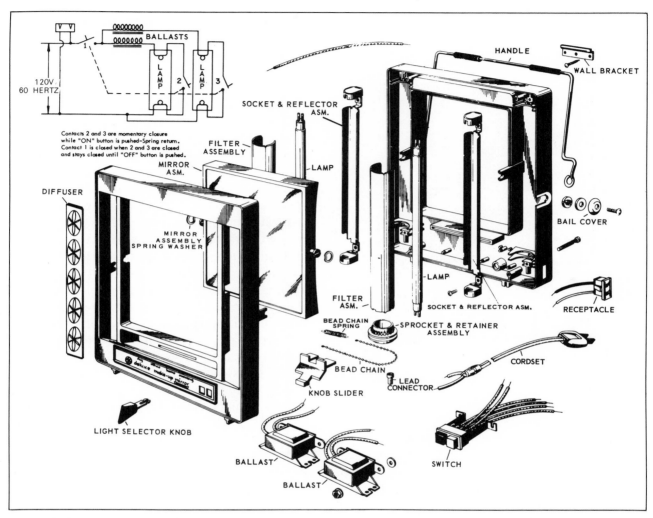

Figure 6-62. Schematic drawing and disassembled view of a typical make-up mirror.

to see if the spring is out of the chain ends or the chain out of the slider or sprockets. Take necessary action to correct.

Table and Floor Lamps

While not considered appliances by most manufacturers, table and floor lamps are often brought into a shop for repair. Such work is usually simple since most of the troubles are found in the line cord. Rewiring is often required also when the lamp socket has loosened and resulting friction has broken the wire. When replacing the lamp cord, determine the number and size of bulbs used in the lamp, and select wire of the correct size and insulation as listed in the table given in Chap. 1.

Disassemble the lamp by loosening any set screws or lock nuts you find. Remember where the parts go, and the order in which they were placed. Disconnect the leads from the socket terminal lugs and remove the old wire. Salvage the plug unless it is broken. Remove and test any switches there may be in the lamp. If they are faulty, replace them with switches of the same type.

Lamps with more than one socket are usually

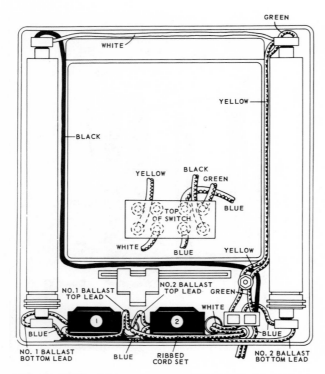

Figure 6-63. Wiring layout of a typical make-up mirror.

wired with No. 18 conductors, and space is left within the lamp where the wires to the various sockets may be joined. They are connected to the one line cord. Secure taping is all the insulation required at splices. No. 18 wire is available in either solid or stranded form. When removing the plug, note how it was fastened to the conductors in the wire, and fasten it to the new wire in exactly the same way. If the plug is broken, it should be replaced. Choose one of similar size, color, and type.

Some table lamps have push-through switches, while others have round knobs which must be turned to light the lamp. If these are broken, purchase another of same size and style at an electrical supply shop. Replace the switch by noting exactly how the broken one was attached, and fastening the new one in the same manner. Where the switch is part of a socket it is usually easier and more economical to replace the whole socket.

The so-called "high-intensity" table lamps often find their way into a service shop. These lamps use bayonet-base 14-V automobile bulbs that work off step-down transformers in the base of the lamp assembly. Some lamps have HIGH-LOW switches that give two intensities of illumination. The lamp voltage in HIGH is slightly above normal, in LOW slightly below.

Troubles with these high-intensity lamps usually occur with a faulty switch, burned-out bulb, defective cordset, open or shorted transformer, and open-circuit wiring. To service, carefully check the lamp and take the necessary action.

Microwave ovens

Since most microwave ovens in the field today are of portable and counter-top design, we have classified them in the small-appliance group. True, some microwave ovens are built into kitchen walls or are a part of electrical ranges (see Chap. 9 in *Refrigerator, Air Conditioning, Range and Oven Servicing*), but they all operate in the same basic manner.

When food is prepared in a conventional oven, the oven is usually preheated before the food is placed in it. During baking, the heat surrounds the food and gradually penetrates to the center to accomplish cooking. When food is prepared on a range surface unit, heat is applied from either an electric coil or gas burner to the food container. Heat is then conducted into the food being cooked by conduction. In the microwave or electronic oven, however, foods are heated by the absorption of microwaves.

HOW THEY WORK

Microwaves are electromagnetic waves of energy, similar to radio, light, and heat waves. This definition is general and covers a number of familiar types of radiation that are not normally considered microwaves, such as radio, TV, and infrared waves. The definition becomes more specific if we add the limitation that the wavelength of the radiant form of energy must lie somewhere between 1 m and 1 mm. Radio and TV waves are thus excluded because their wavelengths are much longer, and infrared waves are excluded because their wavelength is much shorter.

There is a direct relationship between wavelength and frequency: Wavelength is equal to the speed of light divided by the frequency. For example, let us take a radio station such as WMT, which operates at 600 kHz. The wavelength would therefore be the speed of light, 300 million m/s, divided by the frequency 600,000 Hz which is equal to approximately 500 m. The frequency of most microwave ovens is 2,450 million Hz. The wavelength is therefore approximately 0.12 m or 4.7 in. The wavelength of infrared light is much smaller or about 0.0003 m.

The Federal Communications Commission (FCC) limits or controls the design of microwave ovens because a number of communication systems operate at frequencies close to those of microwave ovens. One of these is the police speed radar which operates very close to 2,450 MHz. Another is the telephone companies' relay tower system which carries TV programs and telephone conversations; it operates at approximately 5,000 MHz. The FCC has allocated a number of frequency bands to the operation of microwave ovens and some related types of equipment. The first of these frequencies is 915 MHz, the second is 2,450 MHz, and a still higher one is 5,800 MHz. There are a few allocations at higher frequencies, but these are not generally used in microwave ovens and are therefore not worth mentioning. Most microwave ovens produced today, as previously mentioned, operate at 2,450 MHz.

Microwaves have many of the characteristics of light waves; they travel in a straight line, they can be generated, and they can be reflected, transmitted, and absorbed. They differ in the materials that reflect, transmit, and absorb them and in the way they are generated.

The simplest and the most commonly used generator for lightwaves is a light bulb. In a microwave oven, the generator for the energy is a magnetron, a vacuum tube which operates as an oscillator to produce the microwave energy. Every radio and TV set has an oscillator circuit. This circuit consists of a number of vacuum tubes, resistors, capacitors, and conductors. In the microwave oven, all these oscillator components are built into the tube. Although microwaves can be reflected in the same manner as light, the materials that reflect the two kinds of waves differ. For example, aluminum and stainless steel reflect microwaves while cold-rolled steel to some extent absorbs the microwave power. Another example is the perforated holes in the door of the microwave oven. These holes do reflect the microwave power, but they transmit light. Opaque paper products transmit microwaves while light waves are either absorbed or reflected. Glass and china act much the same as paper but some do absorb power. Water absorbs microwaves while light passes through.

The selective characteristics of microwaves

make it possible to construct an oven where the wall, ceiling, floor, cooking container, and door remain cool. These items do get warm, but this is caused by the transfer of heat from the food. As previously mentioned, foods are heated in microwave ovens by the *absorption* of microwave power. Food is constructed of many millions of molecules per cubic inch. These molecules react to the microwave field in much the same manner as a compass needle reacts to a magnet. If you place a magnet to one side of the compass, the needle will point to the magnet. If you then move the magnet to the other side of the compass, the needle will turn and again point to the magnet. When this process is repeated quickly many times, eventually the friction in the bearing that supports the needle will be heated.

The molecules in food react in a very similar manner to the changing microwave field; that is, they tend to align themselves with the field. The molecules are rotated from their starting position to 180° from their starting position and back to their starting position 2,450 million times a second. This constant and rapid rotation causes the food to heat. As the wave penetrates the food, power is lost to each successive layer of molecules. The center molecules are therefore not rotated a full 180° unless heat is generated toward the center of the food as opposed to the outside of the food.

Contrary to popular belief food prepared in a microwave oven is not cooked from the inside out, but is cooked all the way through at the same time, with more cooking being performed on the exterior of the food. It is, therefore, possible to prepare a rare, medium, or well-done roast in the electronic oven.

The fact that food is heated throughout makes it possible for the microwave oven to cook food fast. Time required to cook an item in the microwave oven is solely dependent upon how much heat is required, and in turn the amount of heat required of the food and the weight of the food. In the conventional process only the surfaces of the food is heated directly by the oven or grill.

The heat required to cook the inside portion has to be conducted from the surface. Three factors govern the time required to cook an item in the more conventional way. A minor one of these is how much heat is required. The major ones are how well the food conducts heat and how much the surface of the food can be heated without causing serious defects. Let us take water, for example. Water is a good conductor of heat, and its surface can be overheated without deterioration; therefore, water can boil fairly quickly on a range. On the other hand, let us take milk. Milk is also a good conductor, but the surface cannot take overheating. If you try to boil milk quickly, the milk will burn; therefore, milk has to be cooked slowly. Cake is an example of a food that conducts heat poorly. Although a small amount of heat is required to bake a cake, it must be cooked rather slowly because the conduction to the center is poor.

Figure 7-1 illustrates how the microwaves are reflected inside the oven. Generated by the magnetron, the microwaves travel in straight lines much the same way as light and follow the waveguide channel into the oven. At the end of the waveguide the microwaves strike the stirrer which is a slowly rotating fan. The fan blades reflect the microwaves, causing them to bounce off the walls, ceiling, back, and bottom of the oven. The microwaves enter the food from all sides, accomplishing cooking inside and outside from all directions.

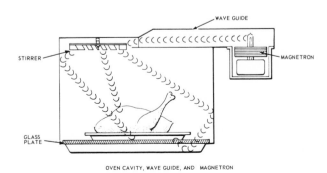

OVEN CAVITY, WAVE GUIDE, AND MAGNETRON

Figure 7-1. How microwaves are reflected inside a microwave oven.

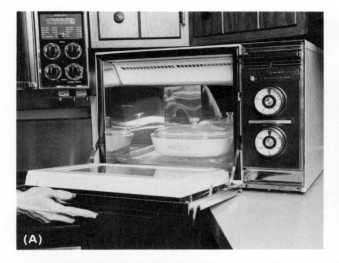

Figure 7-2. Steps in the operation of a microwave oven: (A) Open the door, insert food. (B) Close the door. (C) Set the timer control. (D) Push the START button. (E) To check food, push the STOP button and open the door; the oven shuts off instantly. To resume cooking, close the door and push the START button. In most models, the timer resumes without resetting.

The food, in an appropriate utensil on the glass shelf, is positioned about 1 in above the bottom of the oven. The microwaves pass through the special glass shelf and are reflected off the bottom of the oven to enter the bottom of the food.

Operating a microwave oven, as shown in Fig. 7-2, is simplicity itself. Nothing more is required than to set the desired cooking time on one of the two timers and push the START button. The timer will turn counterclockwise until it reaches OFF, when the oven will shut itself off and the reminder signal will operate. While the oven is operating, the timer in use will be illuminated. Each timer dial lights when turned from OFF.

MICROWAVE OVEN COMPONENTS AND THEIR SERVICING

Let us identify, describe, and find out about the servicing of the major components used in or associated with the electric circuit. While the material is general in nature, it should serve a useful purpose for a better understanding of microwave operation and the servicing of an electronic oven. The components of a microwave oven can be divided into four categories: electronic, control, external enclosure, and convenience.

Electronic Components

The electronic group includes transformers, magnetrons, rectifiers, capacitors, resistors, and chokes.

Power supply. Most portable and counter top models operate on 120 V alternating current. Most built-in or range units and a few counter-top models operate on 240 V alternating current. On some models all the power supply components are located in a removable housing called the *power pack*. Block plugs are provided for easy removal of the complete assembly.

On other models the power supply components are located on various panels in the oven. In either case, the power cord furnishes power to the unit. Remember that when installing a grounded appliance such as a microwave oven in a home that does not have a three-wire grounded receptacle, under no conditions is the grounding prong to be cut off or removed. It is the personal responsibility of the customer to contact a qualified electrician and

have a properly grounded three-prong wall receptacle installed in accordance with the appropriate electrical code. Should a two-prong adapter plug be required *temporarily*, it is the personal responsibility of the customer to have it replaced with a properly grounded three-prong receptacle or the two-prong adapter properly grounded by a qualified electrician in accordance with the appropriate electrical code.

The standard accepted color coding for ground wires, as previously stated, is green or green with a yellow stripe. These grounds leads are *not* to be used as current-carrying conductors. It is extremely important that the service technician replace any and all grounds prior to completion of his service call. Under no conditions should a ground wire be left off, causing a potential hazard to the service technician and the customer.

Transformers. Some microwave ovens employ two transformers in their circuits. They are called the filament and plate (high voltage) transformers.

The filament transformer is the smaller of the two and changes the supply primary voltage (usually 120 V alternating current) to that needed for the magnetron filament (usually 2.5 to 5 V). The plate or high-voltage transformer, on the other hand, is the larger of the transformers used in the unit and steps up the primary supply voltage to a secondary output voltage of approximately 2,400 V alternating current. The two-transformer system enables the magnetron tube filament to be heated prior to application of the high secondary voltage to the tube.

In recent years, there has been a trend by manufacturers to use one transformer of both the step-up and step-down variety. In other words, the one transformer serves to give low filament voltage as well as the necessary high voltage. In either case, the transformer is usually of the saturated type (secondary core operating at magnetic saturation), and the input-line voltage may fluctuate without changing the constant power output of the oven.

To check a transformer, remove the power leads and test for continuity. Although the

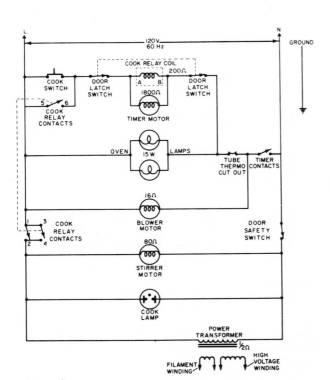

Figure 7-3. Typical microwave oven power circuit.

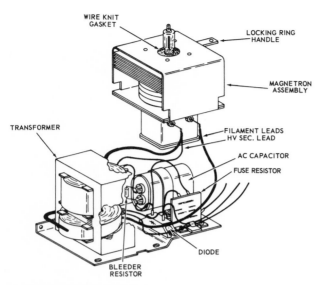

Figure 7-4. Transformer and magnetron assemblies.

resistance may be low, there should be some. If any of the transformer's windings are open or shorted, replace them.

Rectifiers. As mentioned earlier, a rectifier is a solid-state device which can convert the 2,400 V alternating current into 4,000 V direct current. The negative pole of the rectifier is connected to the magnetron, and the positive end is connected to a magnetic-field circuit. The rectifier circuit (Fig. 7-5) is often referred to as a high-voltage doubler, and a simple explanation of its function is as follows. As the unit is started, the transformer supplies approximately 2,400 V to the circuit which charges up the oil-filled capacitor. As the alternating current reverses (flows in the other direction), the diode prevents the energy in the capacitor from being dissipated back through the diode. This charge is then added to the approximately 2,400 V produced in the transformer winding on the positive side of the ac wave, raising the circuit voltage to approximately 4,000 V direct current. The diode used in this circuit is both a one-way current-limiting device and a rectifier, changing the voltage from alternating to direct current. The circuit can be compared with two batteries hooked in series. The resulting voltage of the circuit is not double that of the transformer

output because of the loss in the circuit and the fact that the capacitor does not completely discharge.

As has already been stated, the diode rectifier (as well as any rectifier) will conduct high current in one direction with a low voltage drop but conduct a low current at a high voltage drop in the other direction. If the voltage becomes excessively high, this special single diode will conduct high current for a short time and effectively protect the other components from shorting. Since rectifiers pass current in only one direction, they must be connected properly in the circuit. The symbol used for rectifiers in a circuit diagram is an arrowhead: →|—. The arrowhead indicates the direction of conventional flow through the rectifiers.

To test a typical single diode, proceed as follows.

1. Isolate the diode from the circuit by disconnecting the leads from it.
2. Using the highest ohm scale on the meter, check the resistance between (+) and (−) terminals on the diode board. Then reverse

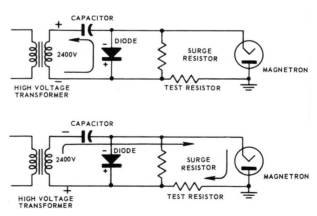

Figure 7-5. How a voltage doubler works: (top) Capacitor charges to peak voltage; (bottom) tube conducts. Note that the diode is connected with opposite polarity from the magnetron in order that the capacitor will charge through the diode during one half-cycle and will discharge through the magnetron on the alternate half-cycle. On the discharge cycle, the capacitor aids the transformer voltage, thereby providing the voltage doubling action.

the meter leads and check again. There should be considerable difference (minimum of one-half of the scale) between the two readings, approximately a 10 : 1 ratio. If reversing the leads does not change the reading, replace the diode. (Be certain to use the highest ohm scale.) Remember that this is only a typical test; consult the service manual for exact details.

In a great many units, the diode board is held in place with plastic-type screws. When making diode replacement in such cases, use only the original screws to fasten it in place; metal screws can cause a short circuit to ground.

As you have seen, high-voltage capacitors play a very important part in the voltage-doubler circuit. If the capacitor fails *open*, no high voltage will be available for the magnetron. A *shorted* capacitor will frequently cause the house circuit fuse to blow. To test a high-voltage capacitor, proceed as follows.

1. Discharge the capacitor (Fig. 7-6). [*Warning*: Never touch or service the high-voltage circuit without discharging the capacitor(s) by shorting across its terminals, to avoid possible electrical shock.]
2. Isolate the capacitor from the circuit by disconnecting the leads. Then check the continuity of the capacitor with the meter on the highest ohm scale.

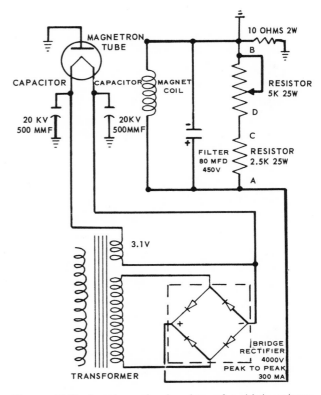

Figure 7-7. A schematic drawing of a high-voltage circuit employing an electrical bridge.

3. A normal capacitor will show continuity for a short time and then indicate open once the capacitor is charged.
4. A shorted capacitor will show a continuous continuity; an open capacitor will show open or infinite resistance.

Another important component in the voltage-doubler circuit is the bleeder resistor(s) which are connected across the high-voltage capacitors. A bleeder resistor is used as a discharge path for the capacitors to ensure that no voltage remains on the capacitor after power is turned off. Since capacitors have the ability to store an electric charge, the bleeder resistor serves as a safety device to discharge the capacitor.

Some voltage-doubler circuits also employ a surge resistor as a current limiter during the initial charging of the capacitors. This resistor is usually of ohmic value and is of the open-coil type. A few models employ a surge current

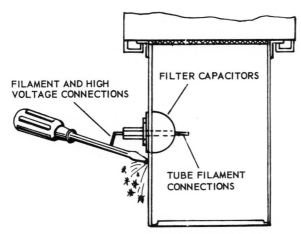

Figure 7-6. How to discharge the high-voltage capacitor.

relay in place of, or in conjunction with, the surge resistor.

While the majority of microwave ovens employ some type of high-voltage doubler circuit, some models use a transformer with a 4,000-V secondary. In such cases, an electrical bridge (Fig. 7-7), usually consisting of silicon rectifiers, is employed to convert the alternating current from the transformer to 4,000 V direct current which in turn supplies the high voltage between the cathode and the plate required for operation.

Magnetic-field circuit. The major components of the magnetic-field circuit are the permanent magnets and the tube housing. They are integral parts of the magnetron tube assembly and may not be disassembled. Because the magnets are permanent, the intensity of the magnetic field imposed on the tube may not be adjusted. But in some models, resistors are used to adjust the strength of the magnetic field supplied to the magnetron. By adjusting the resistor, you regulate the power to the magnet coil, which affects the magnetic field. The strength of the magnetic field affects the power of the magnetron.

To prevent possible demagnetization, tools and other steel objects should not be allowed to come in contact with the magnets or the tube housing.

Magnetron tube. The magnetron tube is the heart of the microwave oven. It converts dc power to radio frequency (rf) power or microwave energy. The magnetron needs a high dc voltage, a low ac voltage, and a magnetic field for normal operation. The tube itself consists of an anode and a filament which also is the cathode.

For the tube to operate, electrons must flow through it from the cathode to the anode. To get the electrons to flow, high voltage has to exist between the cathode and the anode and the cathode must be heated. The heating of the cathode is done by the filament. The filament or heater power is supplied by the 3-V ac winding on the transformer.

The description above is of the operation of a simple diode. To convert a diode to a mag-

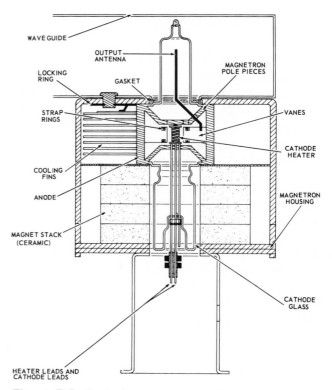

Figure 7-8. Typical magnetron assembly.

netron, a magnetic field is applied parallel to the cathode. The magnetic field causes the electrons to spin around the cathode instead of traveling in a straight line from the cathode to the anode. The spinning action of the electrons and the configuration of the anode cause rf currents to flow on the surface of the anode. An antenna connected to the anode directs the rf power down the waveguide (see p. 231).

Not all energy is transferred by the tube antenna. A small amount of rf current may back-feed down the tube filament leads. If this power were allowed to feed back to the house circuit, it could cause TV and radio interference. To prevent this, ferrite rings and a bypass capacitor are added to the circuit; the choke rings dissipate some of the rf current traveling along the outside of the leads. Then any rf current which passes the rf choke rings is picked up by the capacitor and is dissipated through the capacitor. In checking these rf capacitors, place the ohmmeter leads between the filter capacitor terminal where the filament transformer leads were disconnected and filter box assembly. There should be no continuity reading. If a reading is picked up, it is an indication that the filter capacitor is shorted. Be sure when making this check to use a high-resistance scale on the meter.

Although not too much checking can be done on the magnetron tube, there are two resistance checks. One is to check the filament winding to see if it is open or closed, and the other is to check a ground circuit between the cathode and plate. Some visible checks could be made by looking directly at the magnetron tube, but this would require dissembling the magnetron. To make a check of the filament winding on the magnetron tube, take a resistance reading across the filament leads. The resistance should be between 1 and 2 Ω. In checking the resistance between the cathode of the magnetron tube and ground, there should be no continuity reading. A resistance reading would indicate a high resistive short to ground and a defective magnetron tube.

Magnetrons can fail in one of the following ways and should be changed only if one or more of these conditions exist.

1. *Open filament.* A magnetron with an open

filament will produce no heating power, and an open filament will show no light or glow. An open filament lead or "dead" filament transformer will give the same indication as an open filament. This possibility should be eliminated before the magnetron is considered open. An ohmmeter check across the filter box terminals with the transformer leads removed will disclose an open condition in either the connections to the magnetron or an open magnetron. A visual inspection inside the filter box will determine whether the "open" is in the leads or in the magnetron.

2. *Internal plate cathode short.* A shorted magnetron will usually open the thermal switch or the line fuse. However, occasionally a magnetron will arc internally and not open the fuse. This condition will show up as wide, regularly recurring fluctuations of plate current and also as a varying brilliance in the filaments of the magnetron. A "ticking" sound may also be heard. A shorted plate transformer, filament transformer, or dielectric breakdown of the rf filters in the filter box will also cause the line fuse to open.

3. *Moding.* Moding occurs when the magnetron momentarily operates at a different frequency than the one it was designed for. Moding results in low heating power and is detected by erratic, sharp dips in plate current. A small fluctuation (5 to 10 mA on either side of the desired reading) is normal. "Ticking" noises coinciding with the plate current dips may also be heard.

4. *Low power.* Low emission will result in low cooking power. Food will require a longer-than-normal cooking time. It can be detected by a lower-than-normal steady plate current. Normal plate current in most models is between 50 and 325 mA. Check the service manual for the exact amount.

To visually check the magnetron tube, look for one or more of the following conditions.

1. A vertical crack in the glass envelope on the

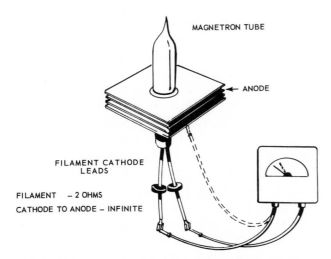

Figure 7-9. Checking the filament of a magnetron.

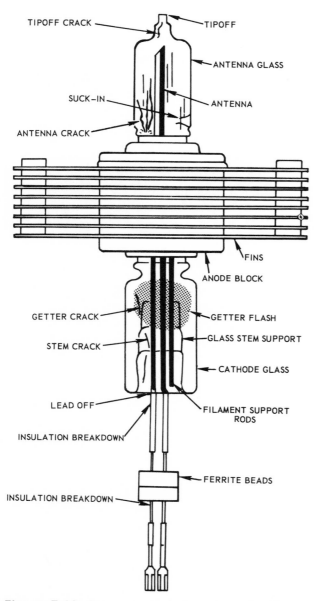

TIPOFF CRACK

TIPOFF

ANTENNA GLASS

SUCK-IN

ANTENNA

ANTENNA CRACK

FINS

ANODE BLOCK

GETTER CRACK

GETTER FLASH

STEM CRACK

GLASS STEM SUPPORT

CATHODE GLASS

LEAD OFF

FILAMENT SUPPORT RODS

INSULATION BREAKDOWN

FERRITE BEADS

INSULATION BREAKDOWN

Figure 7-10. Things to look for when checking a magnetron.

magnetron tube. This indicates that the brass washer or washers were not properly positioned during tube installation.

2. The getter turning white. The getter in the magnetron tube will turn white if the tube has turned to air. (*Turning to air* means that the tube has lost its vacuum.) A poor bonding where the filament leads leave the

tube or a crack in the envelope can cause the tube to turn to air.

3. Sunken place on the glass envelope. If the oven is operated without a load, the microwave energy will peak and cause a high thermal condition on the glass envelope. The temperature can rise to a point where the glass starts to melt, resulting in a sunken place on the envelope because of the high-vacuum condition in the tube.

It is extremely important when replacing the tube to draw up tight all tube housing screws to ensure that there is no air gap between the housing parts. The tube housing carries magnetic flux required for proper operation of the magnetron tube. Any air gap in the housing will interrupt the flow of the magnetic flux lines.

A magnetron tube, like a radio or television tube, must be handled with a reasonable amount of care. Always handle one by the fin area only, and use caution not to touch or strike the glass portion at the top or bottom.

A blower fan arrangement is generally used to cool the magnetron and other electric components. The blower also creates all the airflow necessary to exhaust steam and vapors from the oven capacity. In many units, the blower directs the air across the magnetron by means of a plastic or fabricated duct between the blower and the tube. The blower should turn on as soon as the START-STOP switch is pushed. If 120 V is available at the blower connections and the blower does not work, replace the blower assembly. In a few models, the blower motor coil is internally fused so the unit may be used in built-in installations. The fuse is embedded in the motor winding. If the fuse opens, the entire blower unit must be replaced.

Most microwave ovens have some type of thermal protector to guard the magnetron tube from overheating, such as in the case of a blower failure. Generally, the thermal protector is a normally closed device which opens at approximately 220°F. It resets automatically at approximately 175°F (approximately 3 min off cycle). Cycling of the oven light with no cooking would

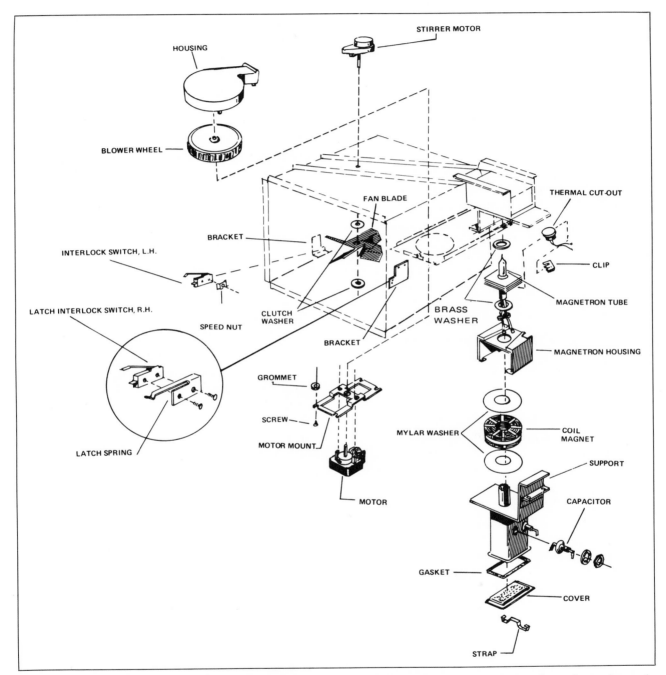

Figure 7-11. Parts of a typical component from the magnetron tube-thermal cut-out portion of an electronic oven.

be an indication of the thermal protector opening and closing. When this occurs, check the blower operation and for obstructions in the filter under the oven.

Some models have an overload relay (usually with a manual reset) to protect the magnetron and field coils in case of excessive plate current. The relay coil may be connected either in the plate or cathode circuit, and the relay contacts are in the power input circuit.

Excessive plate current will energize the overload relay coil and open the contacts, thus interrupting the power supply. But tripping of the overload does not necessarily mean that a failure has occurred in the circuit. Under certain loading conditions of the magnetron, an internal arc may occur and trip the overload. In this case, resetting the overload may be all that is required. If the overload relay does not continue to trip, the oven will function normally. If an electrical failure is present, the relay will trip again.

Waveguide, stirrer, and glass plate. Microwave energy from the magnetron tube must be directed to the food. Furthermore, according to one of the basic design features of these ovens, it must heat the food *evenly*. Several components aid in accomplishing this: the waveguide, stirrer, and glass plate (see Fig. 7-1). Understanding these components is quite easy because microwaves react much the same way as light and travel in a straight line.

The magnetron is connected to a waveguide, which is either a rectangular piece of tubing approximately 2 by 4 in or a round pipe of a diameter of about $2\frac{1}{2}$ in. One end of the waveguide is open, and this end serves as an antenna to radiate the power from the magnetron into the oven cavity. It should be pointed out that with most models the transmission line sections are at ground potential; the coupler and magnetron, however, are operating at high voltage. Because of the high voltage, never service the transmission line when electronic power is applied. When installing the magnetron and waveguide, be certain that the rf gasket is in place and all mounting screws tightened. Failure to do so can result in hazardous levels of microwave leakage. Incidentally, to ensure that the unit does not emit excessive radiation and to meet the guidelines of the United States Department of Health, Education, and Welfare, the oven can be checked for microwave leakage by using an instrument such as the Narda Model 8700 radiation meter. Such instruments should be used as directed in operational instructions.

The stirrer, located near the waveguide or antenna, is usually a fan-shaped blade or blades. In some models, the stirrer blade is rotated by the air circulated through the oven cavity, while in others it is motor-driven at a rather slow rate of revolutions per minute. The stirrer is used to reflect and/or "mix" the microwave energy coming down the waveguide from the tube to ensure that the energy enters the food from all sides to produce an even cooking pattern.

The stirrer motor is energized in some models through the COOK or START relay, while in others through the STOP-START switch. But, in either case, the stirrer is on only during cooking. Testing of the motor is limited to a continuity and voltage check. In units that employ a plastic stirrer cover (see Stirrer Shield below), be certain that it is installed properly so that the stirrer blade does not stall the motor. Failure of the motor and uneven cooking can result.

Food to be cooked is placed on a glass plate suspended approximately 1 in from the bottom of the oven. This plate is of special quality Pyrex glass which is transparent to microwaves. The microwaves go through the glass, strike the bottom of the oven, and are reflected back up into the food from the bottom. This allows the microwave energy to enter the food from all sides. *Note*: Do not operate the oven without a cooking load.

Choke assembly. Once the microwave energy is generated, it must be contained so that only the food is cooked. Oven components are constructed of various materials which reflect, transmit, or absorb power, depending on their intended functions.

Most oven enclosures are constructed of stainless steel, which reflects microwave energy and does not absorb any power. Aluminum will do the same job, but it does not have the fine appearance of stainless steel nor is it as easy to clean.

The primary microwave seal is usually a choke. It is built on the periphery of the door and fits inside the oven cavity. Microwaves, like most everything, follow the path of least resistance; therefore, the microwaves pass by the front edge of the door, strike the back edge of the door, are reflected into the choke cavity, and, in turn, are reflected back into the oven cavity. The choke cavity is filled with material which is transparent to microwaves such as polypropylene. Polypropylene keeps the interior of the choke from becoming filled with dirt and also makes the exterior of the door much easier to clean.

Any power that bypasses the choke, that is not reflected into the oven, is absorbed by the back door gaskets. The gaskets are vinyl but a special grade of vinyl. Ordinary vinyl is transparent to microwaves, so the vinyl in the gasket is loaded with carbon black. This makes the vinyl highly absorbent and gives it the ability to absorb microwave power which is bypassed by the choke. (More on door seals is found below.)

Control Components

The control components include such items as the timer, start switch, latch switches, and door switch.

Timers. Several different timer arrangements are in use on microwave ovens. The three most popular are the following.

1. *Thirty-minute single-speed timer*. The face of this timer is marked in 30 equally spaced minute graduations and subdivided into 30-s intervals. This timer is usually operated by a 120-V motor. A separate buzzer, or bell, is generally used with the 30-min timer. Approximately 5 to 10 s before the end of the timed cycle, a set of contacts close to energize the buzzer or bell. The buzzer or bell continues to sound, even after the cooking has been completed, until the user turns the timer OFF. Some ovens employ 15- or 25-min single-speed timers.

2. *Thirty-five minute two-speed timers*. This timer uses a split-scale arrangement in which the first 5 min are spread out over one-half the dial and subdivided into 15-s intervals. The remaining 30 min are divided into 1-min intervals and spread out over the other half of the dial. The timer motor is generally rated at about 95 V as it usually operates in series with the primary winding of the filament transformer. This circuit is necessary in order to keep the motor energized at the end of the power cycle to operate the continuous buzzer or bell. The buzzer or bell sounds about once per second until the timer is turned off. The mechanism for the buzzer or bell is usually built into the timer and cannot be serviced separately.

3. *Two-timer arrangement*. Many microwave ovens have two separate timers. One allows cooking times up to 5 min for short-time cooking and thawing, while the other allows cooking times up to 30 min. Actually the timer most often used in such an arrangement is the one with a scale from 0 to 5 min. Approximately 70 percent of the items cooked in the oven will be cooked in less than 5 min. Some that can be cooked in this time are hamburgers, hot dogs, and bacon; food can be reheated in 5 min or less, too. The second timer is provided for cooking such food as chicken and roasts. The reason for providing two timers is that for foods that take a very short cooking period, say 45 s, the time must be measured very accurately. It would be almost impossible to set 45 s on a timer which had a full scale of 30 min. The circuit is usually so designed that only one timer will operate at a time. The buzzer or bell system used for the single-speed timer is usually employed in the two-timer arrangement as well.

A microwave oven timer is tested by a continuity check. The schematic diagram in the service manual will tell whether continuity should be established.

A time-delay relay is sometimes used in microwave ovens to provide a warm-up period for the magnetron filament before high voltage is applied to the tube. As a rule, the relay consists of a 120-V coil which operates as soon as the START-STOP switch is pressed. Approximately 5 to 12 s (depending on the model) after the coil is energized, the relay contacts close. This completes a circuit up to the timer switch. In other words,

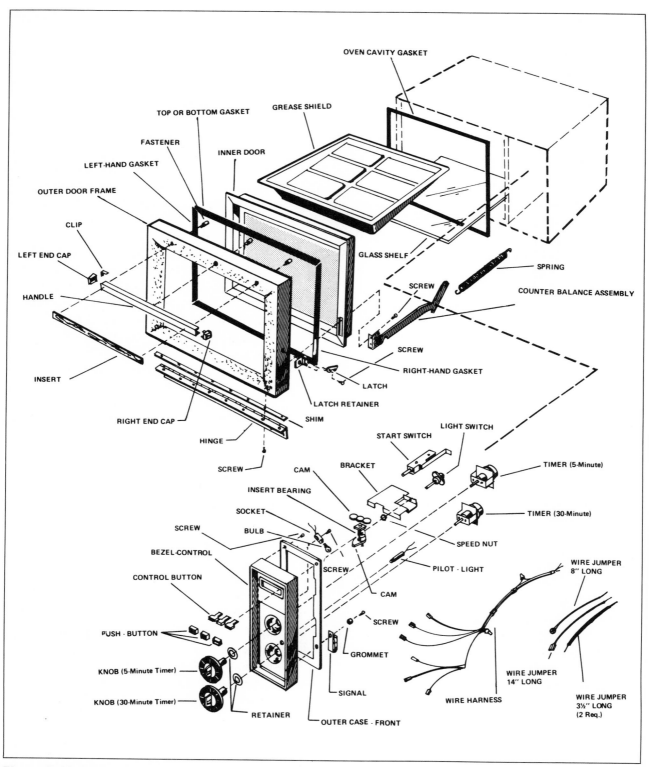

Figure 7-12. Major safety control of a typical microwave oven.

the timer circuit is so arranged that the timing function does not start until after the time-delay period and at the same time that the power circuit is energized.

Most time-delay relays are of the hydraulic type which produce the delay by the time it requires for the magnetic field to lift a metal slug through a fluid-filled sealed chamber. When the slug reaches the top of the chamber, the concentration of the magnetic pull closes the switch contacts.

Door interlock switches. The door interlock switch prevents the user from operating the unit with the door open. It also ensures positive shutoff when the door is open. Most modern microwave ovens have three sets of interlock switches.

The primary and secondary interlocks are the two door latch switches which control the electric system within the oven. In some models, the interlocks control the primary of the power transformer, while in others they control the relay contacts for the magnetron high voltage. But regardless of design, the door latch provides a positive door seal with the door latch (mechanical) and a positive door seal with the latch switch (electrical) before the oven can be put into operation.

The door latch switches are usually located behind the oven front frame beside the oven cavity. They are operated by the door latching catch after the door has reached the closed position. The door latch switch circuit is broken before the door makes any movement when it is opened.

To test door latch switches, make a continuity check of each switch with the door open and closed. Normal reading should indicate closed contacts when the door is closed. Contacts should open as soon as the thumb button or latch is engaged.

The door switch is a backup safety device for the two latch switches. It is connected in series with the high-voltage transformer primary winding, and will break the circuit when the door is opened. This provides secondary protection in case of failure of the primary interlock, the latch switch.

To test a door safety switch, check for continuity. Normal operation will show the switch closed when the door is closed. The contacts should open when the door is opened.

Some oven models have a separate set of switch contacts located in the concealed door safety switch; they monitor the primary and secondary interlock function (latch switch). In case of a primary and secondary function failure, the monitor swich closes and connects a surge resistor across the 120-V line which blows the 15-A built-in oven fuse. This requires service and repair of the oven before it will operate again.

With doors having a monitor, two audible clicks will be heard. The first click is the door safety switch; the second click is the monitor.

START-STOP switch. The ON-OFF switch has many names, among them START-STOP, COOK-STOP, and HEAT-STOP. In most cases, this switch is push-button operated. The START switch will not function if the door is not seated against the reset button.

The light switch, which is also usually of the push-button type, controls the oven light when the door is closed.

Quite a few microwave models have a START or COOK relay which serves two functions.

1. One set of contacts maintains the circuit for the relay coil. This latching effect is required because the initial circuit is established by the momentary START or COOK switch.
2. The other set of contacts provides the circuit for the high-voltage transformer.

The relay coil is energized when all controls are set and the START or COOK switch is pressed.

External Enclosure Components

All the components are enclosed in a well-styled, attractive enclosure. The enclosure materials are chosen not only for their appearance, but also for their cleanability and long life.

Outer case. The microwave oven, as we know, is a source of high voltage. Since the outer case must be removed to service any component, the following safety precautions should be observed to prevent injury by electric shock:

1. Unplug the oven from its electric source.
2. Remove all the screws holding the outer case. Lift the unit from its case.
3. Discharge the high-voltage capacitor by shorting across its terminals with a shorting bar. Two plastic-handled screwdrivers can be used by simultaneously touching each capacitor terminal and the screwdrivers. Also remember to unplug the oven and discharge the capacitor before servicing a circuit each time power has been applied.

Control panel. The control or access panel contains most of the operational controls. It can usually be removed as an assembly for service as follows.

1. Unplug the oven and remove the outer case.
2. Discharge the high-voltage capacitor(s).
3. Disconnect all leads from the controls.

Remove the timer knob. Then remove all necessary screws and lift the panel free.

Door. In most counter-top and portable microwave ovens the door is bottom-hinged, while in built-in and range models it is side-hinged. Either type of door contains a perforated screen window with a tempered glass inside and a full-size plastic or glass window outside. The main body of the door is usually a die-cast picture frame to which the outer window, door trim, handle, and door seals are fastened. To this main door assembly is mounted an inner door assembly, called the choke seal, as was noted earlier. This is the door "plug" which extends inside the oven cavity and consists of the inner window, inner door, polypropylene seal, and seal plate.

As previously mentioned, the primary door seal is accomplished by use of a choke seal. With this type of seal, there is no metal-to-metal contact between the door and the oven. Instead, a scientifically designed gap, between the "plug" and the oven, provides an electronic barrier to microwaves. This gap, and the construction of

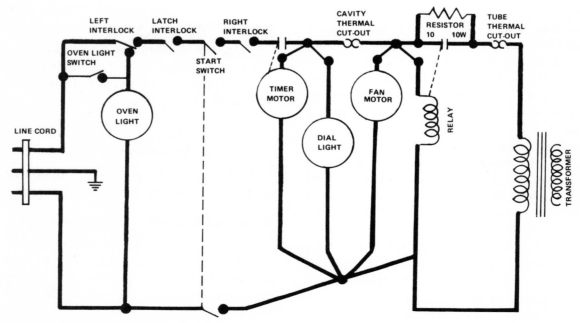

Figure 7-13. Door and control panel parts.

the door in this area, effectively create a short-circuit path for microwaves and thereby prevent their escape. To improve the seal, a secondary absorber seal is used to absorb any leakage which might get past the choke seal. The absorber seal is inside the main part of the door which seals against the front frame just outside the oven cavity. It frequently consists of strips of ferrite-impregnated rubber at the top, sides, and bottom of the door. A plastic cover is usually placed over the ferrite to give longer wear and better appearance.

Door adjustment in most models is limited to a slight side-to-side movement of the hinge. When checking the door, make certain that the hinge-mounting screws are tight. Open and close the door while observing the gap between the door plug and oven cavity as the door approaches full closure. This gap should be about $\frac{1}{8}$ in. The corners of the door can be checked by inserting a narrow paper strip over the corner of the inner door "plug" and pulling out the paper while holding the door open slightly. The paper should move freely, indicating no mechanical interference. If necessary, the door can be moved side to side slightly by loosening the hinge screws on the case frame. Move the door and retighten the screws.

Handle and latch. The handle-and-latch mechanism is fastened to the inside of the door assembly. In some models, when the latch button is pushed, it operates a latch lever which in turn triggers or raises the latch. When the button is released, a spring returns the latch to the lower or latched position. In others, the door latch is an electrically operated mechanical device which prevents the door from being opened when the unit is in operation. The door will remain latched until the electric circuit is interrupted.

Convenience Components

A number of additional items are not included in any of the foregoing groups and can be classified as convenience components. They include the following.

Indicator lights. Most units have an oven light, which is usually located at the rear of the oven,

and dial lights, which show the user that the timer is in operation and also make the control panel more easily readable. In addition, when the dial light is on it usually indicates a complete circuit to the primary of the power transformer.

If either the oven bulb or the dial bulb burns out, be sure to replace it with one of the same type and size as the original bulb.

Stirrer shield. The top of the oven is generally covered by a vacuum-formed plastic shield (usually polypropylene). The top of the oven has a number of sharp corners plus the stirrer. All these areas would be fairly hard to clean; therefore, a shield is provided. Under normal use the shield can be cleaned while it is in the oven. Should the oven ever become exceedingly dirty, the shield can be removed.

Oven rack. Some models feature a round rack which is powered by a 6-r/min motor mounted under the oven. The rack motor is energized at any time the microwave power is on. The purpose of the motor is supposedly to give more even cooking.

Defrost circuit. One of the most popular of the convenience components is the so-called "defrost circuit," which is usually made up of a defrost switch, a defrost timer switch, and a defrost motor and cam.

As shown in Fig. 7-14 in a typical defrost circuit, the defrost switch is a normally closed (NC) switch which allows the unit circuit to be

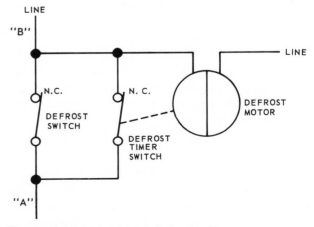

Figure 7-14. A typical defrost circuit.

completed for normal operation when the defrost lever is in the OFF position. If food is to be defrosted, the user places the defrosting latch in the ON position, sets the timer to the recommended time, and waits until the unit shuts off. Placing the defrosting lever in the ON position opens the normally closed (NC) defrost switch and completes the circuit through the defrost timer switch. The defrost timer switch is opened and closed by the cam attached to the motor shaft. Therefore, the circuit through the transformer to the magnetron is interrupted at approximately 30-s intervals which accomplishes the defrosting of the food. As the defrost motor is connected across the line, it functions whenever the unit is in operation. During normal operation of the unit, the circuit is completed from A to B through the defrost switch. During the defrost cycle, the circuit is completed from A to B through the defrost timer switch on an intermittent basis. As the two circuits are parallel, the fact that the motor is attached across the line and functions on a continuous basis has no detrimental effect on the unit.

Oven Power Performance Test

To check whether the oven is performing normally, conduct the following test.

1. Fill a Pyrex glass measuring cup with two cups (1 pt) of tap water.
2. Using an accurate thermometer, check and record the water temperature.
3. Place the filled container in the center of the glass tray in the oven.
4. Close the door.
5. Set the timer to $2\frac{1}{2}$ or 3 min.
6. Push the START button to turn the oven on.
7. Using a stopwatch or a clock with a second hand, time to exactly 2 min and shut the unit off.
8. Open the door and remove the measuring cup.
9. Using the thermometer, stir the water to check the temperature rise of the water. The temperature rise of the water in a *properly* operating unit should be approximately 55 to 60°F when the unit is being operated

at 120 V. But remember that the temperature rise of the water will be greatly affected by the voltage at which the unit is operated. The voltage at the time of the test must be known to properly understand the test results.

Precautions for Proper and Safe Service

All microwave ovens are designed, built, and tested to rigid industry and government standards. If damaged or serviced in a way that will allow microwave energy to escape, a microwave oven can be hazardous. It is important that you observe the following precautions in servicing microwave ovens.

1. Do not attempt to operate the oven with the door open. The oven has been provided with several interlocks to protect you—do not attempt to "fool" them. They are intended to make sure the door cannot be opened while microwave power is on.
2. Do not attempt to operate the oven when any of its components are removed and/or bypassed, when any of the safety interlocks are found to be defective, or when any of the seal surfaces are found to be missing or damaged.
3. Do not operate the oven until after repair, if any of the following conditions exist:
   a. Door does not close firmly against the front frame.
   b. Hinge is broken.
   c. Door seal damaged.
   d. Door is bent or warped.
   e. Any other damage to the oven is visible. Check the choke and gasket area to ensure that this area is clean and free of all foreign matter.
4. Always have the unit disconnected when the rear panel and/or cabinet are removed except when making the "live" tests called for in the service manual. Do not reach into the equipment area while the unit is energized. Make all connections for the test and check them for tightness before plugging the cord into the outlet.

5. Always ground or discharge the capacitors on the filter box with an insulated-handle screwdriver(s) before working in the high-voltage area of the equipment compartment. Some types of failures will leave a charge in these capacitors, and the discharge could cause a reflex action which could make you hurt yourself.

6. Always remember that in the area of the transformer there is *high voltage*. When the unit is operating, keep this area clear and free of anything which could possibly cause an arc or ground.

7. Do not operate the oven until after replacing any defective parts in the interlock, oven door, magnetron, and waveguide.

8. Do not lean on, or place heavy service tools or equipment on, the oven door. Before returning an oven to a customer, be sure that the door spacing is reasonably uniform at the top and sides. Check manufacturers for exact measurements of allowable gap. Also be sure to check for proper door-switch interlock action.

9. Do not operate oven empty.

10. Always operate the unit from an adequately grounded outlet. Do not operate on a two-wire extension cord. Under no circumstances should the round ground prong be removed from the plug. Also never remove any of the required governmental regulatory information from the unit.

The following ten safety or protective devices are usually found in most microwave ovens to ensure maximum safety to the appliance, the customer, and the service technician. While the devices mentioned here may be given slightly different names by various manufacturers and their locations may vary slightly, their operation and function is similar to those listed here:

Device	Usual location	Function	Operation	General conditions requiring protection
1. Magnetron thermal protector (automatic reset)	Rear of magnetron	Protects magnetron from overheating	Connected in series with relay coil with neutral line (lead) to control circuit of power supply. Should magnetron temperature exceed normal conditions, protector opens and interrupts power supply.	a. Magnetron blower failure. b. Dirty filter.
2. Overload relay (manual reset)	Front panel of power pack	Protects magnetron and field coils in case of excessive magnetron current	Relay coil connected in plate current circuit. Excessive magnetron current will energize overload relay coil, and will open contacts, which interrupts power supply.	a. Magnetron arcing —can occur because of misuse of oven; also does not necessarily indicate defective magnetron.
3. Fuse	Front panel of power pack	Protects components and circuit wiring in case of high-voltage short circuit during operation	Connected in series with power supply. Should a short occur, fuse blows and interrupts power supply. (High power only.)	a. Capacitor shorted. b. Silicon rectifiers shorted. c. Defective door shorting switch.
4. Door shorting	Left side of oven on front frame	Disables power supply in event door should be opened during oven operation	Cuts off ac power supply to circuit by blowing 15-A fuse.	a. Personal protection against radiation.

(cont'd)

cont'd.

Device	Usual location	Function	Operation	General conditions requiring protection
5. Door latch switch (primary interlock)	Under cooktop	Prevents operation of entire electronic circuit until door is latched.	Connected in series with other controls and is actuated by a mechanism on the door latch (rear). Door must be completely latched before switch can close and energize circuit to other controls.	a. Personal protection against radiation.
6. Hinge switch (usually the third interlock)	Concealed under left hinge lever	Backup for primary and secondary interlocks.	Hinge-actuated switch which de-energizes main power relay when door is open.	a. Protection against intentional tampering or defeating interlock system.
7. Door gasket (conducting type)	Around oven liner	Prevents significant radiation leakage and prevents arcing between oven and door.	Gasket encased in metallic wire screen. When door is closed, the gasket serves as conductor between oven liner and door as well as a door seal. Good electrical conduction is required to prevent radiation leakage and to prevent arcing. (Door-to-oven liner.)	a. Personal protection against radiation.
8. High-voltage bleeder resistor	Across voltage doubler capacitors	Discharge for high voltage capacitors.	A resistor is connected across each capacitor. The resistors serve as discharge path for the high voltage when power is removed from the range.	a. Personal protection against shock when servicing power pack.
9. Temperature limit switch	Inside oven	Provides over-temperature protection in cook cycle.	Provides direct sensing of oven temperature to accomplish function.	a. Prevent possibility of user opening door above normal cooking temperatures.
10. 5-A fuses (usually two)	Front of power pack	One protects high-voltage circuit on low power. The other one protects 120-V circuit.	Connected in series with each circuit. Should a short occur, fuse blows and interrupts power supply.	a. Shorted capacitor on (low power). b. Silicon rectifiers shorted (low power).

TROUBLE DIAGNOSIS

When attempting to find the trouble with a microwave oven, always discuss problem with the user, who very often will be able to give you a valuable clue as to where to start looking.

Here are some of the more common complaints a service technician will receive about microwave ovens. Also listed are the usual solutions to these problems.

Problem: No power to the oven.

Possible cause

1. No line voltage

Solution

1. Check the line voltage. If absent, inspect house fuse box or circuit breaker and take

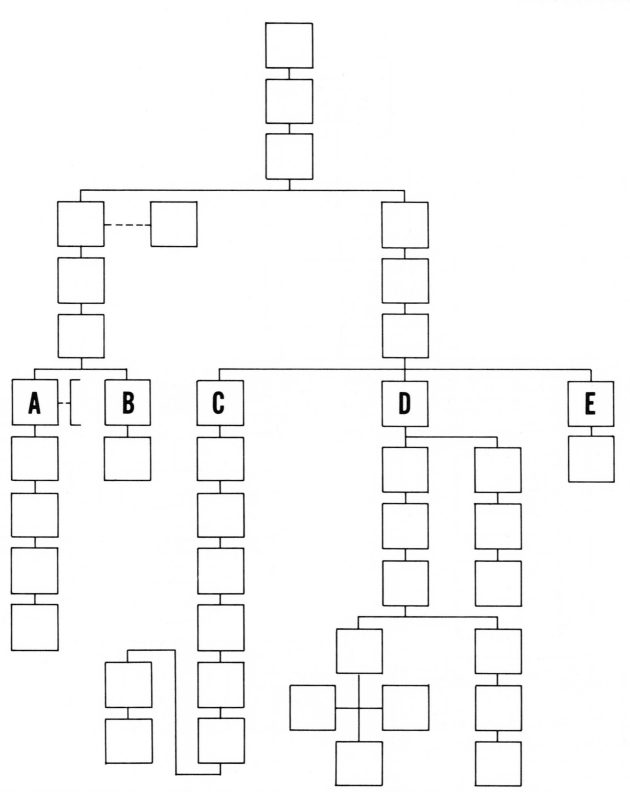

Figure 7-15. Block diagnosis chart: schematic, followed by actual steps in diagnosis; (A) through (E) are specific possible causes of complaints.

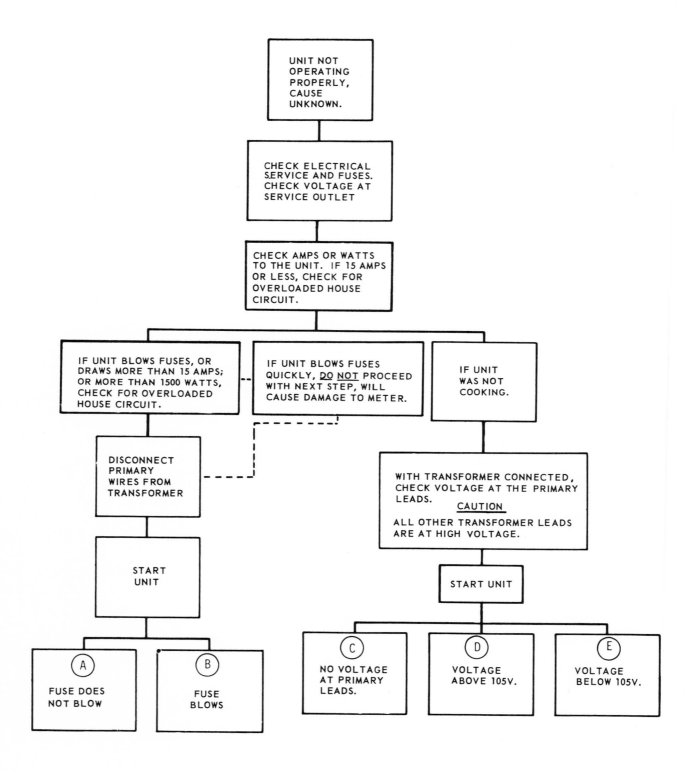

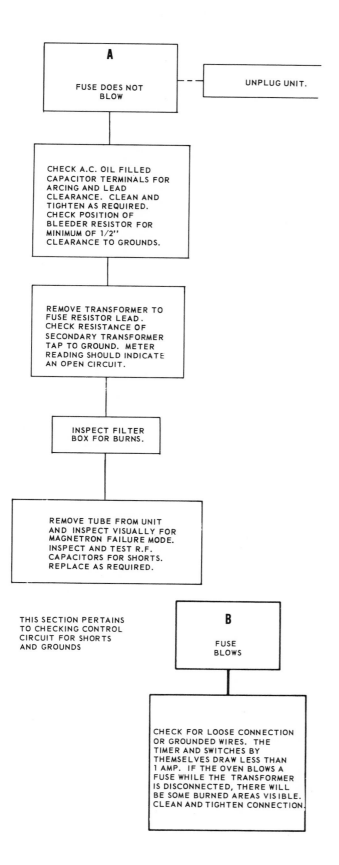

THIS SECTION
PERTAINS TO
CHECKING FOR
OPEN CIRCUIT.

C

NO VOLTAGE AT THE TRANSFORMER PRIMARY LEADS.

WITH UNIT NOT PLUGGED IN. CLOSE DOOR, SET TIMER, PUSH START SWITCH IN.

CHECK CONTINUITY ACROSS TUBE THERMAL SWITCH. NO CONTINUITY, SWITCH OPEN OR DEFECTIVE. ALLOW 3–5 MINUTES TO RESET.

CHECK CONTINUITY ACROSS START–STOP SWITCH. NO CONTINUITY, SWITCH DEFECTIVE

CHECK CONTINUITY ACROSS CAVITY THERMAL SWITCH. NO CONTINUITY, SWITCH OPEN OR DEFECTIVE. REQUIRES MINIMUM OF 20 MINUTES TO RESET.

CHECK CONTINUITY OF LEFT INTERLOCK. NO CONTINUITY, SWITCH DEFECTIVE.

CHECK CONTINUITY ACROSS FUSE RESISTOR. NO CONTINUITY, FUSE OPEN.

CHECK CONTINUITY ACROSS RIGHT INTERLOCK. NO CONTINUITY, SWITCH DEFECTIVE.

CHECK CONTINUITY ACROSS LATCH SWITCH NO CONTINUITY, SWITCH DEFECTIVE.

NOTE: NO CONTINUITY WILL BE SHOWN ACROSS SURGE RELAY CONTACTS WHEN POWER IS OFF.

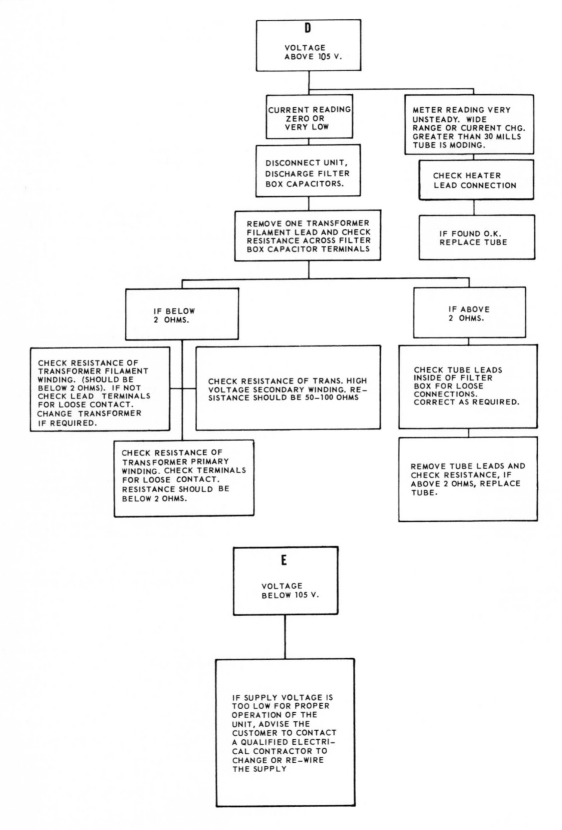

D

VOLTAGE
ABOVE 105 V.

CURRENT READING
ZERO OR
VERY LOW

METER READING VERY
UNSTEADY. WIDE
RANGE OR CURRENT CHG.
GREATER THAN 30 MILLS
TUBE IS MODING.

DISCONNECT UNIT,
DISCHARGE FILTER
BOX CAPACITORS.

CHECK HEATER
LEAD CONNECTION

REMOVE ONE TRANSFORMER
FILAMENT LEAD AND CHECK
RESISTANCE ACROSS FILTER
BOX CAPACITOR TERMINALS

IF FOUND O.K.
REPLACE TUBE

IF BELOW
2 OHMS.

IF ABOVE
2 OHMS.

CHECK RESISTANCE OF
TRANSFORMER FILAMENT
WINDING. (SHOULD BE
BELOW 2 OHMS). IF NOT
CHECK LEAD TERMINALS
FOR LOOSE CONTACT.
CHANGE TRANSFORMER
IF REQUIRED.

CHECK RESISTANCE OF TRANS. HIGH
VOLTAGE SECONDARY WINDING. RE-
SISTANCE SHOULD BE 50–100 OHMS

CHECK TUBE LEADS
INSIDE OF FILTER
BOX FOR LOOSE
CONNECTIONS.
CORRECT AS REQUIRED.

CHECK RESISTANCE OF
TRANSFORMER PRIMARY
WINDING. CHECK TERMINALS
FOR LOOSE CONTACT.
RESISTANCE SHOULD BE
BELOW 2 OHMS.

REMOVE TUBE LEADS AND
CHECK RESISTANCE, IF
ABOVE 2 OHMS, REPLACE
TUBE.

E

VOLTAGE
BELOW 105 V.

IF SUPPLY VOLTAGE IS
TOO LOW FOR PROPER
OPERATION OF THE
UNIT, ADVISE THE
CUSTOMER TO CONTACT
A QUALIFIED ELECTRI-
CAL CONTRACTOR TO
CHANGE OR RE–WIRE
THE SUPPLY

1. (cont'd)

	proper corrective action.
2. START-STOP switch faulty.	2. Conduct a continuity test and, if defective, replace switch.
3. Latch switch or interlock defective.	3. Check continuity and replace if necessary.

Problem: Oven does not heat.

Possible cause — *Solution*

1. Interlock switch inoperative.
1. Readjust switch's position or, if defective, replace.

2. Magnetron defective.
2. Check and replace, if necessary.

3. High-voltage rectifiers faulty.
3. Check and replace, if necessary.

4. Warm-up time delay relay inoperative.
4. Check and replace, if necessary.

Problem: Power output is low during cooking cycle.

Possible cause — *Solution*

1. Low line voltage.
1. Check and take proper corrective action.

2. Diode rectifier defective.
2. Check and replace, if necessary.

3. Magnetron faulty.
3. Check and replace, if necessary.

Problem: Oven will not advance to COOK phase.

Possible cause — *Solution*

1. Dc voltage not available.
1. Check circuit and components responsible for dc voltage, then take appropriate action.

2. Door latch switch inoperative.
2. Check and repair or replace.

3. Inoperative COOK relay (if used).
3. Check relay action and replace if necessary.

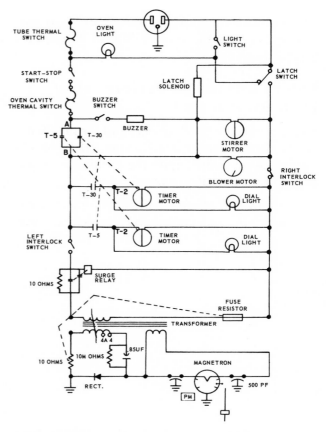

Figure 7-16. Schematic diagram detailed in block diagnosis chart.

Problem: Oven turns off too soon; blower continues to operate.

Possible cause — *Solution*

1. Timer defective.
1. Check timer and replace if necessary.

2. Transformer defective.
2. Check and, if faulty, replace.

3. Thermal protector defective.
3. Check magnetron cooling circuit and then take appropriate action.

The information just given is rather general in nature. More specific troubleshooting details are usually given in the service manual. One method of troubleshooting is to follow a so-called "block" diagnosis chart. Figure 7-15 shows one for a typical microwave oven; as you can see, it presents a very logical method of troubleshooting.

Safety Checks To Perform after Each Service Call

In the interest of good service, and for maximum safety to the user, the following safety checks should be performed upon the conclusion of any service to a microwave oven.

1. Check the electric cord and plug. Advise user if an improperly grounded wall receptacle is being employed, such as a non-grounded adapter.

2. Check the cord-grounding lead to the oven frame. Check for connection and continuity between the ground terminal on the plug and oven frame.

3. Check the plastic stirrer cover for proper installation inside the oven.

4. Make a physical check of the door for buildup of soil on the door seals, or any possible damage. Check the door fit and adjust according to the instructions given in the service manual.

Index